Biology

OCR and Heinemann are working together to provide better support for you

Heinemann is an imprint of Pearson Education Limited, a company incorporated in England and Wales, having its registered office at Edinburgh Gate, Harlow, Essex CM20 2JE. Registered company number: 872828

www.heinemann.co.uk

Heinemann is a registered trademark of Pearson Education Limited

First published 2008

13 12 11 10
10 9

British Library Cataloguing in Publication Data is available from the British Library on request.

ISBN 978 0 435691 80 6

Edited by Jill Bailey, Anne Sweetmore
Designed by Kamae Design
Project managed and typeset by Wearset Ltd, Boldon, Tyne and Wear
Original illustrations © Pearson Education Limited 2008
Illustrated by Wearset Ltd, Boldon, Tyne and Wear
Picture research by Q2AMedia
Cover photo of a mallow pollen grain © Alamy Images
Printed in China (GCC/09)

Acknowledgements
We would like to thank the following for their invaluable help in the development and trialling of this course: Alan Cadogan, Rob Duncan, Richard Fosbery, Amanda Hawkins, Dave Keble, Maggie Perry, Maggie Sindall and Jenny Wakefield-Warren.

The authors and publisher would like to thank the following for permission to reproduce photographs:

p3 Riccardo Cassiani-Ingoni/Science Photo Library; **p4** Photo Researchers, Inc./Photolibrary.com; **p5 L** BSIP LECA/Science Photo Library; **p5 M** Science Photo Library/Photolibrary.com; **p5 R** Simon Fraser/Newcastle General Hospital/Science Photo Library; **p8** Science Photo Library/Photolibrary.com; **p9 T** Colin Cuthbert/Science Photo Library; **p9 TR** Simon Fraser/Newcastle General Hospital/Science Photo Library; **p9 BL** Science Photo Library/Photolibrary.com; **p9 BR** Phototake Inc./Photolibrary.com; **p11 T** Phototake Inc./Photolibrary.com; **p11 R** Biophoto Associates/Science Photo Library; **p17** Photo Researchers, Inc./Photolibrary.com; **p19** Roger Ressmeyer/Corbis; **p21** Science Photo Library/Photolibrary.com; **p29 L** JJP/Philippe Plailly/Eurelios/Science Photo Library; **p29 R** Photo Researchers, Inc./Photolibrary.com; **p30** M.I. Walker/Photo Researchers/Dinodia Photo Library; **p32 M** Photo Researchers, Inc./Photolibrary.com; **p32 B** Phototake Inc./Alamy; **p33 B** Photo Researchers, Inc./Photolibrary.com; **p33 M** RMF/Visuals Unlimited; **p33 T** CMEABG-LYON-1, ISM/Science Photo Library; **p35** Phototake Inc./Photolibrary.com; **p36 T** Phototake Inc./Photolibrary.com; **p36 B** Science Photo Library/Photolibrary.com; **p37 T** Science Photo Library/Photolibrary.com; **p37 B** Steve Gschmeissner/Science Photo Library; **p43** Dr. Keith Wheeler/Science Photo Library; **p44 T** Phototake Inc./Alamy; **p44 M** Phototake Inc./Alamy; **p44 B** Phototake Inc./Photolibrary.com; **p45** Steve Gschmeissner/Science Photo Library; **p46** Science Photo Library/Photolibrary.com; **p48 T** Astrid & Hanns-Frieder Michler/Science Photo Library; **p48 B** Phototake Inc./Photolibrary.com; **p52** Michael Abbey/Science Photo Library; **p54** Photo Researchers, Inc./Photolibrary.com; **p61** Science Photo Library/Photolibrary.com; **p68** Photolibrary.com; **p69 TR** Phototake Inc./Photolibrary.com; **p69 TL** Phototake Inc./Photolibrary.com; **p69 B** Phototake Inc./Photolibrary.com; **p70 T** Hugh A. Johnson/Visuals Unlimited; **p70 B** Phototake Inc./Photolibrary.com; **p71 L** Dr. George Wilder/Visuals Unlimited; **p71** J.C. Revy/Science Photo Library; **p79** Phototake Inc./Photolibrary.com; **p87** Dr. Mark J. Winter/Science Photo Library; **p94 L** Dr. Jeremy Burgess/Science Photo Library; **p94 R** Science Source/Science Photo Library; **p97 L** Andrew Syred/Science Photo Library; **p97 M** Susumu Nishinaga/Science Photo Library; **p97 R** Russell Knightley/Science Photo Library; **p101** Coneyl Jay/Science Photo Library; **p104 T** Yakobchuk Vasyl/Shutterstock; **p104** Dr. Tim Evans/Science Photo Library; **p105 T** J. Gross, Biozentrum/Science Photo Library; **p105 B** Astra productions/Zefa/Corbis; **p108** Tom McHugh/Science Photo Library; **p109** John Bavosi/Science Photo Library; **p111 T** Pacific Stock/Photolibrary.com; **p111 MT** Hermann Eisenbeiss/Science Photo Library; **p111 M** Chin Kit Sen/Shutterstock; **p111 MB** Carlos Caetano/Shutterstock; **p111 B** Hinrich Baesemann/dpa/Corbis; **p114** Martyn F. Chillmaid/Science Photo Library; **p117** Phototake Inc./Photolibrary.com; **p118** Phototake Inc./Alamy; **p121 TL** Bettmann/Corbis; **p121 TR** Bettmann/Corbis; **p121 BL** Bettmann/Corbis; **p121 BR** Photo Researchers, Inc./Photolibrary.com; **p124 T** Doug Allan/Naturepl.com; **p124 M** Ferenc Cegledi/Shutterstock; **p124 B** David Tipling/Alamy; **p125** Blickwinkel/Alamy; **p136** A.E. Eriksson, T.A. Jones, A. Liljas/Protein Data bank; **p137** N. Ramasubbu/Protein Data Bank; **p139** David A. Northcott/Corbis; **p140** Martyn F. Chillmaid/Science Photo Library; **p151** Ianni Dimitrov/Alamy; **p158** Geoffrey Kidd/Alamy; **p159** Gustoimages/Science Photo Library; **p160 L** Eye of Science/Science Photo Library; **p160 R** Andrew Syred/Science Photo Library; **p161 T** CNRI/Science Photo Library; **p161 MR** BSIP, PR Bouree/Science Photo Library; **p161 B** Roger Harris/Science Photo Library; **p162** Eye of Science/Science Photo Library; **p163** Dr. John Brackenbury/Science Photo Library; **p167** Photo Researchers/Dinodia Photo Library; **p176** St. Mary's Hospital Medical School/Science Photo Library; **p177** CNRI/Science Photo Library; **p179 T** Science Photo Library/Photolibrary.com; **p179 B** Biophoto Associates/Science Photo Library; **p181** John Bavosi/Science Photo Library; **p185** AJ Photo/Science Photo Library; **p191** Istockphoto; **p192 L** Leslie J. Borg/Science Photo Library; **p192 R** Yanta/Shutterstock; **p195 L** Frank Sochacki; **p195 R** Frank Sochacki; **p196 T** Frank Sochacki; **p196 B** Frank Sochacki; **p202 R** A.B. Dowsett/Science Photo Library; **p202 L** M.I. Walker/Science Photo Library; **p203 T** Eye of Science/Science Photo Library; **p203 M** Norm Thomas/Photo Researchers/Dinodia Photo Library; **p203 B** Mike Wilkes/Naturepl.com; **p205 L** Andrew Syred/Science Photo Library; **p205 R** Ingo Arndt/Naturepl.com; **p208** Alex Rakosy, Custom Medical Stock Photo/Science Photo Library; **p210** Steve Vowles/Science Photo Library; **p212** Eye of Science/Science Photo Library; **p213** Mike Norton; **p216 L** Charles Darwin Online University of Cambridge; **p216 M** Malcolm Schuyl/Alamy; **p216 R** Colin Purrington; **p217** Interfoto Pressebildagentur/Alamy; **p220** Mary Evans Picture Library/Alamy; **p221** G.P. Bowater/Alamy; **p225** Annie Haycock/Science Photo Library; **p226** Mark Carwardine/Naturepl.com; **p227 T** Anthony Cooper/Science Photo Library; **p227 B** Ascal Goetgheluck/Science Photo Library; **p228 T** Associated Press; **p228 B** Convention on Biological Diversity; **p231** Interfoto Pressebildagentur/Alamy

The authors and publisher would like to thank the following for permission to use copyright material:

Figure 1, p164 reproduced with permission from www.traveldoctor.co.uk

Every effort has been made to contact copyright holders of material reproduced in this book. Any omissions will be rectified in subsequent printings if notice is given to the publisher.

Websites
There are links to websites relevant to this book. In order to ensure that the links are up-to-date, that the links work, and that the links are not inadvertently made to sites that could be considered offensive, we have made the links available on the Heinemann website at www.heinemann.co.uk/hotlinks. When you access the site, the express code is 1806P.

Exam Café student CD-ROM

Original illustrations, screen designs and animation by Michael Heald
Photographs © iStock Ltd
Developed by Elektra Media Ltd

Technical problems
If you encounter technical problems whilst running this software, please contact the Customer Support team on 01865 888108 or email software.enquiries@pearson.com

OCR Biology

OCR and Heinemann are working together to provide better support for you

Pete Kennedy and Frank Sochacki
Series Editor: Sue Hocking

Official Publisher Partnership

Contents

Introduction

How to use this book

In this book you will find a number of features planned to help you.

- **Module openers** – these introductions set the context for the topics covered in the module. They also have a short set of questions that you should already be able to answer from your previous science courses.
- **Double-page spreads** are filled with information and questions about each topic.
- **End-of-module summary and practice** pages help you link together all the topics within each module.
- **End-of-module practice examination questions** have been selected to show you the types of question that may appear in your examination.

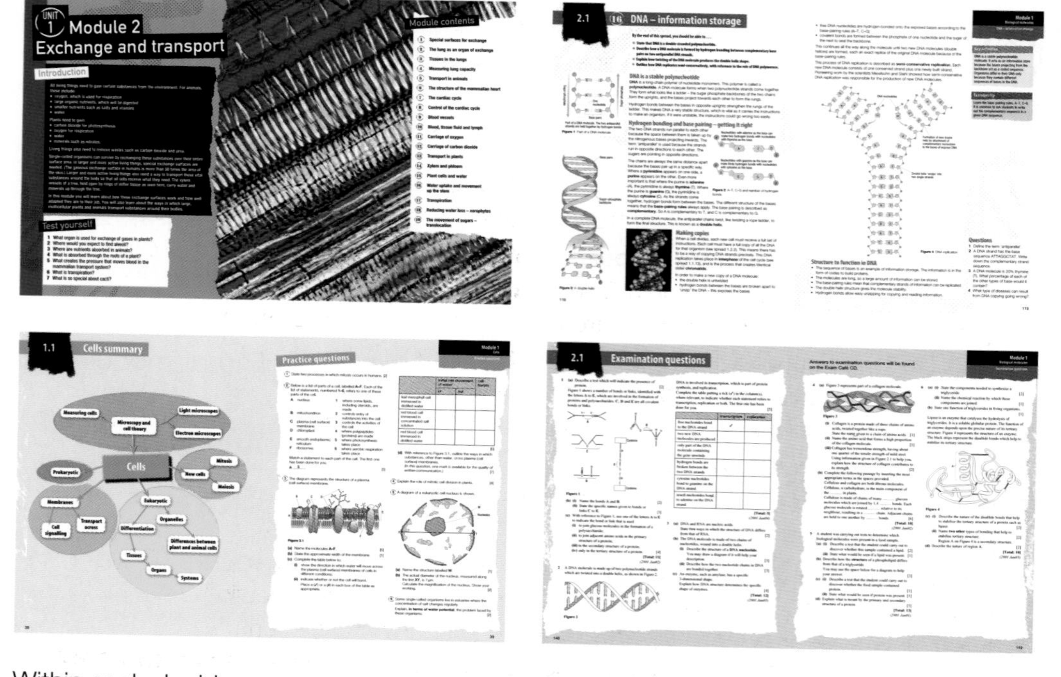

Within each double-page spread you will find other features to highlight important points.

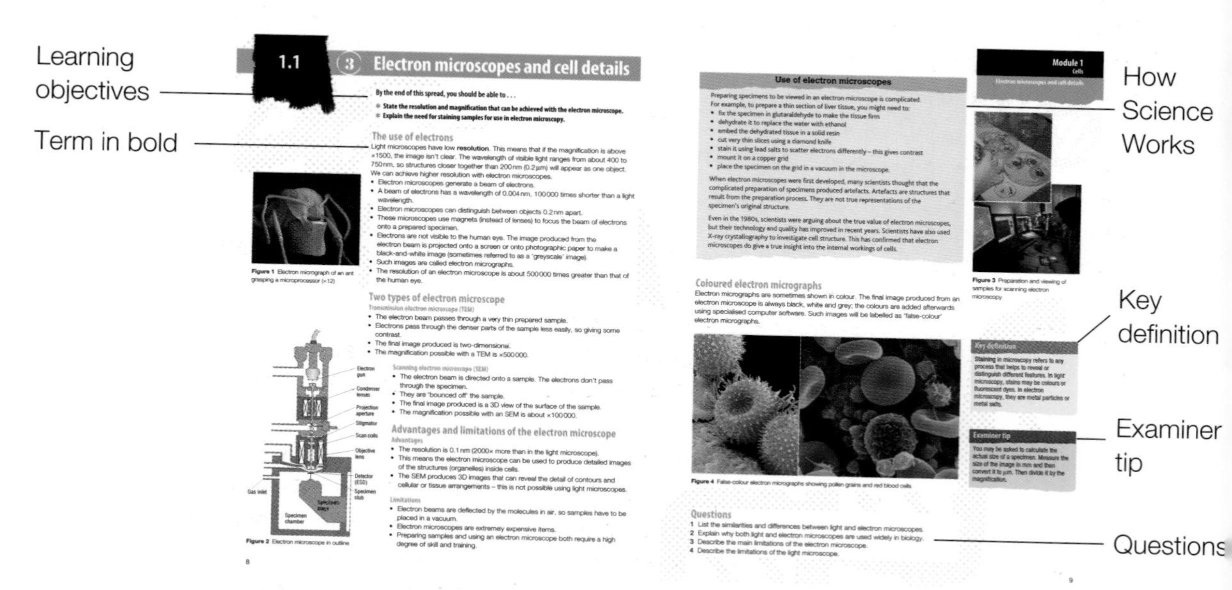

- **Learning objectives** – these are taken from the Biology AS specification to highlight what you need to know and to understand.
- **Key definitions** – these are terms in the specification. You must know the definitions and how to use them.
- **Terms in bold** are important terms, used by biologists, that you are expected to know. You will find each term in bold listed in the glossary at the end of the book.
- **Examiner tips** are selected points to help you avoid common errors in the examinations.
- **Worked examples** show you how calculations should be set out.
- **How Science Works** – this book has been written to reflect the way that scientists work. Certain sections have been highlighted as good examples of how science works.
- **Questions** – at the end of each spread are a few questions that you should be able to answer after studying that spread.

In addition, you'll find an Exam Café CD-ROM in the back of the book, with more questions, revision flashcards, study tips, answers to the examination questions in the book and more.

The examination

It is useful to know some of the language used by examiners. Often this means little more than simply looking closely at the wording used in each question on the paper.

- Look at the number of **marks allocated** to a part question – ensure you write enough points down to gain these marks. The number of marks allocated for any part of a question is a guide to the depth of treatment required for the answer.
- Look for words in **bold**. These are meant to draw your attention.
- Look for words in *italics*. These are often used for emphasis.

Diagrams, tables and equations often communicate your answer better than trying to explain everything in sentences.

Each question will have an **action word**. Some action words are explained below.

- *Define*: Specify the meaning of the word or term.
- *Explain*: Set out reasons or purposes using biological background. The depth of treatment should be judged from the marks allocated for the question.
- *State*: A concise answer is expected with no supporting argument.
- *List*: Provide a number of points with no elaboration. If you are asked for **two** points, then give only two!
- *Describe*: Provide a detailed account (using diagrams/data from figures or tables where appropriate). The depth of the answer should be judged from the marks allocated for the question.
- *Discuss*: Give a detailed account that addresses a range of ideas and arguments.
- *Deduce*: Draw conclusions from information provided.
- *Predict*: Suggest possible outcome(s).
- *Outline*: Restrict the answer to essential detail only.
- *Suggest*: Apply your biological knowledge and understanding to a situation which you may not have covered in the specification.
- *Calculate*: Generate a numerical answer, with working shown.
- *Determine*: The quantity cannot be measured directly but can be obtained by calculation. A value can be obtained by following a specific procedure or substituting values into a formula.
- *Sketch*: Produce a simple, freehand drawing. A single, clear sharp line should be used. In the context of a graph, the general shape of the curve would be sufficient.

When you first read the question, do so very carefully. Once you have read something incorrectly, it is very difficult to get the incorrect wording out of your head! If you have time at the end of the examination, read through your answers again to check that you have answered the questions that have been set.

UNIT 1

Module 1

Cells

Introduction

All living organisms consist of cells. These are the fundamental units of life. In this module you will learn how all cells have features in common and how cells can differ dramatically from each other. Most of the organisms that you are familiar with consist of cells with a nucleus. These are eukaryotic cells, like the neural stem cells shown here differentiating into neurones (red) and nerve support cells (green). You will also learn about another type of cell, called a prokaryotic cell.

Cells are very small and much of our knowledge of them comes from the use of microscopes. You will learn how microscopes work and how the information gathered from microscopy is of value in biology.

One of the features common to all cells is the cell surface membrane. You will learn how this is structured, what it does and why it is so important to living organisms.

The characteristics of living things are:

- movement
- respiration
- sensitivity
- nutrition
- excretion
- reproduction
- growth.

Throughout this unit, these themes show how the cellular nature of life relates to the living processes common to all organisms. You will learn about the process of generating new cells from old ones in growth, repair and reproduction. This module introduces you to the fundamental basis of life on Earth.

Test yourself

1. How many different types of cell do you know?
2. What are the limitations of the light microscope and the electron microscope in observing cells?
3. How big are cells?
4. Why can we observe chromosomes in a dividing cell but not in a non-dividing cell?
5. What happens when cell division gets out of control?
6. What does 'partially permeable' mean?
7. What is found inside cells?

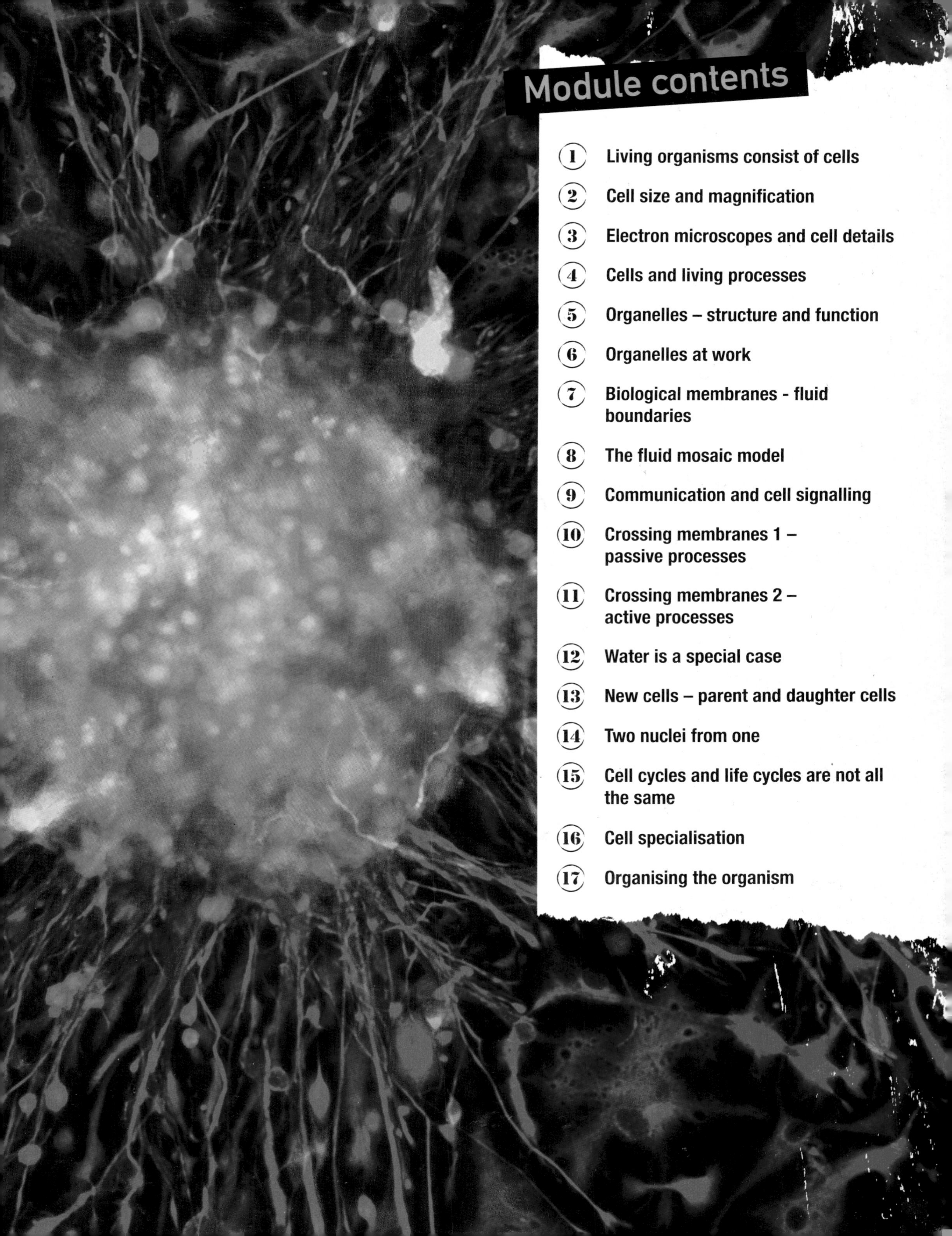

Module contents

1.1 ① Living organisms consist of cells

By the end of this spread, you should be able to . . .

- * **State the resolution and magnification that can be achieved by a light microscope.**
- * **Explain the difference between magnification and resolution.**
- * **Explain the need for staining samples in light microscopy.**

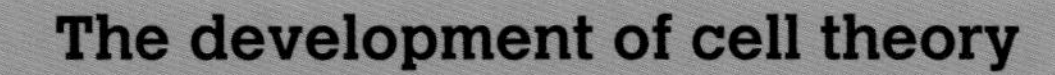

The development of cell theory

In the 1660s, Robert Hooke developed a compound microscope using several lenses. He used it to examine slices of cork taken from under the bark of an oak tree. Hooke noted that the slices were made up of a lot of tiny chambers. They resembled the rooms, or cells, in which monks lived, so he called the tiny chambers in the cork 'cells'. With better microscopes, other scientists studied biological material and saw that all plant and animal material was made up of many cells. By the 1840s the Cell Theory – developed by two scientists, Schleiden and Schwann – was accepted. The Cell Theory, as extended by the work of Virchow in 1855 and Weisman in 1880, states:

- All living things consist of cells.
- New cells are formed only by the division of pre-existing cells.
- The cell contains information that acts as the instructions for growth. This information can be passed to new cells.

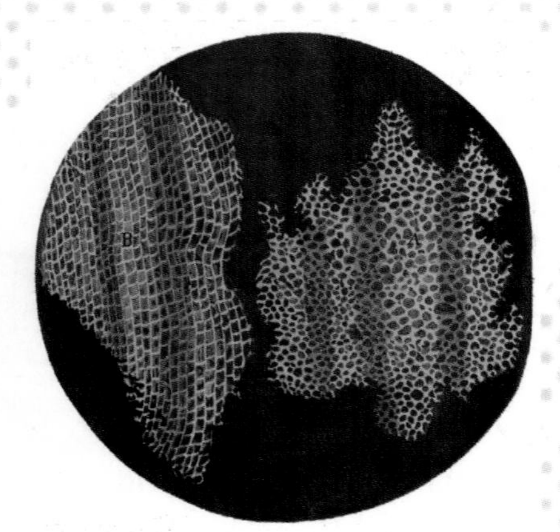

Figure 1 Micrograph showing cork cell section, as from Hooke's work

Investigating cell make-up

Cells are very small. Most are too small to be seen with the naked eye, and certainly not in any detail. In order to investigate cells, we need to be able to produce images that are both enlarged and more detailed. During the past 60 years, light microscopes have improved and electron microscopes have been developed. This has allowed scientists to study cells in detail. Other scientists have studied the chemical reactions going on in the different parts of cells. The results of these investigations have enabled us to understand how the structure of cell parts allows them to carry out their functions.

Magnification and resolution

In the dark, a car's headlights at some distance away appear as one light source – if you take a photograph it will also show one light source. You could enlarge the photograph many times but it would still show only one light source. It would be increasingly blurred as the **magnification** increased. This is because magnification on its own does not increase the level of detail seen.

The term **resolution** refers to the ability to see two distinct points separately. In the example above, as the car moves closer, the one light source you saw in the distance 'resolves' into two when the car is close enough for your eye to see the two headlights as separate points.

In order to investigate cells and their component parts, you need both high resolution and high magnification.

Key definitions

Magnification is the degree to which the size of an image is larger than the object itself. Numerically, it is the image size divided by the actual size of the object, measured using the same units. It is usually expressed as ×10, ×1.5, etc.

Resolution is the degree to which it is possible to distinguish between two objects that are very close together. The higher the resolution, the greater the detail you can see.

The light microscope

- Light microscopes use a number of lenses to produce an image that can be viewed directly at the eyepieces.
- Light passes from a bulb under the stage, through a condenser lens, then through the specimen.
- This beam of light is focused through the objective lens, then through the eyepiece lens.

- To view specimens at different magnifications, light microscopes have a number of objective lenses that can be rotated into position.
- Usually four objective lenses are present: ×4, ×10, ×40 and ×100. The ×100 objective is an oil immersion lens.
- The eyepiece lens then magnifies the image again. This is usually ×10.

The total magnification of any specimen is given by multiplying the objective magnification by the eyepiece magnification. In the example above, the microscope would be capable of producing images that are magnified by ×40, ×100, ×400 and ×1000.

If you change the eyepiece lens to a ×15 one, then the total magnification will be ×60, ×150, ×600 and ×1500.

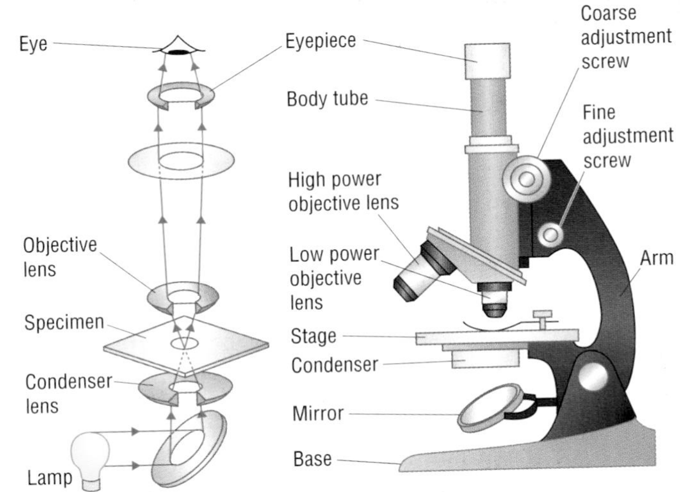

Figure 2 Light microscope showing lenses and light paths

Advantages and limitations of the light microscope

Magnification – Most light microscopes are capable of magnification up to a maximum of ×1500.

Resolution – The maximum resolving power using light is 200 nm. This means that if two objects are closer together than 200 nm, they will be seen as one object. This limit is due to the magnitude of the wavelength of light. Two objects can be distinguished only if light waves can pass between them.

Specimens – A wide range of specimens can be viewed using a light microscope. These include living organisms such as *Euglena* and *Daphnia*. You can also view thin sections of larger plants and animals, and smear preparations of blood or cheek cells.

The light microscope is used widely in education, laboratory analysis and research. But because it doesn't have high resolution, it can't give detailed information about internal cell structure.

Preparation of specimens for the light microscope

You can view some specimens directly. However, a lot of biological material is not coloured, so you can't see the details. Also, some material distorts when you try to cut it into thin sections.

Preparation of slides to overcome these problems involves the following steps:

1 Staining – Coloured stains are chemicals that bind to chemicals on or in the specimen. This allows the specimen to be seen. Some stains bind to specific cell structures. Acetic orcein stains DNA dark red. Gentian violet stains bacterial cell walls.

2 Sectioning – Specimens are embedded in wax. Thin sections are then cut without distorting the structure of the specimen. This is particularly useful for making sections of soft tissue, such as brain.

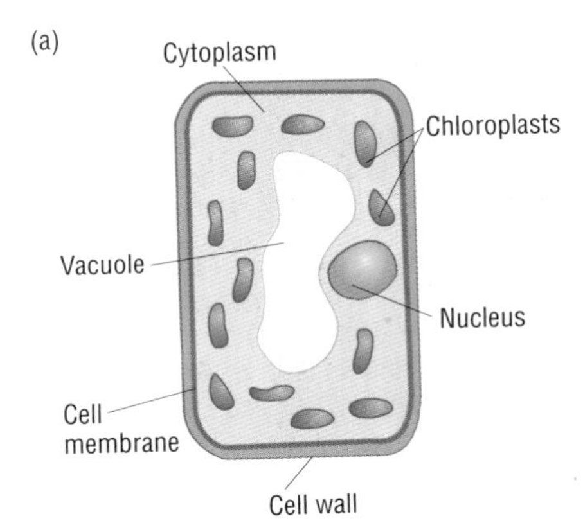

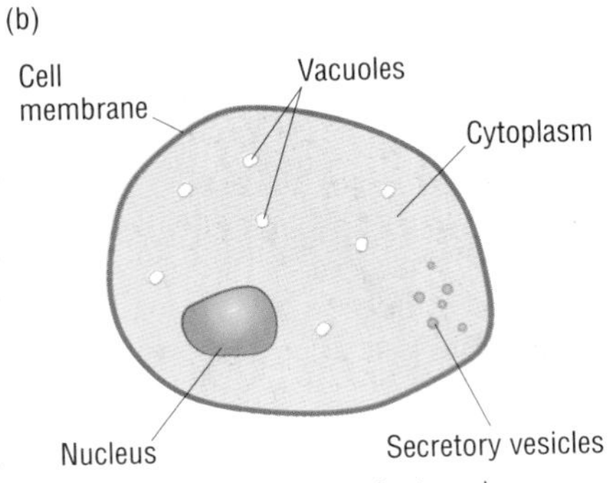

Figure 3 **a** Generalised plant and **b** animal cells as seen under the light microscope

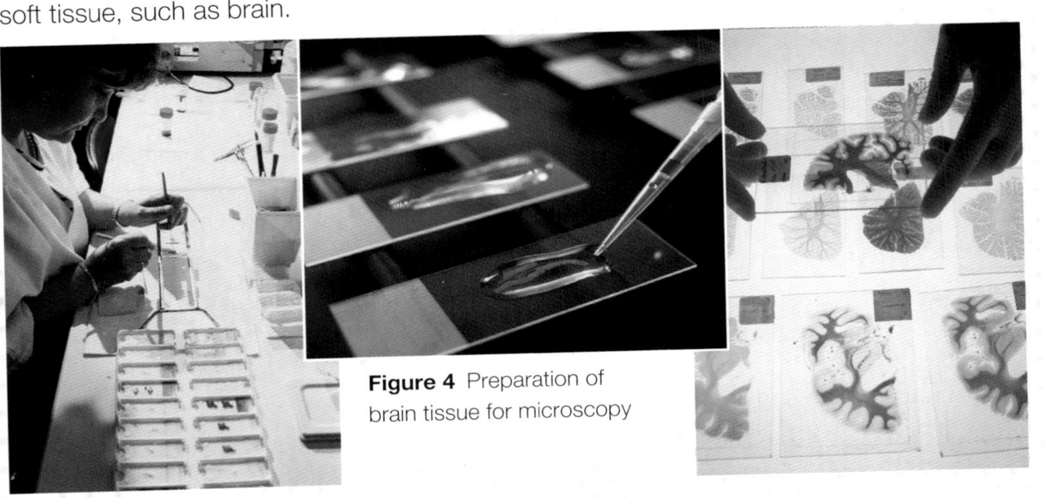

Figure 4 Preparation of brain tissue for microscopy

Questions

1 Why did Hooke use the term 'cell'?

2 Why do sections of tissue need to be cut into thin slices for examination under a microscope?

3 Suggest why light microscopes are so useful in biology.

1.1 How Science Works

2 Cell size and magnification

By the end of this spread, you should be able to . . .

* **Calculate the linear magnification of an image such as a photomicrograph or electron micrograph.**

Unit	Symbol	Equivalent in metres	Fraction of a metre
metre	m	1	One
decimetre	dm	0.1	one tenth
centimetre	cm	0.01	one hundredth
millimetre	mm	0.001	one thousandth
micrometre	μm	0.000 001	one millionth
nanometre	nm	0.000 000 001	one thousand millionth

Table 1 Relationships between units of measurement

1mm × 100,000 = converting to nm

Key definitions

A **micrometre** is equal to one millionth (10^{-6}) of a metre. It is the standard unit for measuring cell dimensions.

A **nanometre** is one thousandth (10^{-3}) of a micrometre. It is therefore one thousand millionth (10^{-9}) of a metre. It is a useful unit for measuring the sizes of organelles within cells and for measuring the size of large molecules.

Suitable units of measurement

Many structures that biologists need to study, such as cells and organelles, are very small, so magnified images of those structures have to be made, as described in spreads 1.1.1 and 1.1.3.

The metre

The international (SI) unit for length is the metre (m). People are familiar with this and can visualise it. It is divided into 1000 mm. However, cells and organelles are so small that these units are too big to be of any real use. Most animal cells are around 20–40 millionths of a metre (0.000 02–0.000 04 m) long, so when measuring cells the unit μm (micrometre) is used.

The micrometre

One mm is divided into 1000 equal divisions, each of one **micrometre** (μm). One μm is one millionth of a metre (10^{-6} m). Animal cells are usually 20–40 μm long.

The nanometre

Some biological structures are so small that even the μm unit is of little use.

One μm is divided into a thousand equal parts, each of one nm (**nanometre**). One nm is one thousand millionths of a metre. Cell surface membranes are about 10 nm wide and ribosomes are about 20 nm in diameter. Viruses are about 40–100 nm in diameter.

Limits of resolution

Resolution is the degree to which it is possible to distinguish clearly between two objects that are very close together (spread 1.1.1). It is the smallest distance apart that two separate objects can be seen clearly as two objects.

The resolution for the
- human eye is 100 μm.
- light microscope is 200 nm.
- electron microscope is 0.20 nm.

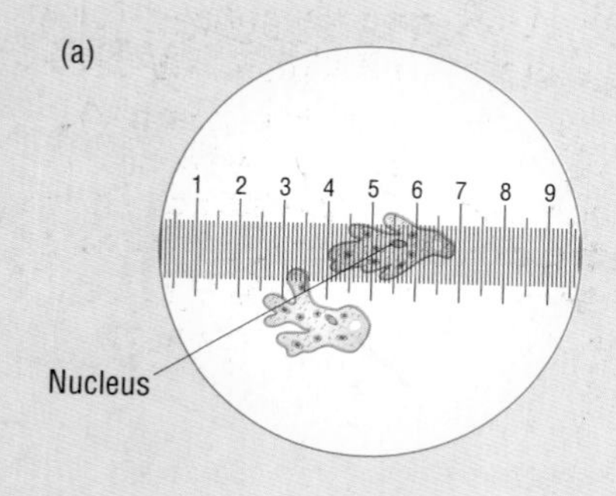

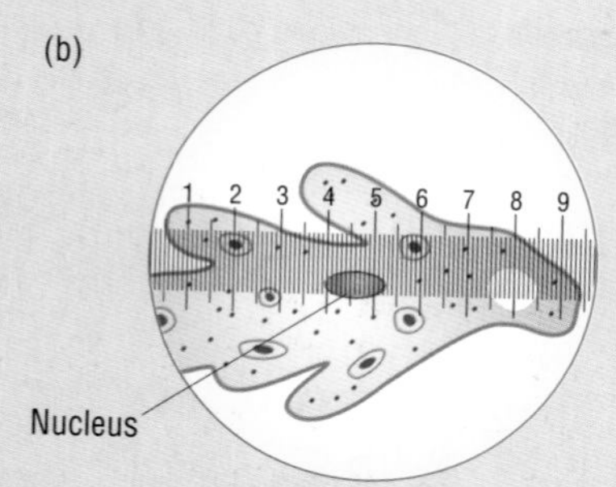

Figure 1 An amoeba seen under a light microscope **a** with a ×10 objective (total magnification = ×100) and **b** with a ×40 objective (total magnification ×400)

Measuring cells and organelles seen with a light microscope

- A microscope eyepiece can be fitted with a graticule.
- This is transparent with a small ruler etched on it.
- As the specimen is viewed the eyepiece graticule scale is superimposed on it and the dimensions of the specimen can be measured (just as you can measure a large object by placing a ruler against it) in eyepiece units (Figure 1).

The scale of the eyepiece graticule is arbitrary – it represents different lengths at different magnifications. The image of the specimen looks bigger at higher magnifications but the actual specimen has not increased in size. The eyepiece scale has to be calibrated (its value worked out) for each different objective lens.

Calibration of the eyepiece graticule

- A microscopic ruler on a special slide, called a *stage micrometer*, is placed on the microscope stage.
- This ruler is 1 mm long and divided into 100 divisions.
- Each division is 0.01 mm or 10 μm
- With a ×4 objective lens and ×10 eyepiece (magnification = ×40), 40 epu (eyepiece units) = 1 mm (1000 μm). Therefore 1 epu = 1000/40 = 25 μm.
- With a ×10 objective lens (total magnification = ×100), 100 epu = 1000 μm. So 1 epu = 10 μm.

For modern microscopes used in schools the values of the eyepiece divisions at different magnifications are:

Magnification of eyepiece lens	Magnification of objective lens	Total magnification	Value of one eyepiece division/μm
×10	×4	×40	25
×10	×10	×100	10
×10	×40	×400	2.5
×10	×100 (oil immersion lens)	×1000	1.0

Table 2 Values of eyepiece divisions at different magnifications

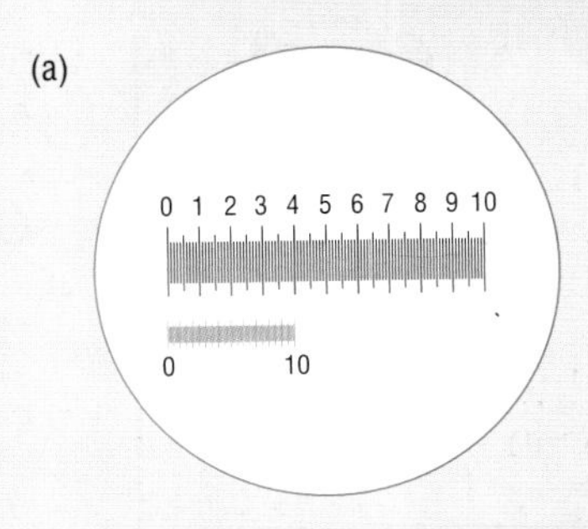

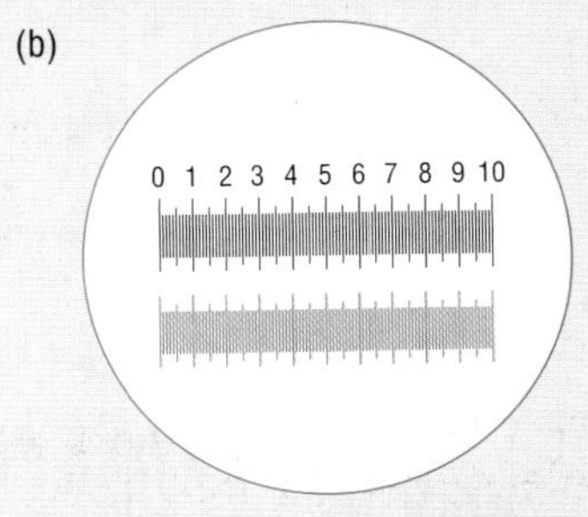

Figure 2 Eyepiece graticule and stage micrometer at **a** ×40 magnification and **b** ×100 magnification. (Note the spelling of stage *micrometer* – it is a measuring device not a unit of measurement)

Worked example

In Figure 1a the length of the nucleus is 3.2 epu.
At ×100 magnification, one epu = 10 μm
so the length of this nucleus is 3.2 × 10 = 32 μm.

In Figure 1b the length of the same nucleus is 13 epu.
At ×400 magnification, one epu = 2.5 μm
so the length of the nucleus is 13 × 2.5 = 32.5 μm.

Examiner tip

You must show working and units when answering exam questions about measurements. Methodically working through calculations means that even if the final answer is incorrect, you may receive some credit for showing the right idea.

Magnification and micrographs

There is a relationship between actual size, magnification and image size where:

actual size = image size/magnification.

This means that it is possible to work out either the magnification of a micrograph or drawing, or the actual size of the cell or part of the cell shown in the micrograph or drawing.

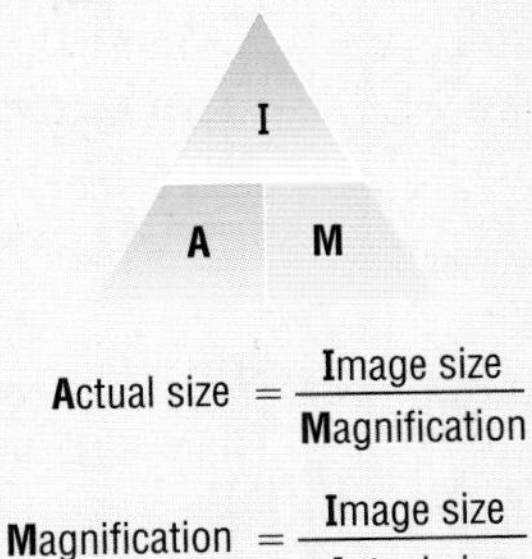

$$\text{Actual size} = \frac{\text{Image size}}{\text{Magnification}}$$

$$\text{Magnification} = \frac{\text{Image size}}{\text{Actual size}}$$

Figure 3 IMA triangle

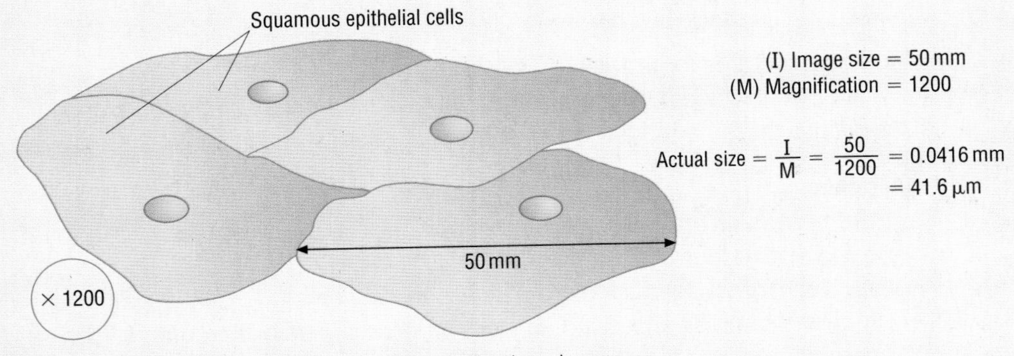

(I) Image size = 50 mm
(M) Magnification = 1200

$$\text{Actual size} = \frac{I}{M} = \frac{50}{1200} = 0.0416\ \text{mm} = 41.6\ \mu\text{m}$$

Figure 4 Calculations of actual size from magnification given

Questions

1. If a nucleus measures 100 mm on a diagram, with a magnification of ×10 000, what is the actual size of the nucleus?
2. Draw up a table to show each of the following measurements in metres (m), millimetres (mm) and micrometres (μm): 5 μm, 0.3 m, 23 mm, 75 μm.

1.1 (3) Electron microscopes and cell details

By the end of this spread, you should be able to . . .

* **State the resolution and magnification that can be achieved with the electron microscope.**
* **Explain the need for staining samples for use in electron microscopy.**

The use of electrons

Light microscopes have low **resolution**. This means that if the magnification is above ×1500, the image isn't clear. The wavelength of visible light ranges from about 400 to 750 nm, so structures closer together than 200 nm (0.2 μm) will appear as one object. We can achieve higher resolution with electron microscopes.

- Electron microscopes generate a beam of electrons.
- A beam of electrons has a wavelength of 0.004 nm, 100 000 times shorter than a light wavelength.
- Electron microscopes can distinguish between objects 0.2 nm apart.
- These microscopes use magnets (instead of lenses) to focus the beam of electrons onto a prepared specimen.
- Electrons are not visible to the human eye. The image produced from the electron beam is projected onto a screen or onto photographic paper to make a black-and-white image (sometimes referred to as a 'greyscale' image).
- Such images are called electron micrographs.
- The resolution of an electron microscope is about 500 000 times greater than that of the human eye.

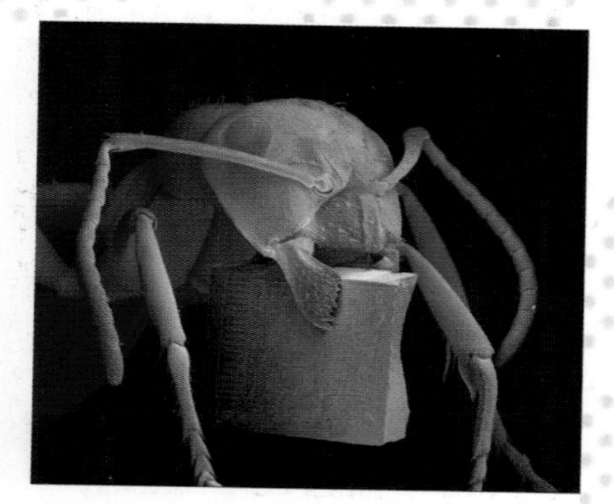

Figure 1 Electron micrograph of an ant grasping a microprocessor (×12)

Two types of electron microscope

Transmission electron microscope (TEM)

- The electron beam passes through a very thin prepared sample.
- Electrons pass through the denser parts of the sample less easily, so giving some contrast.
- The final image produced is two-dimensional.
- The magnification possible with a TEM is ×500 000.

Scanning electron microscope (SEM)

- The electron beam is directed onto a sample. The electrons don't pass through the specimen.
- They are 'bounced off' the sample.
- The final image produced is a 3D view of the surface of the sample.
- The magnification possible with an SEM is about ×100 000.

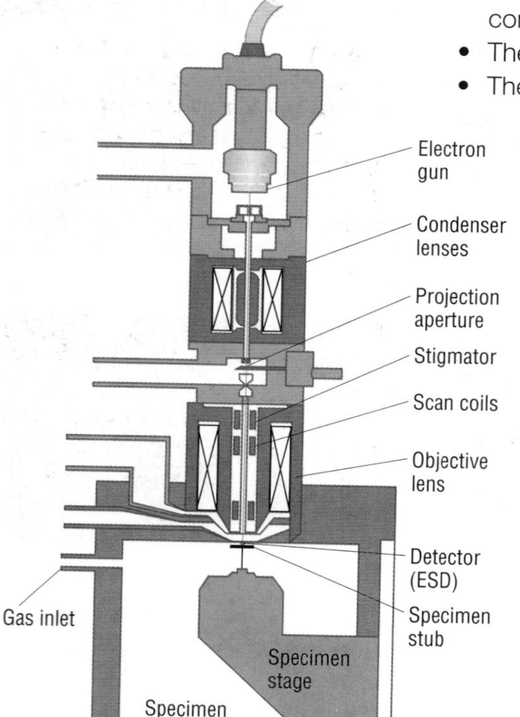

Figure 2 Electron microscope in outline

Advantages and limitations of the electron microscope

Advantages

- The resolution is 0.1 nm (2000× more than in the light microscope).
- This means the electron microscope can be used to produce detailed images of the structures (organelles) inside cells.
- The SEM produces 3D images that can reveal the detail of contours and cellular or tissue arrangements – this is not possible using light microscopes.

Limitations

- Electron beams are deflected by the molecules in air, so samples have to be placed in a vacuum.
- Electron microscopes are extremely expensive items.
- Preparing samples and using an electron microscope both require a high degree of skill and training.

Use of electron microscopes

Preparing specimens to be viewed in an electron microscope is complicated. For example, to prepare a thin section of liver tissue, you might need to:

- fix the specimen in glutaraldehyde to make the tissue firm
- dehydrate it to replace the water with ethanol
- embed the dehydrated tissue in a solid resin
- cut very thin slices using a diamond knife
- stain it using lead salts to scatter electrons differently – this gives contrast
- mount it on a copper grid
- place the specimen on the grid in a vacuum in the microscope.

When electron microscopes were first developed, many scientists thought that the complicated preparation of specimens produced artefacts. Artefacts are structures that result from the preparation process. They are not true representations of the specimen's original structure.

Even in the 1980s, scientists were arguing about the true value of electron microscopes, but their technology and quality has improved in recent years. Scientists have also used X-ray crystallography to investigate cell structure. This has confirmed that electron microscopes do give a true insight into the internal workings of cells.

Figure 3 Preparation and viewing of samples for scanning electron microscopy

Coloured electron micrographs

Electron micrographs are sometimes shown in colour. The final image produced from an electron microscope is always black, white and grey; the colours are added afterwards using specialised computer software. Such images will be labelled as 'false-colour' electron micrographs.

Figure 4 False-colour electron micrographs showing pollen grains and red blood cells

Key definition

Staining in microscopy refers to any process that helps to reveal or distinguish different features. In light microscopy, stains may be colours or fluorescent dyes. In electron microscopy, they are metal particles or metal salts.

Examiner tip

You may be asked to calculate the actual size of a specimen. Measure the size of the image in mm and then convert it to μm. Then divide it by the magnification.

Questions

1. List the similarities and differences between light and electron microscopes.
2. Explain why both light and electron microscopes are used widely in biology.
3. Describe the main limitations of the electron microscope.
4. Describe the limitations of the light microscope.

1.1 (4) Cells and living processes

By the end of this spread, you should be able to . . .

* **Explain the importance of the cytoskeleton in providing mechanical strength to cells, aiding transport within cells and enabling cell movement.**
* **Recognise the structures undulipodia (flagella) and cilia, and outline their functions.**
* **Compare and contrast the structure and ultrastructure of plant cells and animal cells.**

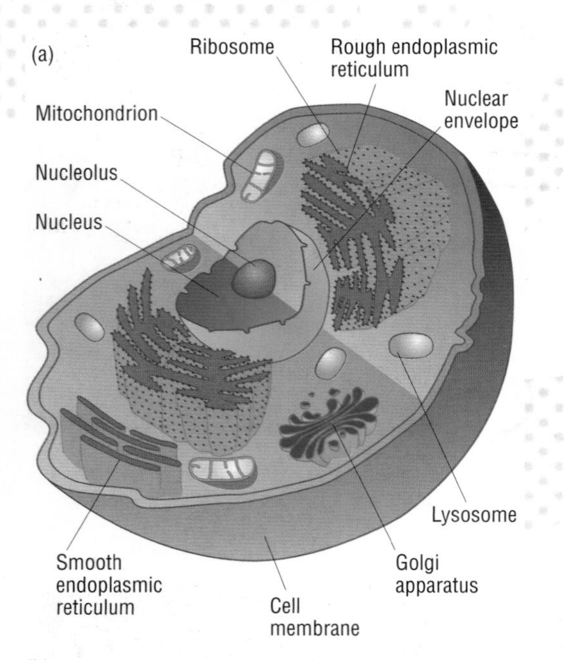

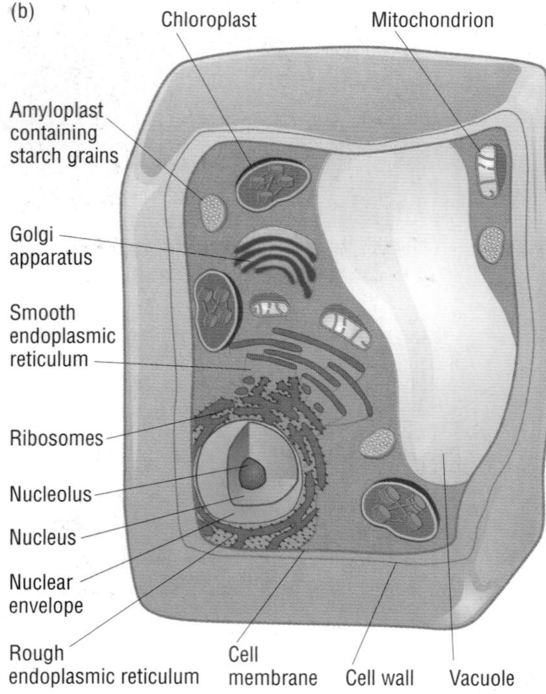

Figure 1 Generalised **a** animal cell; **b** plant cell under the EM

Characteristics of living things

All living organisms are said to share seven characteristics – movement, respiration, sensitivity, nutrition, excretion, reproduction and growth. (Some bacteria do not respire, but use other chemical reactions to obtain energy.)

It is easy to see growth in plants or reproduction in animals. But many organisms consist of only a single living cell. Such cells exhibit all the seven characteristics of living things. Living cells contain special structures that each carry out a particular function.

Organelles and ultrastructure

When you look at a cell under the light microscope, the most obvious feature you will see is the very large nucleus. You may also see other structures, such as chloroplasts and large vacuoles in plant cells. These structures are organelles. Using an electron microscope, it is possible to identify a range of organelles in plant and animal cells. The detail of the inside of cells, as revealed by the electron microscope, is termed the cell's **ultrastructure** (sometimes also called fine structure).

Division of labour

Most organelles are found in both plant and animal cells. They have the same functions in each type of cell. Each type of organelle has a specific role within the cell. This is called **division of labour**. The different organelles work together in a cell, each contributing its part to the survival of the cell.

Movement and stability in cells

Cytoskeleton

Cells contain a network of fibres made of **protein**. These fibres keep the cell's shape stable by providing an internal framework called the cytoskeleton.

Some of the fibres (called actin filaments) are like the fibres found in muscle cells. They are able to move against each other. These fibres cause the movement seen in some white blood cells. They also move some organelles around inside cells.

There are other fibres known as **microtubules**. These are cylinders about 25 nm in diameter. They are made of a protein called tubulin. Microtubules may be used to move a microorganism through a liquid, or to waft a liquid past the cell. Other proteins present on the microtubules move organelles and other cell contents along the fibres. This is how **chromosomes** are moved during **mitosis** (see spread 1.1.7). It is how **vesicles** move from the **endoplasmic reticulum** to the **Golgi apparatus** (see spreads 1.1.5 and 1.1.6). These proteins are known as **microtubule motors**. They use **ATP** to drive these movements.

Flagella (undulipodia) and cilia

In eukaryotes (organisms that have cells with nuclei) flagella – correctly called undulipodia – and cilia are structurally the same. They are hair-like extensions that stick out from the surface of cells. Each one is made up of a cylinder that contains nine microtubules arranged in a circle. There are also two microtubules in a central bundle. Undulipodia are longer than cilia.

The undulipodium that forms the tail of a sperm cell can move the whole cell. The long, whip-like undulipodium on the protoctist (see spread 2.3.6) *Trichomonas* (which causes urinary tract infections) enables it to move. In ciliated epithelial tissue, the sweeping movements of the cilia move substances such as mucus across the surface of the cells.

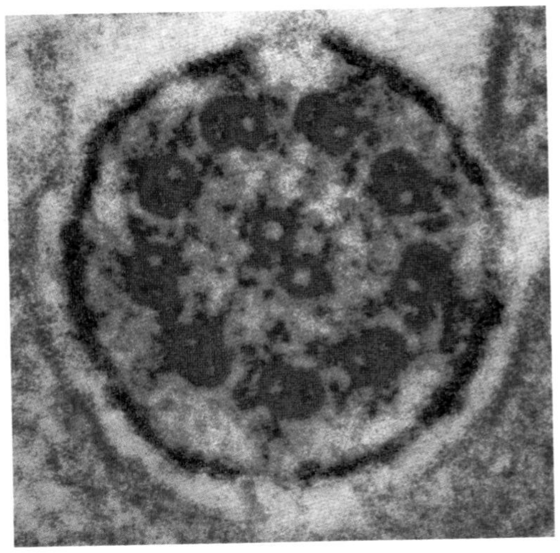

Figure 2 A transmission electron micrograph (TEM) of a cross-section through a cilium, showing the 9+2 arrangement of microtubules

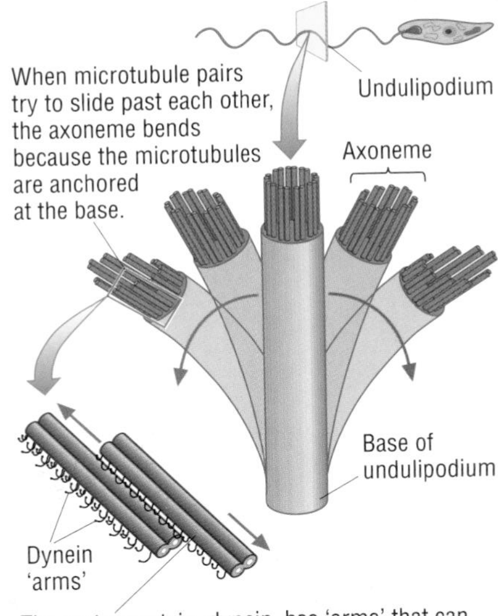

Figure 3 Cilia microtubules and movements

Undulipodia and cilia can move because the microtubules can use energy from ATP.

Cilia and undulipodia were first seen under the light microscope. Cilia are short (less than 10 μm long) and undulipodia are long. Scientists at the time didn't know that they had the same internal structure, so they gave them different names.

- Undulipodia usually occur in ones or twos on a cell.
- Cilia often occur in large numbers on a cell.

Some bacteria have flagella. These look like eukaryotic undulipodia, but their internal structure is very different. These are true 'motors' – they are made of a spiral of protein (called flagellin) attached by a hook to a protein disc at the base. Using energy from ATP, the disc rotates, spinning the flagellum.

Vesicles and vacuoles

- Vesicles are membrane-bound sacs found in cells. They are used to carry many different substances around cells.
- In plant cells, the large cell vacuole maintains cell stability. It is filled with water and solutes so that it pushes the cytoplasm against the cell wall, making the cell **turgid**. If all the plant cells are turgid, this helps to support the plant. This is especially important in non-woody plants.

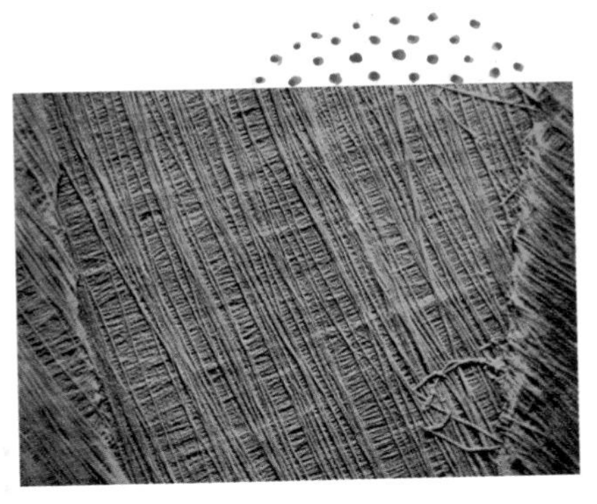

Figure 4 Cell wall structure showing bundles of cellulose fibres

Plant cell walls

- These are on the outside of the plant cell plasma membranes (cell surface membranes).
- Plant cell walls are made of cellulose, a carbohydrate polymer made up of glucose subunits.
- The cellulose forms a sieve-like network of strands that makes the wall strong.
- Because it is held rigid by the pressure of the fluid inside the cell (turgor pressure), it supports the cell and so helps to support the whole plant.

Key definition

The **cytoskeleton** refers to the network of protein fibres found within cells that gives structure and shape to the cell, and also moves organelles around inside cells.

Examiner tip

Remember that movements of cilia outside cells, and of organelles inside cells, require energy input from the cell. Stating that the energy is derived from ATP is often worth marks.

Questions

1 Define the term 'division of labour'.

2 List the characteristics of living things and describe how the organelles allow cells to show those characteristics (you may need to use this spread and later pages).

3 Suggest why:
(a) chloroplasts are moved around plant cells
(b) white blood cells need to be able to move.

1.1 (5) Organelles – structure and function

By the end of this spread, you should be able to . . .

* **Recognise structures as seen under the electron microscope, e.g. nucleus, nucleolus, nuclear envelope, rough and smooth endoplasmic reticulum, Golgi apparatus, ribosomes, mitochondria, lysosomes and chloroplasts.**
* **Outline the functions of these structures.**

Organelles surrounded by membranes

Many organelles are membrane-bound. They form separate compartments within the cell.

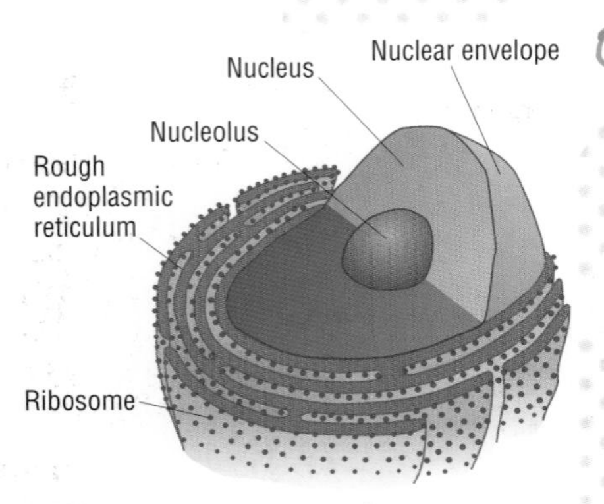

Figure 1 Nucleus and endoplasmic reticulum

The nucleus

Structure	Function
The nucleus is the largest organelle. When stained, it shows darkened patches known as chromatin. It is surrounded by a nuclear envelope. This is a structure made of two membranes with fluid between them. A lot of holes, called *nuclear pores*, go right through the envelope. These holes are large enough for relatively large molecules to pass through. There is a dense, spherical structure, called the nucleolus, inside the nucleus.	The nucleus houses nearly all the cell's genetic material. The chromatin consists of DNA and proteins. It has the instructions for making proteins. Some of these proteins regulate the cell's activities. When cells divide, chromatin condenses into visible chromosomes (see spread 1.1.14). The nucleolus makes RNA and ribosomes. These pass into the cytoplasm and proteins are assembled at them.

Endoplasmic reticulum (ER)

Structure	Function
ER consists of a series of flattened, membrane-bound sacs called *cisternae*. They are continuous with the outer nuclear membrane. *Rough endoplasmic reticulum* is studded with ribosomes. *Smooth endoplasmic reticulum* does not have ribosomes.	Rough ER transports proteins that were made on the attached ribosomes. Some of these proteins may be secreted from the cell. Some will be placed on the cell surface membrane. Smooth ER is involved in making the lipids that the cell needs.

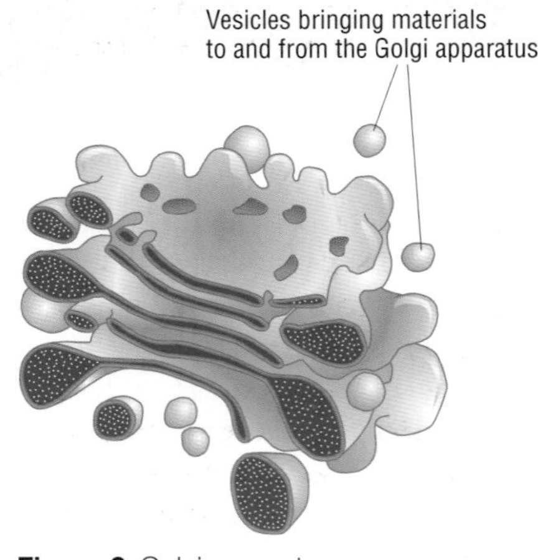

Figure 2 Golgi apparatus

Golgi apparatus

Structure	Function
A stack of membrane-bound, flattened sacs. (It looks a bit like a pile of pitta bread.)	Golgi apparatus receives proteins from the ER and modifies them. It may add sugar molecules to them. The Golgi apparatus then packages the modified proteins into vesicles that can be transported. Some modified proteins may go to the surface of the cell so that they may be secreted.

Mitochondria (singular: mitochondrion)

Structure	Function
These may be spherical or sausage-shaped. They have two membranes separated by a fluid-filled space. The inner membrane is highly folded to form **cristae**. The central part of the mitochondrion is called the **matrix**.	Mitochondria are the site where **adenosine triphosphate (ATP)** is produced during respiration. ATP is sometimes called the universal energy carrier because almost all activities that need energy in the cell are driven by the energy released from ATP.

Chloroplasts

Structure	Function
These are found only in plant cells and the cells of some protoctists. Chloroplasts also have two membranes separated by a fluid-filled space. The inner membrane is continuous, with an elaborate network of flattened membrane sacs called **thylakoids**. These look like piles of plates. A stack of thylakoids is called a granum (plural: *grana*). Chlorophyll molecules are present on the thylakoid membranes and in the intergranal membranes.	Chloroplasts are the site of photosynthesis in plant cells. Light energy is used to drive the reactions of photosynthesis, in which **carbohydrate molecules** are made from carbon dioxide and water.

Lysosomes

Structure	Function
These are spherical sacs surrounded by a single membrane.	Lysosomes contain powerful digestive enzymes. Their role is to break down materials. For example, white blood cell lysosomes help to break down invading microorganisms; the specialised lysosome (acrosome) in the head of a sperm cell helps it penetrate the egg by breaking down the material surrounding the egg.

Organelles without membranes surrounding them

Ribosomes

Structure	Function
These are tiny organelles. Some are in cytoplasm and some are bound to ER. Each ribosome consists of two subunits.	Ribosomes are the site of protein synthesis in the cell (where new proteins are made). They act as an assembly line where coded information (mRNA) from the nucleus is used to assemble proteins from amino acids.

Centrioles

Structure	Function
These are small tubes of protein fibres (microtubules). There is a pair of them next to the nucleus in animal cells and in the cells of some protoctists.	Centrioles take part in cell division. They form fibres, known as the **spindle**, which move chromosomes during nuclear division (see spread 1.1.4).

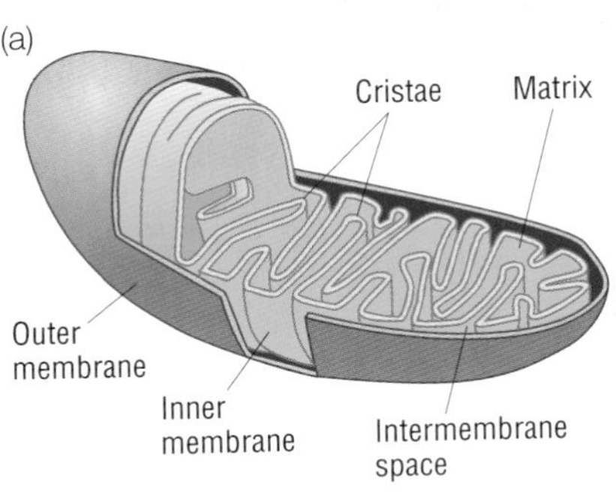

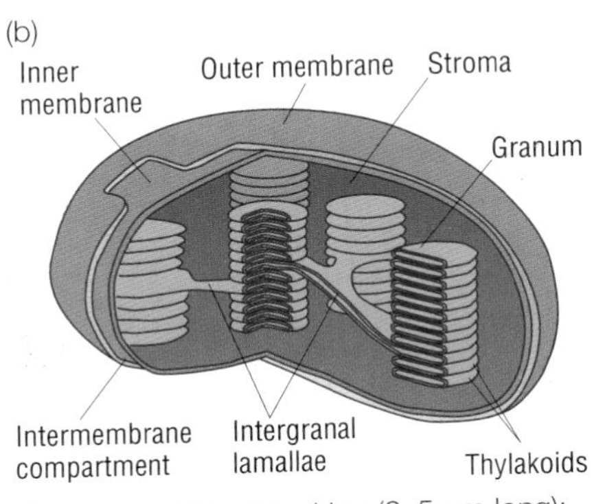

Figure 3 **a** Mitochondrion (2–5 μm long); **b** chloroplast (4–10 μm long)

Key definition

The term **organelle** refers to a particular structure of a cell that has a specialised function. Some organelles are membrane-bound, others are not. All perform a particular role in the life processes of the cell.

Examiner tip

It is best to use the term **nuclear envelope**, not 'nuclear membrane'.

Questions

1. List all the cell organelles above in order of their size.
2. Which cells in a plant are the ones that contain most chloroplasts?
3. Suggest why muscle cells contain a lot of mitochondria, whereas most fat storage cells do not.
4. What type of cells in plants and animals would contain extensive Golgi apparatus?

Organelles at work

By the end of this spread, you should be able to . . .

* **Outline the interrelationship between the organelles involved in the production and secretion of proteins.**
* **Compare and contrast the structure of prokaryotic cells and eukaryotic cells.**

Division of labour

The synthesis of a protein illustrates division of labour within a cell.

Some of the cells within some organisms produce hormones. Hormones are chemical messenger molecules that help to coordinate the activities of the whole organism, including auxins in plants, and insulin and growth hormone in animals. Plant 'hormones' are more usually referred to as *plant growth regulators* or *plant growth substances*.

- The instructions to make the hormone are in the **DNA** in the nucleus.
- The specific instruction to make the hormone is known as the **gene** for that hormone. A gene is on a **chromosome**.
- The nucleus copies the instructions in the DNA into a molecule called **mRNA**.
- The mRNA molecule leaves the nucleus through a nuclear pore and attaches to a **ribosome**. In this case, the ribosome is attached to rough **endoplasmic reticulum** (ER).
- The ribosome reads the instructions and uses the codes to assemble the hormone (protein).
- The assembled protein inside the rough ER is pinched off in a **vesicle** and transported to the Golgi apparatus.
- The Golgi apparatus packages the protein and may also modify it so that it is ready for release. The protein is now packaged into a vesicle and moved to the cell surface membrane, where it is secreted outside.

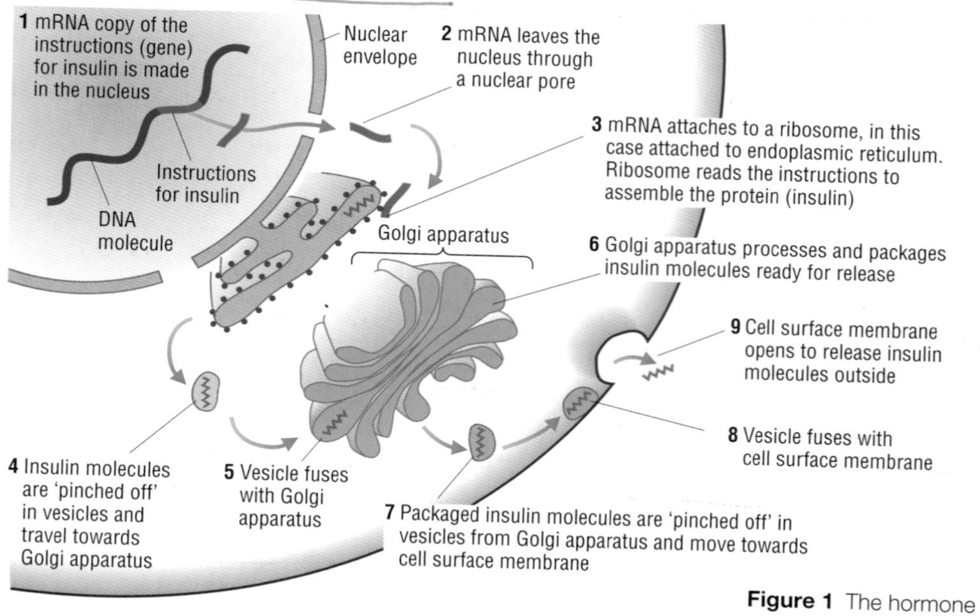

Figure 1 The hormone production sequence

Prokaryotes and eukaryotes

Eukaryotic cells have a nucleus

The cells described so far all contain organelles, some of which are bound by membranes. This gives these cells a complicated internal structure, where each organelle performs a specific role. Such cells are called **eukaryotic cells**. Eukaryotic means *having a true nucleus*. All organisms except prokaryotes have this cell structure.

Prokaryotic cells do not have a nucleus

Cells of the kingdom *Prokaryotae* are very different. They are bacteria and, at 1–5 μm, are much smaller than eukaryotic cells.

It is important to remember that **prokaryotic cells** show all the characteristics of living organisms.

Their features include:

- They have only one membrane, the cell surface membrane around the outside (they do not contain any membrane-bound organelles such as mitochondria and chloroplasts).
- They are surrounded by a cell wall, but this is usually made of peptidoglycan (also called murein), not cellulose.
- They contain ribosomes that are smaller than eukaryotic ribosomes.
- Their DNA is in the cytoplasm in the form of a single loop sometimes called a 'circular chromosome' or 'bacterial chromosome', unlike the linear chromosomes (separate strands) of eukaryotes. Many prokaryotic cells also contain very small loops of DNA called plasmids.
- The DNA is not surrounded by a membrane, as in the nuclear envelope of eukaryote cells. The general area in which the DNA lies is called the *nucleoid*.
- ATP production takes place in specialised infolded regions of the cell surface membrane called *mesosomes*.
- Some prokaryotic cells have **flagella**. These function like eukaryotic undulipodia, but they have a different internal structure.

Examiner tip

If you are comparing eukaryote and prokaryote cells and you refer to the cell walls, you must describe what each type of cell wall is made of.

Prokaryotes and disease

Some prokaryotic (or bacterial) cells are well known because of the diseases they cause. Some strains of bacteria are resistant to antibiotics. MRSA (methicillin-resistant *Staphylococcus aureus*) is one such strain. These resistant strains can cause problems because the resistance is coded on plasmid DNA. Bacteria can share plasmids with each other, so they can pass resistance between cells. They can also pass on resistance to daughter cells during binary fission.

A great deal of research goes on into looking for new antibiotics that are capable of stopping the growth of bacterial cells, while at the same time not causing too much damage to eukaryotic cells.

Prokaryotes that help

Many prokaryotic species are important to humans:

- The food industry uses particular bacterial species, for example for cheese and yoghurt.
- In mammalian intestines, bacterial cells help with vitamin K production and help digest some foods.
- Skin is covered with a 'normal flora' of bacteria. These help to prevent harmful microorganisms getting into the body.
- Sewage treatment and natural recycling rely on bacterial cells digesting and respiring dead and waste material.

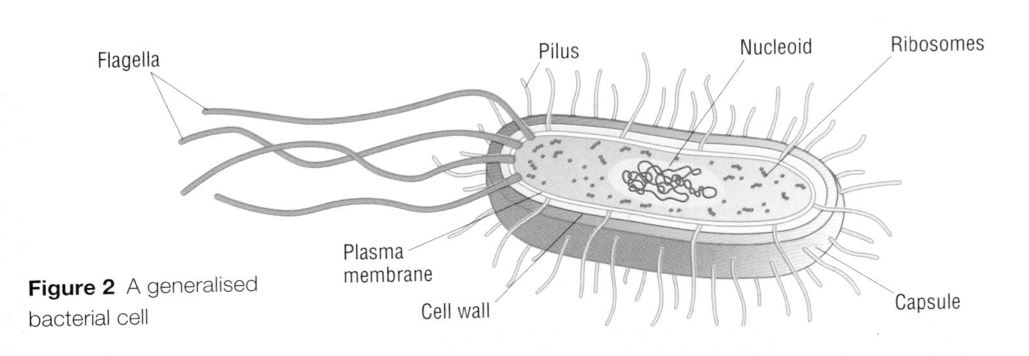

Figure 2 A generalised bacterial cell

Questions

1. List all the features that prokaryotic and eukaryotic cells have in common.
2. List the differences between prokaryotic and eukaryotic cells.
3. Mitochondria and chloroplasts contain small loops of DNA. They also contain ribosomes that are the same size as prokaryotic ribosomes. Suggest an explanation for these features.

1.1 (7) Biological membranes – fluid boundaries

By the end of this spread, you should be able to . . .

* **State that plasma (cell surface) membranes are partially permeable barriers.**
* **Outline the roles of membranes within cells and at the surface of cells.**

Key definition

The **phospholipid** bilayer is the basic structural component of plasma membranes (cell surface membranes). It consists of two layers of phospholipid molecules. Proteins are embedded in this layer.

Surrounding all cells, separating the cell contents from the outside world, are membranes. In **eukaryotic cells**, membranes are also found around many organelles. As well as separating the cell components from each other, membranes carry out a number of different functions. The basic structure of all cell surface membranes is the same.

The roles of membranes

The major roles of membranes include:

- separating cell contents from the outside environment
- separating cell components from cytoplasm
- cell recognition and signalling
- holding the components of some metabolic pathways in place
- regulating the transport of materials into or out of cells.

Examiner tip

Do not use the term *semi-permeable* when referring to cell membranes that allow water and other solutes to cross them. Refer to them as *partially permeable.*

The nature of phospholipids

To understand how membranes perform their various functions, you need to understand phospholipids. The basic structure of a phospholipid molecule is shown in the diagram. The phosphate 'head' is **hydrophilic** – water-loving – while the two fatty acid 'tails' are **hydrophobic** – water-hating. These properties come from the way charges are distributed across the molecule.

Molecules with charges that are evenly distributed around the molecule do not easily dissolve or mix with water, and in fact repel water molecules. You can see this if you try to mix oil and water. The oil quickly separates from the water, forming a layer on the surface, which can easily be skimmed off. Molecules with more unevenly distributed charges can interact with water molecules quite easily.

If phospholipid molecules are mixed with water, they form a layer at the water surface. The phosphate heads stick into the water, while the fatty acid tails stick up out of the water.

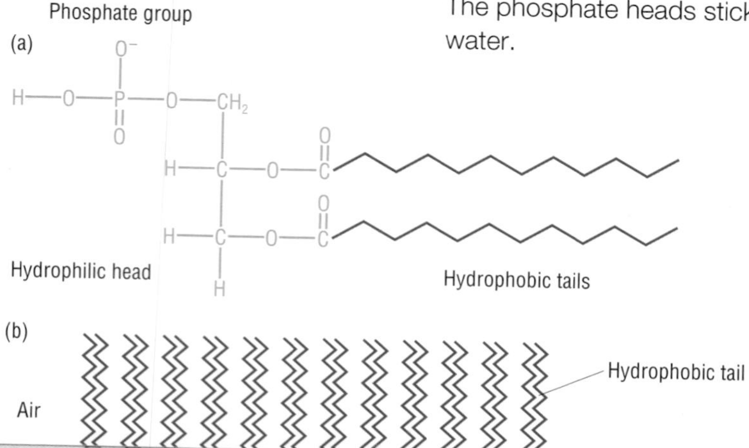

Figure 1 a Phospholipid molecule enlarged and **b** layer of phospholipids at the surface of water

Layers and bilayers

If phospholipid molecules are completely surrounded by water, a bilayer can form. Phosphate heads on each side of the bilayer stick into the water, while the hydrophobic fatty acid tails point towards each other in a sort of mirror image.

This means the hydrophobic tails are held away from the water molecules. In this state, the phospholipid molecules can move freely, just as fluid molecules do, within the plane of the membrane. Very rarely some may 'flip-flop' from one monolayer to the other. The hydrophilic head group cannot easily pass through the hydrophobic region in the middle of the bilayer. This gives the bilayer some stability, despite the fact that the phospholipid molecules are not actually bonded together.

All membranes are basically the same

The phospholipid bilayer is the basic structural component of all biological membranes. Essentially, the hydrophobic layer formed by the phospholipid tails creates a barrier to many molecules and separates the cell contents from the outside world. This thin layer of oil is ideal as a boundary in living systems, where most metabolic reactions take place in a water-based environment.

You cannot see membranes through the light microscope, but they can be seen using an electron microscope. Here the membranes appear as two dark 'tramlines' (the phospholipid heads) separated by a pale region (the fatty acid tails). The electron microscope has revealed that membranes are about 7–10 nm thick.

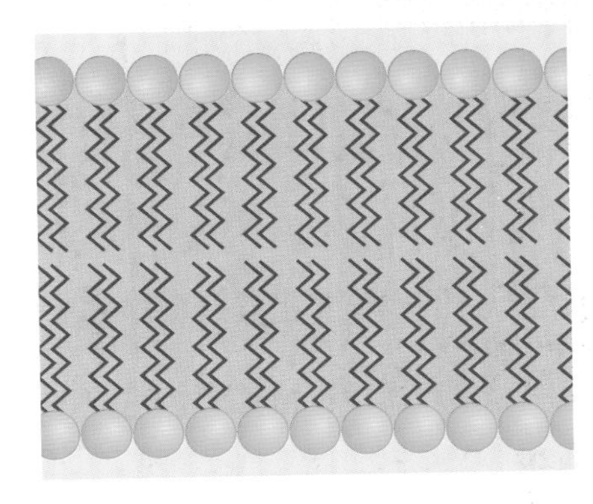

Figure 2 A phospholipid bilayer

Performing the different roles

A simple phospholipid bilayer would be incapable of performing all the functions of biological membranes. It would also be too fragile to function as a barrier within or around cells. Other components are needed in order to make a functioning biological membrane.

The number and type of these other components varies according to the function of the particular membrane. This specialisation of cell membranes is a part of the process of **differentiation** (see spread 1.1.16).

Some examples are listed below.

- The plasma membranes (cell surface membranes) of the cells in a growing shoot contain receptors that allow them to detect the molecules that regulate growth.
- Muscle cell membranes contain a large number of the channels that allow rapid uptake of glucose to provide energy for muscle contraction.
- The internal membranes of chloroplasts contain chlorophyll and other molecules needed for photosynthesis.
- The plasma membranes of white blood cells contain special proteins that enable the cells to recognise foreign cells and particles.

In a multicellular organism, this means the cell membranes of specialised cells have different properties, as do the membranes of specific organelles.

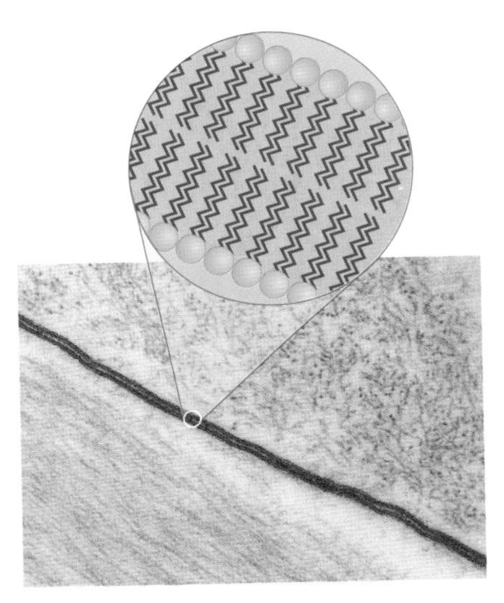

Figure 3 A phospholipid bilayer seen under the electron microscope

Permeability

All membranes are permeable to water molecules because water molecules can diffuse through the lipid bilayer. Some membranes are up to 1000 times more permeable to water because they contain aquaporins (protein channels that allow water molecules through them). Cell membranes that are permeable to water and some solutes are described as **partially permeable membranes**.

Questions

1 Why can phospholipid molecules in a bilayer move only in the plane of the bilayer?
2 Why do muscle cells need to be able to take up glucose rapidly?
3 Why do we describe cell membranes as partially permeable rather than semi-permeable?

1.1 (8) The fluid mosaic model

By the end of this spread, you should be able to . . .

* **Describe the fluid mosaic model of membrane structure.**
* **Describe the roles of the components of the cell membrane, including phospholipids, cholesterol, glycolipids, proteins and glycoproteins.**
* **Outline the effects of changing temperature on membrane structure and permeability.**

Key definition

Fluid mosaic refers to the model of cell membrane structure. The lipid molecules give fluidity and proteins in the membrane give it a mosaic (patchwork) appearance. The molecules can move about.

The fluid mosaic model

In 1972 Singer and Nicholson proposed a model of how all the components found in cell membranes might be arranged to form a biological membrane. Their model revised the one proposed by Davson and Danielli some years before, and is now widely accepted as describing the way biological membranes are formed and function.

The term **fluid mosaic** is used to describe the molecular arrangements in membranes. The main features of the fluid mosaic model are:

- a bilayer of **phospholipid** molecules forming the basic structure
- various protein molecules floating in the phospholipid bilayer, some completely freely, some bound to other components or to structures within the cell
- some (extrinsic) proteins partially embedded in the bilayer on the inside or the outside face; other (intrinsic) proteins completely spanning the bilayer.

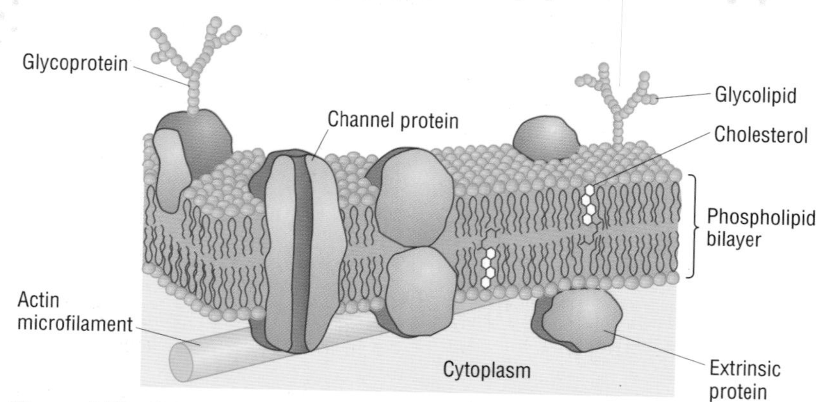

Figure 1 The fluid mosaic model

Glycoproteins and glycolipids

Some of the phospholipid molecules making up the bilayer, and some of the proteins found in the membrane, also have small carbohydrate parts attached to them. Where phospholipid molecules have a carbohydrate part attached they are called **glycolipids**. Where protein molecules have a carbohydrate part attached they are called **glycoproteins**.

Membrane components and their roles

Membrane stability and fluidity

- **Cholesterol** gives the membranes of some **eukaryotic cells** mechanical stability. This steroid molecule fits between fatty acid tails and helps make the barrier more complete, so substances like water molecules and ions cannot pass easily and directly through the membrane.

Membrane transport functions

- **Channel proteins** allow the movement of some substances across the membrane. Molecules of sugars such as glucose are too large and too **hydrophilic** (water-loving) to pass directly through the phospholipid bilayer. Instead, they enter and leave cells using these protein channels.
- **Carrier proteins** actively move some substances across the membrane. In plant roots, for example, magnesium ions are actively pumped (using ATP energy) into the root hair cells from the surrounding soil, so that the plant ensures a supply of magnesium for the manufacture of chlorophyll. When mineral ions are actively transported into root hair cells, they lower the water potential of those cells. This makes water enter by osmosis. Nitrate ions are actively transported into xylem vessels to lower the water potential and cause water uptake from the surrounding root cells.

Recognition and communication

- **Receptor sites** Some allow hormones to bind with the cell so that a cell 'response' can be carried out. A cell can respond to a hormone only if it has a receptor for that hormone on its cell surface membrane. Cell membrane receptors are also important in allowing drugs to bind, and so affect cell metabolism.
- Glycoproteins and glycolipids may be involved in cells signalling that they are 'self', to allow recognition by the immune system. Glycoproteins can also bind cells together in tissues. Note that some hormone receptors are glycoprotein and some are glycolipid.

Metabolic processes

- **Enzymes** and **coenzymes**. Some reactions in **photosynthesis** take place in membranes inside chloroplasts. Some stages of **respiration** take place in membranes of mitochondria. Enzymes and coenzymes may be bound to these membranes. The more membrane there is, the more enzymes and coenzymes it can hold. This helps to explain why mitochondrial inner membranes are folded to form **cristae**, and why chloroplasts contain many stacks of membranes called **thylakoids**.

Membranes and temperature

Increasing temperature gives molecules more **kinetic energy**, so they move faster. This increased movement of phospholipids and other components makes membranes leaky, which allows substances that would normally not do so to enter or leave the cell.

A good illustration of this feature uses tissue samples from beetroot. Sections of beetroot tissue exposed to increasing temperatures release more and more of the red pigment found in the cells, as the cell surface membrane and the vacuole membrane are damaged by heat.

Organisms that live in very hot or very cold environments need differently adapted molecular components of their membranes, for example the cholesterol content, so that their membranes can perform the functions needed to maintain life.

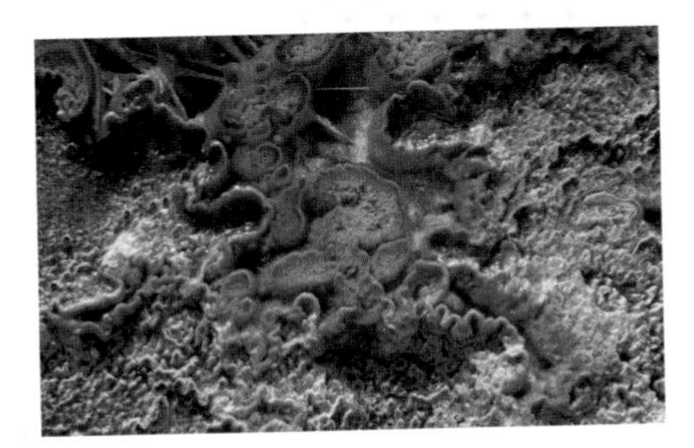

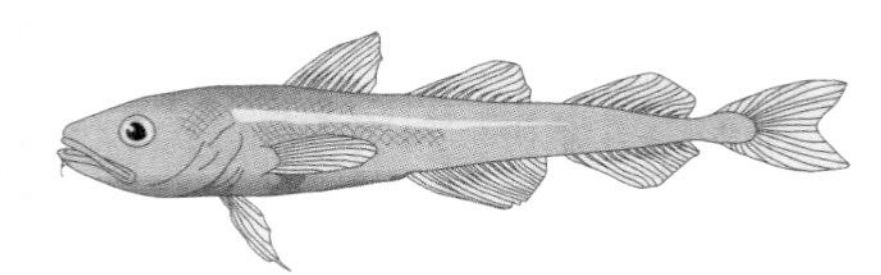

Figure 2 Top: Hot springs bacteria can tolerate temperatures of 80°C; bottom: Arctic cod live in waters that never rise above 0°C.

Questions

1 What types of molecule are likely to be involved in:
   (a) cell signalling and recognition?
   (b) allowing small charged molecules to pass through the cell membrane?
   (c) driving metabolic reactions?
2 If a protein spans the membrane, what property must the part of the protein embedded in the central part of the bilayer have?
3 What would happen to the beetroot cell membranes described above if the temperature continued to increase?

9 Communication and cell signalling

By the end of this spread, you should be able to . . .

* **Explain the term 'cell signalling'.**
* **Explain the role of membrane-bound receptors as sites where hormones and drugs can bind.**

Key definition

Cell signalling: cells communicate with one another by signals. Many molecules act as signals – some signal during processes taking place inside cells; others signal from one cell to others. Cytokines are an example of cell signals.

Examiner tip

Signal molecules fit into receptor molecules on cell surface membranes. This is because their shapes are **complementary** to each other. *Don't* say they have 'the same' shape as each other.

Sensitivity and response

One of the characteristics of living organisms is that they can sense what is present in their environment and respond appropriately to changes in that environment.

On a simple level, a single-celled organism, such as *Amoeba*, must be able to detect nutrient molecules in the water around it. It must then move towards the nutrient and take the molecules into the cell. If the amoeba cannot detect that there are nutrients present, cannot move towards the nutrients, or cannot take in the nutrients, it will be unable to survive.

In multicellular organisms, the survival of the whole organism requires each cell to play its part. So cells must be able to detect the various internal and external signals used to coordinate and carry out the processes involved in growth, development, movement and excretion. Cells must then be able to carry out reactions or functions in response to the signals. The processes involved in communication between the cells of multicellular organisms are extremely complex.

In order to detect signals, cells must have on their surface 'sensors' capable of receiving signals. These sensors are known as **receptors**. They are often protein molecules or modified protein molecules.

Hormone receptors

In multicellular organisms, communication between cells is often mediated by hormones. Hormones are chemical messengers, produced in specific tissues and then released into the organism. Any cell with a receptor for the hormone molecules is called a **target cell**.

A hormone molecule binds to a receptor on a target cell surface membrane because the two have complementary shapes (like two adjacent pieces of a jigsaw). Binding of the hormone and receptor causes the target cell to respond in a certain way.

The insulin receptor

An example of a hormone receptor is the insulin receptor. Insulin is released from special cells (beta-cells) in islets of Langerhans in the pancreas, in response to increased blood sugar levels.

Insulin is a protein molecule that attaches to the insulin receptors on the plasma membranes (cell surface membranes) of many cells, including muscle cells and liver cells.

When insulin attaches to its receptor, it triggers internal responses in the cell that lead to more glucose channels being present in the plasma membrane. This allows the cell to take up more glucose from the blood, so reducing the blood glucose level.

Medicinal drugs – interfering with receptors

A number of medicinal drugs have been developed that are complementary to the shape of a type of receptor molecule.

Such drugs are intended to block receptors. Beta-blockers are used to prevent heart muscle from increasing the heart rate in people for whom such an increase could be dangerous. Some drugs mimic a natural neurotransmitter that some individuals cannot produce – drugs used to treat schizophrenia work in this way.

Hijacking receptors

Viruses enter cells by binding with receptors on the cell's plasma membrane that normally bind to the host's signalling molecules. Human immunodeficiency virus (HIV), which causes AIDS, can infect humans because it can enter the cells of the immune system. It has a shape that fits into one of the receptors on the cell surface of some important types of immune cells such as helper T-lymphocytes. Once HIV enters such a cell, after a period of inactivity it may reproduce inside the cell and eventually destroy it.

Some poisons also bind with receptors. The toxin extracted from the bacterium *Clostridium botulinum* binds with receptors on muscle fibres and prevents them from working properly, causing paralysis. This toxin is lethal, but it is used in small quantities in cosmetic surgery under the name BOTOX®, to paralyse small muscles in the face and reduce wrinkling of the skin.

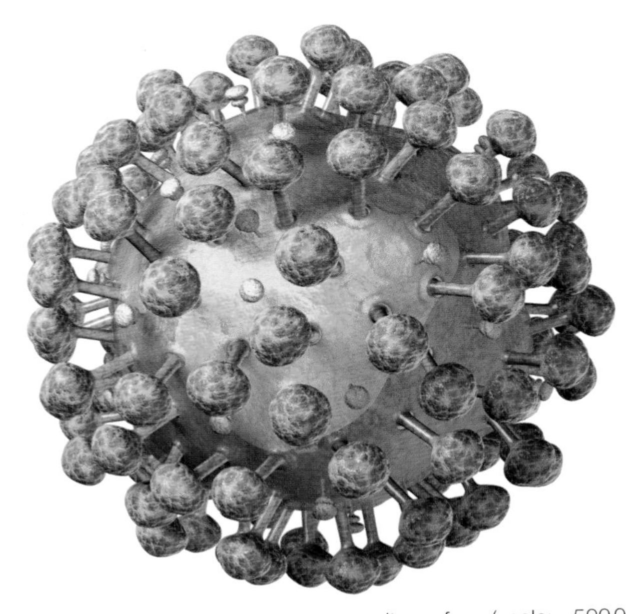

Figure 1 HIV particle with receptors on its surface (scale: ×500 000)

Questions

1. Suggest how flu viruses infect human cells of the respiratory tract.
2. Name two target tissues for insulin.
3. Botulinum toxin binds to the ends of nerves and prevents them from releasing chemicals that normally cause muscles to contract. There are eight different botulinum toxins, some stronger than others. Suggest why some of the toxin molecules are more potent than others.

1.1 (10) Crossing membranes 1 – passive processes

By the end of this spread, you should be able to . . .

* **Describe and explain what is meant by passive transport (diffusion and facilitated diffusion).**
* **Describe the role of membrane proteins in passive transport.**

Key definition

Diffusion is the movement of molecules from a region of high concentration of that molecule to a region of lower concentration of that molecule down a concentration gradient.

Examiner tip

Some questions need you to identify clues from a diagram about the type of transport that is happening. Look to see if the movement is against the **concentration gradient** and whether **ATP** is being used.

Transport and survival

In order to survive, cells need a supply of nutrient molecules. Most cells also need a supply of oxygen for aerobic respiration. The reactions in living cells (known collectively as **metabolism**) generate waste products that must be removed from the cell. Any molecules that need to enter or leave a cell (or enter or leave a membrane-bound organelle) will usually have to cross a membrane to do so.

Diffusion – molecules even out

In a gas or a liquid, the molecules (or ions) move around. Even if the gas or liquid is not mixed up or stirred, molecules will continue to move freely and randomly in all directions. This is because the molecules are not held together as they are in a solid. They possess **kinetic energy** that keeps them moving. Processes such as diffusion that depend only on this energy are termed passive processes.

If molecules (or ions) are packed and held together – like the scent molecules in a piece of wrapped fish – they can move around only within the packing. But if the packing is unwrapped, scent molecules spread around the available space until they are evenly distributed. This is because molecules in a gas or liquid bump into each other and spread away from where they occur in high concentration. This tendency to even out is **diffusion**.

Each molecule diffuses down its own concentration gradient

It is important to remember that each different type of molecule or ion in a gas or a liquid will diffuse down its own concentration gradient. Diffusion of one type of molecule can take place in one direction while, at the same time, a different type of molecule can diffuse in the opposite direction.

In a plant cell carrying out photosynthesis in bright light, oxygen gas (a waste product of photosynthesis) will be diffusing out of the cell as its concentration builds up. At the same time, carbon dioxide gas will be diffusing in, because it is used up in photosynthesis.

Diffusion and net movement

When diffusion has taken place, molecules are distributed evenly. This does not mean their movement stops – the molecules continue to move around. We refer to this state, where there is no overall movement of molecules in one direction, as equilibrium – there is no net movement.

In living organisms, a number of activities and features ensure that equilibrium is not reached. Cells use up carbon dioxide in photosynthesis, so the level within the cell remains much lower than that outside the cell. This maintains a concentration gradient for carbon dioxide.

The rate of diffusion

Diffusion is a passive process. This means that molecules continue to diffuse down their concentration gradient without using energy from the cell. They have their own kinetic energy.

The rate of diffusion is affected by a number of factors.

- Temperature – increasing temperature gives molecules more kinetic energy. The rate of random movement increases and so the rate of diffusion increases.
- Concentration gradient – having more molecules on one side of a membrane increases the concentration gradient. This increases the rate of diffusion.
- Stirring/moving – stirring a liquid, or the movement of air currents in a gas, increases the movement of molecules and thus the rate of diffusion.
- Surface area – diffusion across membranes occurs more rapidly if there is a greater surface area to diffuse across. Cells are adapted to increase the surface area for diffusion. Red blood cells are biconcave. Epithelial cells in the small intestine have folds called microvilli. Alveoli increase the surface area of the lungs.
- Distance/thickness – diffusion is slowed down by thick membranes. There is a greater distance for molecules to travel.
- Size of molecule – smaller molecules or ions diffuse more quickly than larger ones.

Taking advantage of diffusion in cells

Lipid-based molecules

As the membrane is made of phospholipids, fat-soluble molecules can simply pass through the bilayer. They diffuse down a concentration gradient. Steroid hormones are lipid-based and so diffuse through membranes into cells.

Very small molecules and ions

Carbon dioxide and oxygen molecules are small enough to pass through the bilayer between the phospholipid molecules. Water molecules are very small. Some water molecules will pass directly through the membrane even though they are polar (charged).

Larger or charged molecules need to be carried across

Small, charged particles such as sodium ions, or larger molecules such as glucose, cannot pass through the lipid bilayer. Two types of protein molecule are involved in allowing such substances to pass through membranes. Because these proteins allow substances to pass through the membrane, the diffusion of these molecules and ions is known as **facilitated diffusion**.

The two protein types are:

- **channel proteins** – these basically form pores in the membrane, which are often shaped to allow only one type of ion through; many are also 'gated', meaning they can be opened or closed – gated sodium ion channel proteins are involved with the working of the nervous system
- **carrier proteins** – these are shaped so that a specific molecule (e.g. glucose) can fit into them at the membrane surface – when the specific molecule fits, the protein changes shape to allow the molecule through to the other side of the membrane.

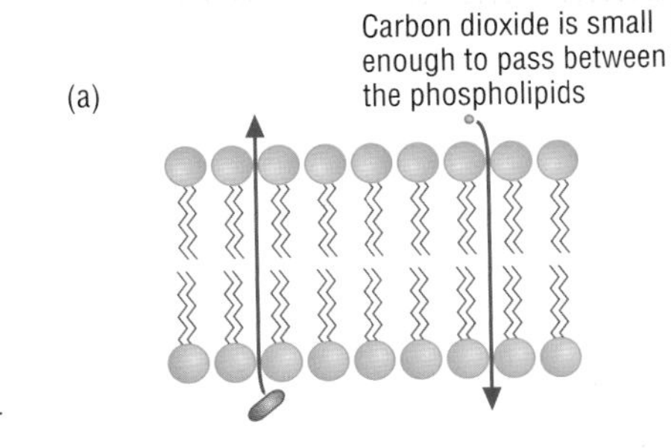

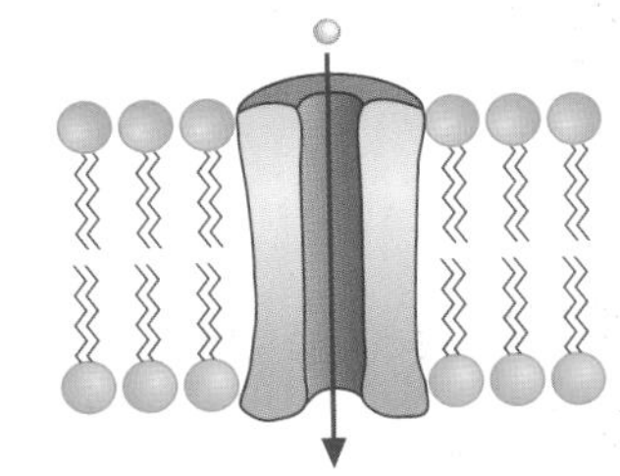

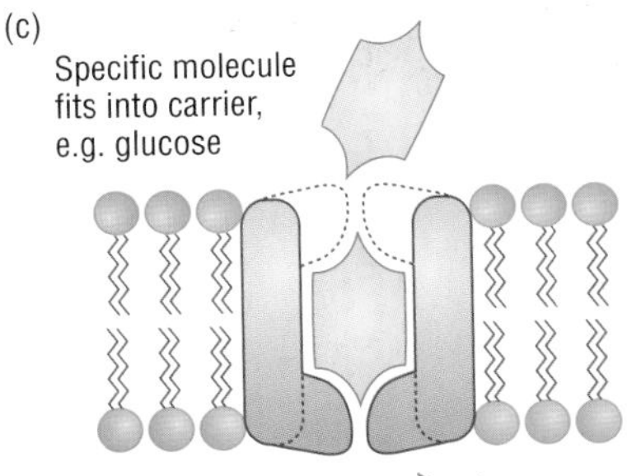

Figure 1 **a** Simple diffusion; **b** facilitated diffusion using a channel protein; **c** facilitated diffusion using a carrier protein

Membranes take some control

Different membranes can have different carrier and channel proteins. This means that cells and organelles have some control over the types of substance that are allowed to pass in or out across the membrane.

Substance moved by:	Examples
Simple diffusion	Gases like oxygen and carbon dioxide Lipid-based molecules like steroid hormones
Facilitated diffusion using channel proteins	Ions like sodium ions and calcium ions
Facilitated diffusion using carrier proteins	Larger molecules like glucose and amino acids

Table 1 Substances moved by diffusion and facilitated diffusion in most cells

Questions

1. List two substances that diffuse through a cell surface membrane in a plant cell of a leaf and explain how the direction of diffusion will change in daylight compared with darkness.
2. Explain why a single-celled organism such as *Amoeba* can gain enough oxygen for its respiration through simple diffusion across its membrane.
3. Give a definition for 'facilitated diffusion'. Compare and contrast simple diffusion and facilitated diffusion.

1.1 (11)

Crossing membranes 2 – active processes

By the end of this spread, you should be able to . . .

* **Describe and explain what is meant by active transport, endocytosis and exocytosis.**

Concentration gradients don't always help

The needs of cells cannot always be met by the process of diffusion. Sometimes, for a cell to function properly it may need more of a particular substance in the cytoplasm than is present outside the cell. In other cases, cells may need to move materials into or out of a cell more quickly than simple diffusion allows.

- Magnesium ions are often in very short supply in soil. Plant cells need magnesium ions to make chlorophyll. The plant cell must be able to move magnesium ions into the cell against a concentration gradient.
- Active transport helps us absorb glucose from our intestines.

Key definition

Active transport refers to the movement of molecules or ions across membranes, which uses ATP to drive protein 'pumps' within the membrane.

Active transport – 'pumping' molecules across membranes

Some of the **carrier proteins** found in membranes act as pumps. These proteins are similar to the protein carriers used for facilitated diffusion. They are shaped in a way that fits (is **complementary** to) the molecule they carry. They carry larger or charged molecules and ions through membranes. These are the molecules and ions that cannot pass through the lipid bilayer by diffusion.

These protein pumps differ significantly from the proteins used in facilitated diffusion:

- they carry specific molecules one way across the membrane
- in carrying molecules across the membrane, they use metabolic energy in the form of **ATP**
- they can carry molecules in the opposite direction to the concentration gradient
- they can carry molecules at a much faster rate than by diffusion
- molecules can be accumulated either inside cells or organelles, or outside cells.

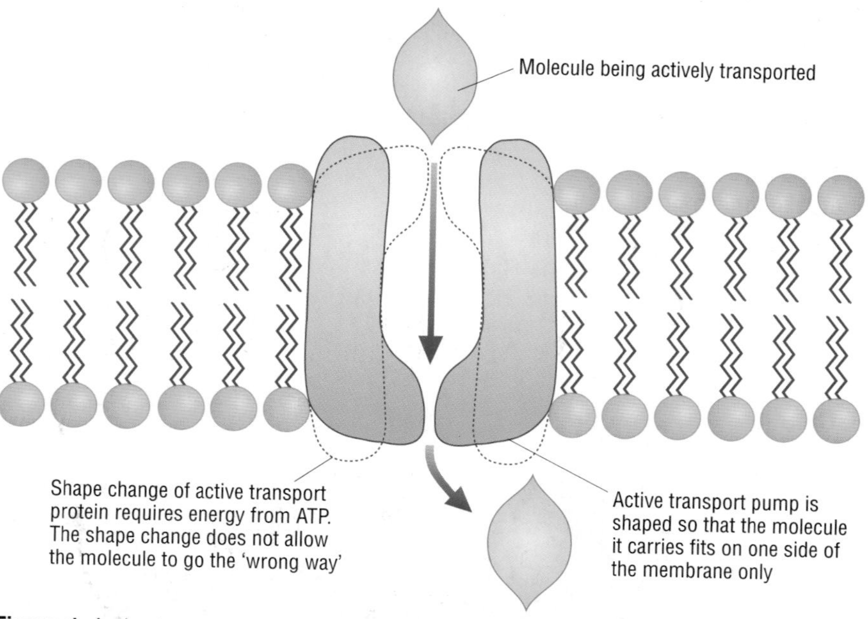

Figure 1 Active transport

Ensuring one-way flow

The energy used in pumping molecules across membranes by active transport changes the shape of the carrier protein. The shape change means that the specific molecule to be transported – or pumped – fits into the carrier protein on one side of the membrane only. As the molecule is carried through, the carrier uses energy from ATP. This changes its shape so that the molecule being carried across now leaves the carrier protein. The molecule cannot enter the transport protein, because the protein is now a different shape so it will not fit.

Calcium ion movement in muscles – an example of active transport

Muscle fibres can contract only if calcium ions are present. When a muscle is stimulated to contract, calcium ions are released from membrane-bound stores (specialised **endoplasmic reticulum**), where they are in very high concentration. When the muscle needs to relax again, the calcium ions are pumped rapidly back into the stores by the many calcium ion pumps found on the membrane of the specialised endoplasmic reticulum.

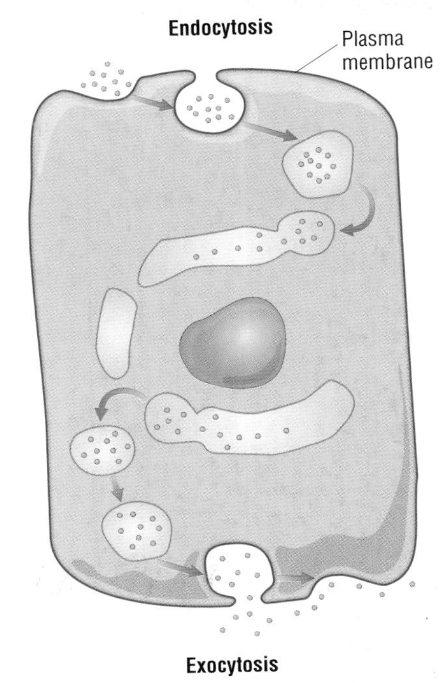

Figure 2 Exocytosis and endocytosis

Moving large amounts – bulk transport

Some cells need to move large quantities of material either in or out. The process is described as **endocytosis** when it involves bringing materials into the cell, and **exocytosis** when it involves moving material out of the cell. This bulk transport is possible because membranes can easily fuse, separate and 'pinch off'. Like active transport, bulk transport requires energy in the form of ATP. In this case, the energy is used to move the membranes around to form the **vesicles** that are needed, and to move the vesicles around the cell.

Some examples of bulk transport are listed below.

- Hormones – pancreatic cells make insulin in large quantities. The insulin is processed and packaged into vesicles in the Golgi apparatus. These vesicles fuse with the outer membrane to release insulin into the blood.
- In plant cells, materials required to build the cell wall are carried outside in vesicles.
- Some white blood cells engulf invading microorganisms by forming a vesicle around them. This vesicle then fuses with lysosomes so that the enzymes from the lysosomes can digest the microorganism. Such cells are called phagocytes.

In or out – solid or liquid?

Different names are given to the movements of materials in bulk transport:

- *endo* – inwards
- *phago* – solid material
- *exo* – outwards
- *pino* – liquid material.

So the bulk movement of liquid material out of a cell would be described as 'exopinocytosis'.

Table 1 Summary of transport methods

Passive processes (i.e. no energy input from ATP required)	Diffusion	Down a concentration gradient lipid soluble or very small molecules through lipid bilayer
	Facilitated diffusion	Down a concentration gradient charged or hydrophilic molecules or ions via channel or carrier proteins
	Osmosis	Down a water potential gradient through bilayer or protein pores
Active processes (i.e. energy input in the form of ATP is required)	Active transport	Against a concentration gradient via carrier proteins that use energy from ATP in order to change shape
	Endocytosis and exocytosis	Bulk transport of materials via vesicles that can fuse with or break from the cell surface membrane

Examiner tip

Questions on active transport often involve interpretation of data in tables or graphs. Look for differences in molecule or ion concentrations that change if ATP production is affected in some way.

Questions

1. Compare and contrast active transport and facilitated diffusion.
2. Explain why active transport allows substances to be accumulated in an area, whereas facilitated diffusion doesn't.
3. What term would describe the bulk movement of materials shown by phagocytes?

Water is a special case

By the end of this spread, you should be able to . . .

- * **Explain what is meant by osmosis, in terms of water potential.**
- * **Recognise and explain the effects of solutions of different water potentials on plant and animal cells.**

Water diffuses

As with any other molecule or ion in a liquid or a gas, water molecules that are free to move around will diffuse from a region where there are a lot of them to a region where there are fewer of them. But if there is any other substance dissolved in the water, this will affect the concentration of 'free' water molecules. A substance that can dissolve is called a **solute**. The liquid it dissolves in is called a **solvent**. The two together form a **solution**.

For example, if you place a brick in a bucket of water, the brick will not change the water other than moving it around. If the brick were made of sugar and dissolved in the water, the water would become thicker in consistency and form a solution very different from pure water.

Both buckets contain the same number of water molecules, but in the second bucket many of the water molecules are clustered around the dissolved sugar molecules. In effect, these molecules are not 'free' to diffuse.

Water potential – the concentration of 'free' water

Water potential is a measure of the tendency of water molecules to diffuse from one place to another.

Water always moves from a region of high water potential to a region of lower water potential – it moves from a region of high concentration of 'free' water molecules to a region of lower concentration.

In our example above, the first bucket contains pure water. All the water molecules are free to diffuse, so pure water has the highest concentration of freely moving water molecules – it has the highest water potential. As solutes are dissolved, water molecules cluster around them forming a solution. This lowers the concentration of free water molecules, and so lowers the water potential. The more solute dissolved, the lower the water potential of the solution.

Examiner tip

Always describe the movement of water by osmosis in terms of **water potential** and **water potential gradients**, and learn the differences between these effects in plant and animal cells.

Figure 1 Osmosis and membrane

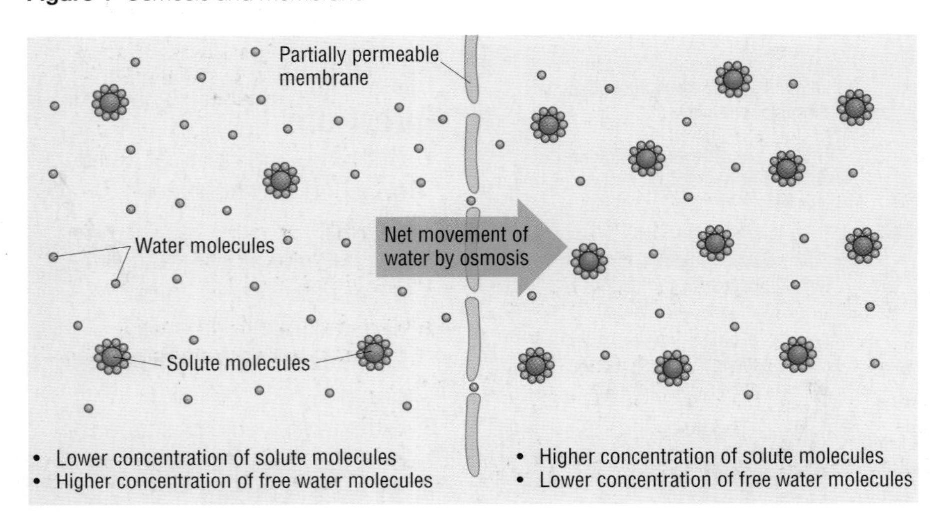

Osmosis and water potential

Osmosis is a special kind of **diffusion**. It refers *only* to the movement of water molecules:

- by diffusion, and
- across a **partially permeable membrane**.

Water potential is a measure of the concentration of water molecules that are able to diffuse. Osmosis is the movement of water molecules from a region of higher water potential to a region of lower water potential (down a water potential gradient) across a partially permeable membrane.

As with diffusion, net movement of molecules occurs until the concentrations are evened out. So osmosis will occur until the water potential is the same on both sides of the membrane.

Osmosis, water potential and cells

The water potential of cells is lower than that of pure water, because of the sugars, salts and other substances dissolved in the cytoplasm. In plant cells, the large vacuole also contains dissolved substances.

Cells in solutions of high water potential

The cell membrane is a partially permeable membrane. Placing plant or animal cells in pure water (or in any solution with a water potential *higher* than the cell contents) means there is a water potential gradient from outside to inside the cells. Water molecules will move down the water potential gradient *into* the cells by osmosis. The cells will swell. In animal cells, the membrane will eventually burst open.

In a plant cell, the swelling cytoplasm and vacuole will push the membrane against the cell wall. The cell will not burst, because the wall will eventually stop the cell getting any larger. Osmosis will stop at this point, even though there may still be a water potential gradient.

Cells in solutions of low water potential

Placing plant or animal cells in a concentrated salt or sugar solution (with a water potential *lower* than the cell contents) means there is a water potential gradient from inside to outside the cell. So water molecules will move *out of* the cell by osmosis. The cell will shrink. In animals, the cell contents will shrink and the membrane will wrinkle up. In plant cells, the cytoplasm and vacuole will shrink as they lose water, and the cell surface membrane will pull away from the cell wall. This is called **plasmolysis**.

Units of measuring water potential

Water potential is measured in kiloPascals (kPa). Confusingly, pure water – which has the highest water potential – has a value of 0 kPa. Dissolving solute in water reduces the water potential, so water potential values go down from 0 to negative figures. The larger the negative figure, the more solute dissolved and the lower the water potential.

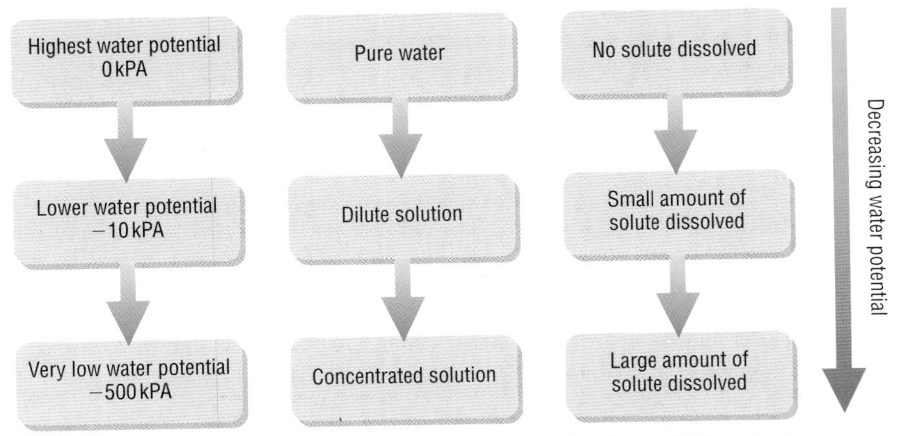

Figure 4 Ladder showing water potential, free water molecules and kPa values

Questions

1. List the factors that affect the rate of diffusion and explain how osmosis is affected by these factors.
2. Explain why dissolving more solute decreases the water potential of a solution.
3. Freshwater amoebae must continually move large volumes of water out of the cell. Explain why this is the case and suggest how they might carry this out.
4. As water moves into a cell placed in a hypotonic (more watery) solution, what happens to the water potential in the cell?

Key definitions

Solute – a solid that dissolves in a liquid.

Solvent – a liquid that dissolves solids.

Solution – a liquid containing dissolved solids.

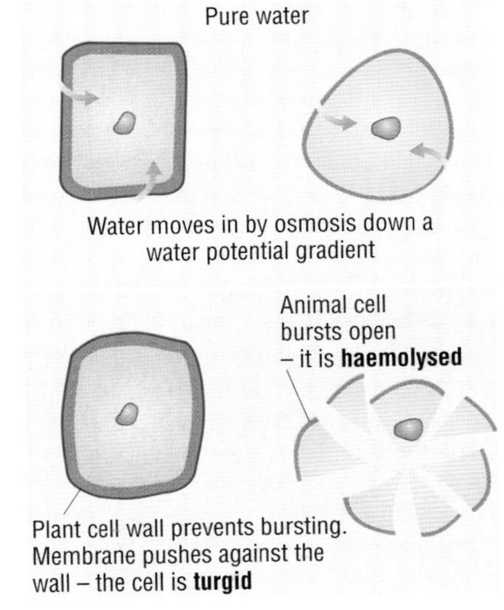

Figure 2 Plant and animal cells in solutions of high water potential – before and after

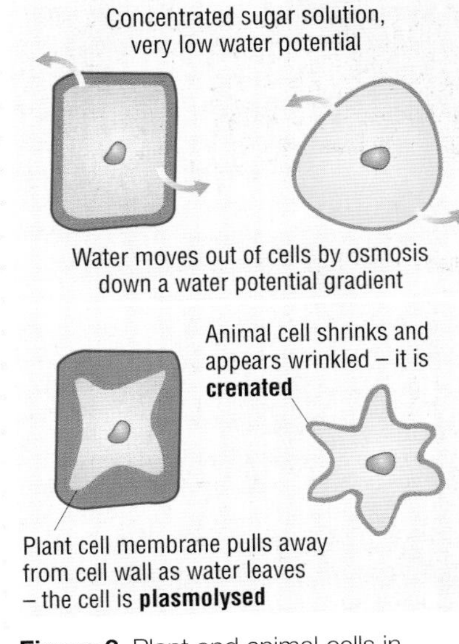

Figure 3 Plant and animal cells in solutions of low water potential – before and after

New cells – parent and daughter cells

By the end of this spread, you should be able to . . .

* **State that mitosis occupies only a small percentage of the cell cycle and that the remaining percentage includes the copying and checking of genetic information.**

Key definition

The **cell cycle** describes the events that take place as one parent cell divides to produce two new daughter cells which then each grow to full size.

For some organisms, the cell cycle is the life cycle, and each daughter is a new single-celled organism.

Examiner tip

When you are asked about the products of the cell cycle, you need to say that the daughter cells are *genetically identical* both to each other *and* to the parent cell.

Parents and daughters

Some organisms, e.g. *Amoeba*, consist of only one cell. Others, such as oak trees, have many millions of cells. New daughter cells form from parent cells in a series of events called the cell cycle. The daughter cells produced must be able to carry out the same functions as the parent cell. This is true in all cases – whether the daughter cell is a new organism, or a new cell in a growing organism. It is also true if the new cell is replacing a worn-out cell, such as a skin cell.

Chromosomes have the instructions

Chromosomes are in the nucleus of **eukaryotic cells**. Each chromosome contains one molecule of **DNA**, which includes specific lengths of DNA called **genes**. So the chromosomes hold the instructions, sometimes called the blueprint, for making new cells. The daughter cells produced during the cell cycle must contain a copy of all of these instructions, so they must contain a full set of chromosomes, copied exactly from the chromosomes in the parent cell.

In humans there are 46 chromosomes in the nucleus of each cell; in onion cell nuclei there are 12. Chimpanzees have 48 and dogs have 78. Some ferns have over 1000.

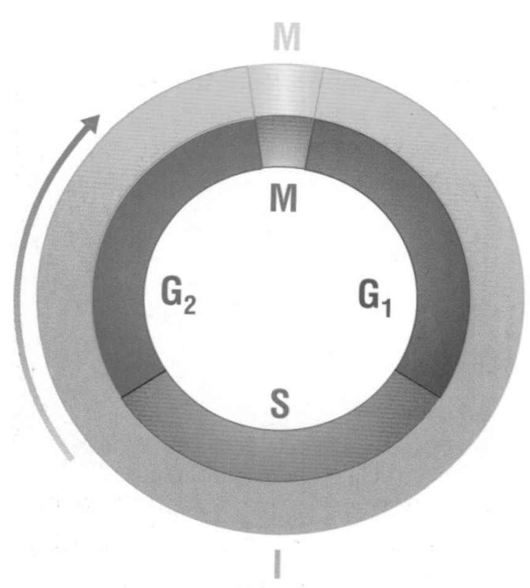

G phases indicate growth of organelle numbers and cell size overall
S indicates synthesis of new DNA (ie DNA replication)

Figure 1 M is nuclear division (mitosis) and cytokinesis (cleavage of cytoplasm). G_1 is biosynthesis (e.g. proteins made, organelles replicate). S is synthesis of new DNA (replication of chromosomes). G_2 is growth

Copying and separating

In eukaryotes, the molecules of DNA that make up each chromosome are wrapped around proteins called histones. The DNA and the histone proteins together are called **chromatin**. Before a cell can divide to produce two new daughter cells, the DNA of each chromosome must be replicated. Two replicas are produced. Each is an exact copy of the original, and they remain held together at a point called the **centromere**. This plays an important role in the process of nuclear division.

At this stage, you can't see the chromosomes under a light microscope. Each chromosome now consists of two replica DNA strands. These replicas are called a pair of sister **chromatids**. When they are separated from each other, each one will end up in a different new daughter cell.

Before this happens, the chromatin must be coiled up (supercoiled) to form visible chromosomes. Each one is then short and sturdy enough to be moved around more easily. In this state, chromosomes can take up stains and be seen under a light microscope. Chromatin threads are about 30 nm thick, but after supercoiling a chromosome is about 500 nm thick.

Supercoiled chromosomes can't perform their normal functions in the cell, so the length of time they spend coiled up needs to be as short as possible.

Checks and balances

As chromosomes are being replicated, proof-reading enzymes move along the new DNA strands and check that the copying has been done properly. If the genes are not copied precisely, the resulting **mutations** may mean the new cells fail to function.

Copying the information carried by the DNA in a human each time the cell divides is roughly equivalent to copying out a full 30-volume set of *Encyclopedia Britannica* 20 times over – without making any mistakes!

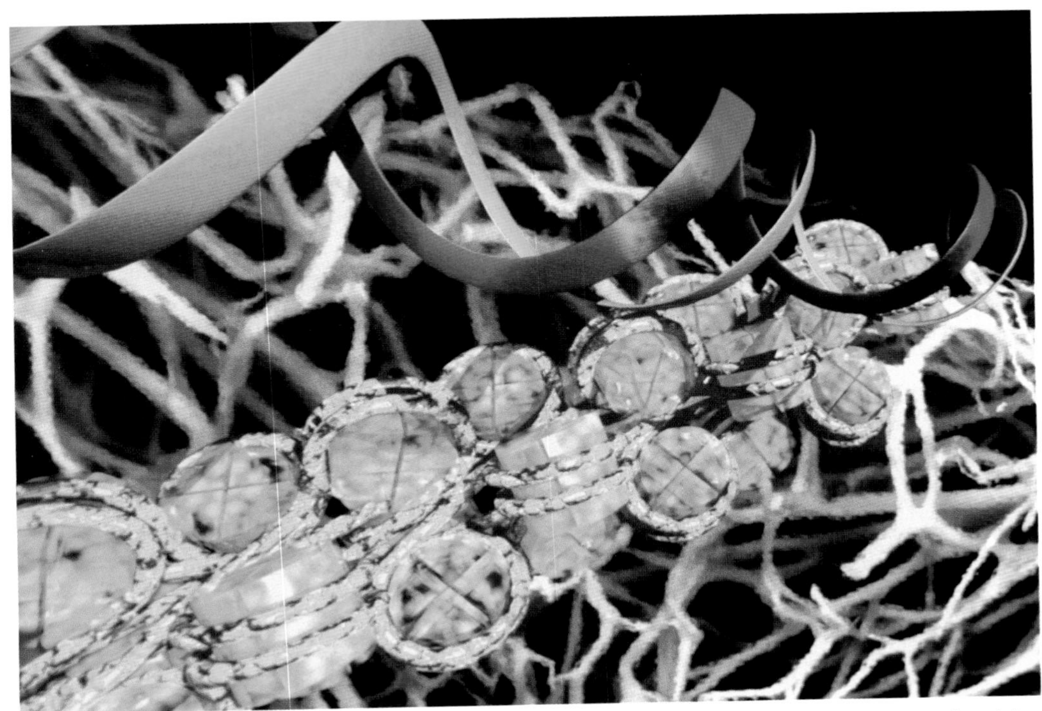

Figure 2 The DNA double helix (top left) is coiled and wrapped in loops (yellow) around cores of proteins called histones (blue). This packaged form of DNA and protein, called chromatin (background), fits into the cell nucleus

Figure 3 Electron micrograph showing fully condensed chromosome consisting of a pair of sister chromatids

How fast is the cycle?

The length of time taken for a parent cell to divide into two daughter cells, and for each to grow to full size, varies between species and cell type. It is also affected by the availability of nutrients for the cells. Some prokaryotic cells can go through the whole process in around 30 minutes, whereas yeast cells (which are single-celled eukaryotic organisms) take about 4 hours.

The cell cycle is divided into stages:

- **interphase** – DNA replicates in this stage
- **mitosis** – the nucleus divides and chromatids separate
- **cytokinesis** – the cytoplasm divides or cleaves
- growth phase – each new cell grows to full size.

Mitosis occupies only a small proportion of the cell cycle and the remaining larger portion includes copying and checking genetic information on the DNA and processes associated with growth.

It's not *just* about the chromosomes

Daughter cells need more than just a set of chromosomes. Each new cell, like its parent, must carry out a number of metabolic functions. In order to survive, it needs its own membranes, cytoplasm, organelles, enzymes and other proteins.

Questions

1. A horse skin cell contains 64 chromosomes. How many chromosomes are there in
   (a) a horse brain cell,
   (b) a horse liver cell?
2. How many chromosomes are there in a human red blood cell?
3. How many DNA molecules are there in the nucleus of an onion cell that is not dividing?

Two nuclei from one

By the end of this spread, you should be able to . . .

* **Explain the significance of mitosis for growth, repair and asexual reproduction in animals and plants.**
* **Describe, with the aid of diagrams, the stages of mitosis.**

Key definition

Mitosis refers to the process of nuclear division where two genetically identical nuclei are formed from one parent cell nucleus.

What's so important about making new cells?

All organisms need to produce genetically identical daughter cells.

- Asexual reproduction – single-celled organisms, such as *Paramecium*, divide to produce two daughter cells that are separate organisms. Some multicellular organisms, such as *Hydra*, produce offspring from parts of the parent.
- Growth – multicellular organisms grow by producing new extra cells. Each new cell is genetically identical to the parent cells and so can perform the same functions.
- Repair – damaged cells need to be replaced by new ones that perform the same functions and so need to be identical (as with growth).
- Replacement – red blood cells and skin cells are replaced by new ones.

Examiner tip

Learn the main events at each stage, and be able to recognise how a cell would look at each stage. You could also make up a mnemonic to help you remember the sequence: PMAT.

Remember that bacteria don't do mitosis – they don't have the linear chromosomes or a centriole or spindles.

Mitosis in four stages

Mitosis refers to the process of nuclear division where two genetically identical nuclei are formed from one parent cell nucleus. To help describe what is happening as one nucleus divides to become two, the sequence of events in mitosis is divided into four named stages – **prophase**, **metaphase**, **anaphase** and **telophase**.

In reality, it is a continuous process involving:

- prophase – replicated chromosomes supercoil (shorten and thicken)
- metaphase – replicated chromosomes line up down the middle of the cell
- anaphase – the replicas of each chromosome are pulled apart from each other towards opposite poles of the cell
- telophase – two new nuclei are formed.

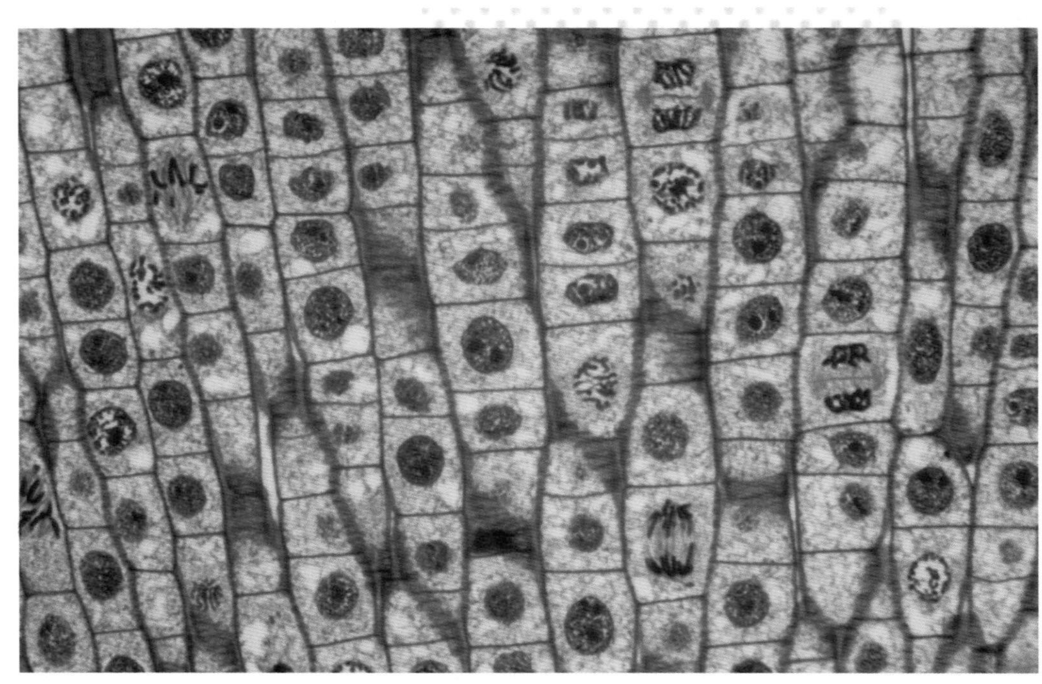

Figure 1 A light micrograph of onion root showing cells at various stages of the cell cycle

Chromosomes become visible in prophase

Each chromosome has already replicated in interphase. In prophase, the chromosomes shorten and thicken (supercoil), and, using a light microscope, you can see that they consist of a pair of sister chromatids.

At this time the **nuclear envelope** breaks down and disappears. An organelle called a **centriole** divides into two, and each daughter centriole moves to opposite ends (poles) of the cell to form the **spindle**. This is a structure made of protein threads that look rather like the lines of longitude on a globe.

Chromosomes line up in metaphase

The chromosomes move to the central region of the spindle (the equator) and each becomes attached to a spindle thread by its **centromere**.

Replicated chromosomes separate in anaphase

The replicated sister chromatids that make up the chromosome are separated from each other when the centromere that holds them together splits. At this point, each of the 'sisters' effectively becomes an individual chromosome. Each one is identical to the original chromosome in the parent cell from which it was copied.

The spindle fibres shorten, pulling the sister chromatids further and further away from each other towards the poles. They assume a V-shape because the centromeres, attached to the spindle fibres, lead.

Two new nuclei form in telophase

As the separated sister chromatids (now chromosomes) reach the poles of the cell, a new nuclear envelope forms around each set.

The spindle breaks down and disappears. The chromosomes uncoil, so you can no longer see them under the light microscope.

Two new cells can now be made

The whole cell now splits to form two new cells, each one containing a full set of chromosomes identical to that found in the original parent cell. This splitting in two is called **cytokinesis**.

Having identical genetic information, in the form of identical chromosomes, means each daughter cell is capable of doing everything the parent cell could do.

A time and a place

There are some important differences between when, where and how plant and animal cells go through the cell cycle.

- In animals most cells are capable of mitosis and cytokinesis, whereas in plants only special cells, known as **meristem cells**, can divide in this way.
- Plant cells do not have centrioles – the tubulin protein threads are made in the cytoplasm.
- In animal cells cytokinesis starts from the outside ('nipping in' the cell membrane), but in plant cells cytokinesis starts with the formation of a cell plate where the spindle equator was. New cell membrane and new cell wall material is laid down along this cell plate.

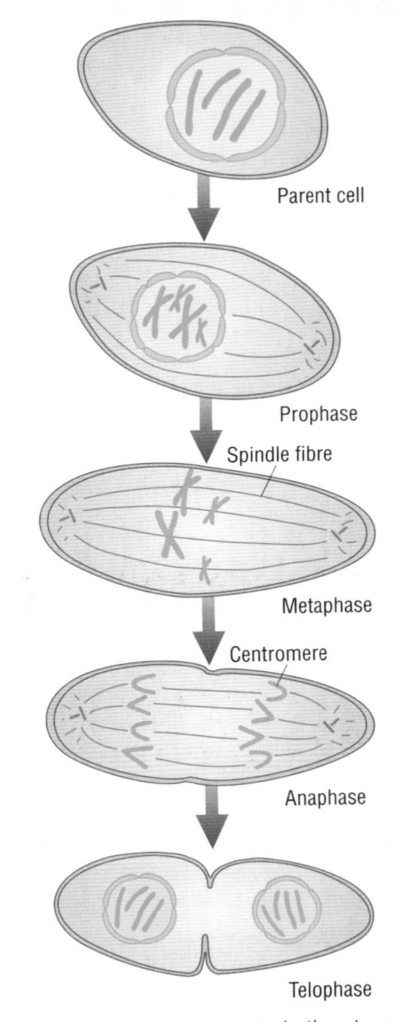

Figure 2 The main events in the stages of mitosis

Questions

1. In an onion cell nucleus, there are normally 12 chromosomes. In which stages of mitosis are there 24 chromosomes?
2. Describe the events of mitosis that help ensure each daughter cell receives one copy of each chromosome.
3. Sketch a graph showing how the DNA content of a cell changes against the phases of the cell cycle.
4. Suggest why most plants cells (except meristem cells) are not capable of undergoing mitosis and cytokinesis.

1.1 (15) Cell cycles and life cycles are not all the same

By the end of this spread, you should be able to . . .

* **Define the term stem cell.**
* **Explain the meaning of the term homologous pair of chromosomes.**
* **Outline the process of cell division by budding in yeast.**
* **State that cells produced as a result of meiosis are not genetically identical.**

Key definition

Clones: genetically identical cells or organisms derived from one parent.

Examiner tips

It is important to spell some words correctly. One such example is mitosis and meiosis. If you write 'meitosis' or 'miosis' you will not get a mark because the examiner won't know which process you are referring to.

When you are talking about clones or mitosis, you should always say '*genetically* identical cells', not just 'identical cells'.

Natural and artificial clones

Some new cells produced by **mitosis** and **cytokinesis** are new and separate organisms, for example *Amoeba* daughter cells. The two new organisms are genetically identical to each other and to the parent cell. Both can survive and reproduce. Genetically identical cells are called **clones**. All the bacteria in a single colony have been produced by one cell dividing (by **binary fission**), and they are also clones.

Many plants undergo asexual reproduction using specialised parts of the plant that are derived from adult plant cells. Examples include potato tubers and strawberry plant runners. These specialised parts can produce many new individual organisms that are genetically identical to the original parent. They are also called clones. This form of asexual reproduction is known as vegetative propagation.

Bacteria are **prokaryotes**. They have a single, naked (not associated with **histone** proteins) strand of **DNA** that is in the cytoplasm, not in a nucleus. They may also have small **plasmids** of DNA. These may have genes for antibiotic resistance. Because bacteria can swap plasmids, they are used in genetic engineering. Bacteria divide by binary fission, *not* by mitosis. (The term mitosis refers only to cell division involving chromosomes.)

Artificial cloning has been practised by farmers and growers for many years. For example, cuttings taken from some plants can be made to grow into adult plants. They are all genetically identical to the parent plant from which the cuttings were taken, so they are clones of the parent. They are also all genetically identical to each other.

The artificial cloning of animals became big news in 1997. Scientists placed a nucleus from an adult sheep's udder cell into a sheep egg cell that had had its nucleus removed. The resulting cell was placed into the uterus of another female sheep (a surrogate). At the end of the pregnancy this lamb was born, and was named Dolly. Dolly the sheep was a clone of another animal. This technique continues to raise many ethical issues.

Figure 1 Strawberry plant showing overground runners between individuals

Stem cells

Mitosis produces genetically identical daughter cells that carry a full set of genetic information. They are potentially capable of becoming any one of the different cell types found in the fully grown organism. Such cells are known as stem cells. They can be described as omnipotent or totipotent.

It is important to know that stem cells occur in small numbers in an adult animal. Bone marrow contains stem cells that can divide to produce all the blood and bone cell types required by an adult human.

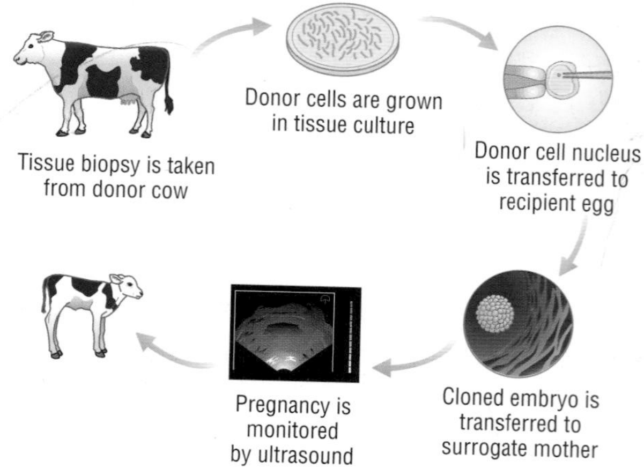

Figure 2 Steps in cloning animals

There is great interest in stem cells at present because scientists hope that one day they may become widely used in medical treatments. Using an individual's own cells to repair damaged or diseased tissues and organs would have great benefits.

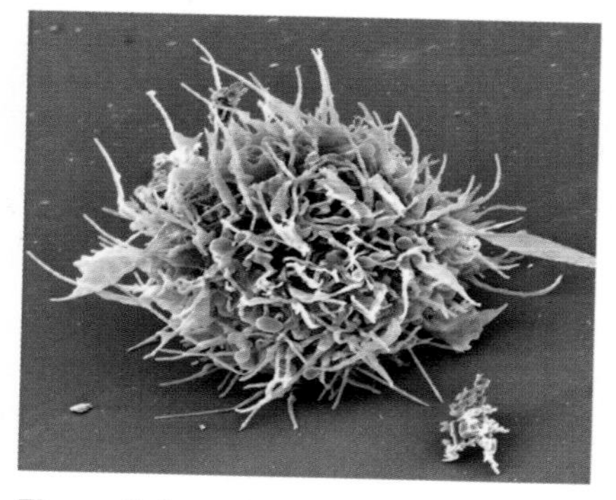

Figure 3 Scanning electron micrograph of a mammalian stem cell in culture

A time and a place

There are some important differences between when, where and how the cells of different organisms go through the cell cycle.

When and where cells divide

In animals, many cells are capable of mitosis and cytokinesis. This means organisms can repair damage to most organs by producing new cells to replace damaged ones.

In plants, only the cells of special growing regions (meristems), known as **meristem cells**, can divide in this way. Meristems are located at the root and shoot tips, and in a ring of tissues in the stem or trunk (which allows for an increase in girth). These relatively tiny parts of the plant are responsible for the growth of the whole organism.

Cytokinesis

In animal cells, cytokinesis starts from the outside – 'nipping in' the cell membrane and cytoplasm along what is termed a cleavage furrow.

In plant cells, cytokinesis starts with the formation of a cell plate where the spindle equator was. The cell then lays down new membrane and cell wall material along this plate. The cell plate is not a solid structure, but a single plane along which the new cell wall forms.

Cells of yeast (a single-celled fungal organism, e.g. *Candida albicans*) undergo cytokinesis by producing a small 'bud' that nips off the cell, in a process called budding.

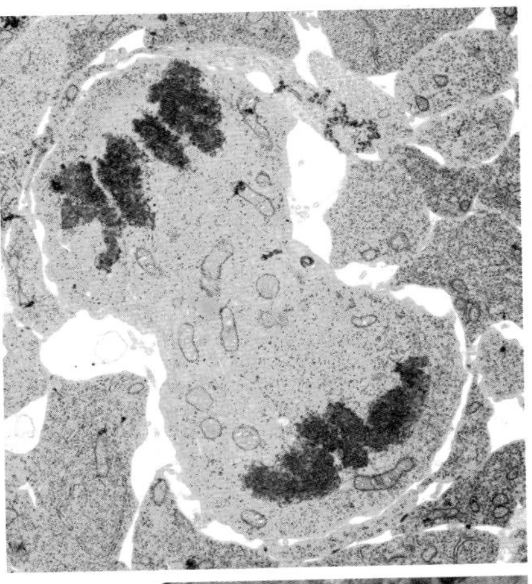

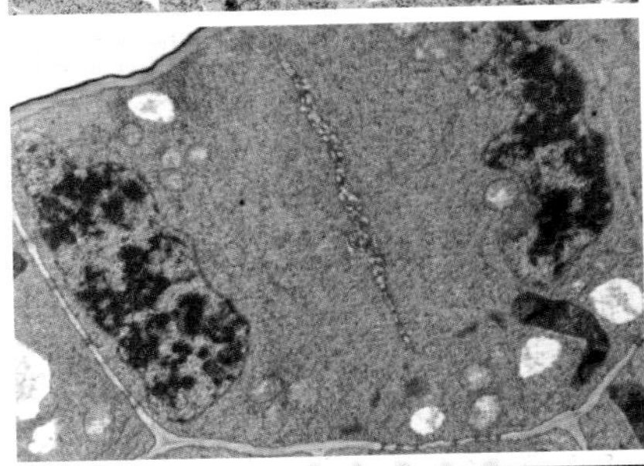

Figure 4 Cell with cleavage furrow, plant cell plate and yeast cell budding (top to bottom)

Producing cells that are genetically different

Sexual reproduction involves the fusing of two cell nuclei, usually from two different individuals, in order to produce offspring.

Each cell contributes half of the total genetic information (**genome**) required by the offspring. This means that special cells containing half the adult number of **chromosomes** must be produced. Such cells are called **gametes**. Fusion of two gametes – male and female – produces a **zygote**, which can then divide by mitosis to grow into a new individual organism.

The type of cell division usually responsible for producing gametes is called **meiosis** and it takes place only in specific regions of the adult organism known as the sex organs or gonads.

Normal adult cells of most eukaryotes contain two sets of chromosomes – they are diploid. Their genome consists of pairs of homologous chromosomes, each containing the same genes, but not necessarily the same versions (alleles) of each gene. During meiosis, one member of each homologous pair goes into each daughter cell.

The resulting daughter cells are haploid – each contains only one set of chromosomes. The haploid cells are not genetically identical to each other because each pair of homologous chromosomes separates into haploid cells independently of all others. The haploid daughter cells produced will differ because they contain the particular alleles of each gene found on the members of the homologous pairs they receive. Independent assortment and segregation of chromosomes is covered in more detail in A2.

So meiosis differs from mitosis in two important ways:

- meiosis produces cells containing half the number of chromosomes
- meiosis produces cells that are genetically different from each other, and from the parent cell.

These features, together with the fusion of gametes from different individuals, means that the offspring of sexually reproducing organisms are always different from each other (apart from identical twins, who are natural clones).

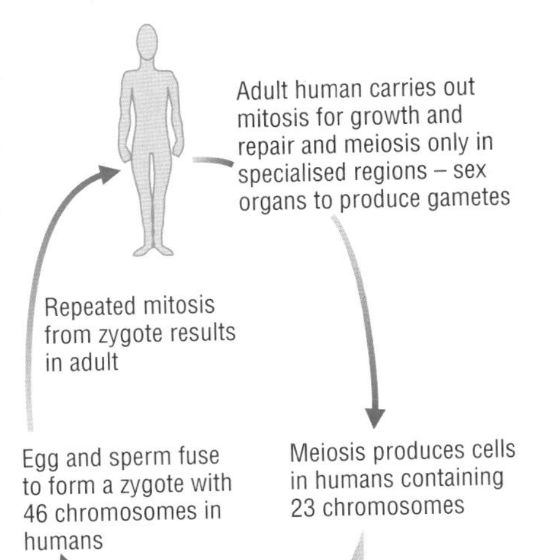

Figure 5 The human life cycle

Questions

1. Define the term 'clone'.
2. Suggest the advantages and disadvantages to farmers of crops that are genetically identical.
3. Many plants can reproduce both sexually and asexually. What are the advantages of this feature?

1.1 (16) Cell specialisation

By the end of this spread, you should be able to . . .

* **Define the term differentiation with respect to the production of erythrocytes and neutrophils derived from stem cells in bone marrow, and the production of xylem vessels and phloem sieve tubes from cambium.**
* **Describe and explain how cells become specialised for different functions, with reference to erythrocytes, neutrophils, epithelial cells, sperm cells, palisade cells and root hair cells.**
* **Explain the meaning of the terms tissue, organ, and organ system.**

Key definition

Differentiation refers to the changes occurring in cells of a multicellular organism so that each different type of cell becomes specialised to perform a specific function.

Examiner tip

Never refer to energy being made or produced in a cell. You can say that **ATP** is *made*. Energy is *released*.

Size matters

Within any **eukaryotic cell** there are different **organelles**, each performing a particular function. These organelles all contribute to the survival of the cell. From the **mitochondria** releasing energy for the cell's needs to the **ribosomes** assembling cell proteins, each organelle contributes to the processes needed to sustain life.

There is a physical limit to the size that a single cell can reach. This is governed by the need to support structures within the cell, and by the increasing difficulty of getting enough oxygen and nutrients into a cell to support its needs as its size increases.

Single-celled organisms have a large surface-area-to-volume ratio. They can receive oxygen and remove carbon dioxide by diffusion through the membrane. Multicellular organisms have a smaller surface-area-to-volume ratio, and not all cells are in contact with the external medium. This means they need specialised cells, forming tissues and organs, to carry out particular functions. These functions include delivery of oxygen and nutrients, and removal of waste.

Differentiation and specialisation

When organisms consist of many cells, some cells will be different from others. In this way, some cells perform one role very well. They become specialised in that role, while other cells are specialised for other roles (see Figure 1).

We refer to cells becoming specialised to carry out a particular role or function as differentiation.

Cells can differentiate in a number of ways, with changes to:
- the number of a particular organelle
- the shape of the cell
- some of the contents of the cell.

In some cases, differentiation may involve all three types of change.

Erythrocytes and neutrophils

Erythrocytes (red blood cells) and neutrophils (a type of white blood cell) play different roles (see Figure 2). Both are human cells and each began with the same set of chromosomes, so each is potentially capable of carrying out the same functions.

All blood cells are produced from undifferentiated stem cells in the bone marrow.

The cells destined to become erythrocytes lose their nucleus, mitochondria, Golgi apparatus and rough endoplasmic reticulum. They are packed full of the protein **haemoglobin**. The shape of the cells changes so that they become biconcave discs, and they are then capable of transporting oxygen from lungs to tissues.

(a) Sperm cell

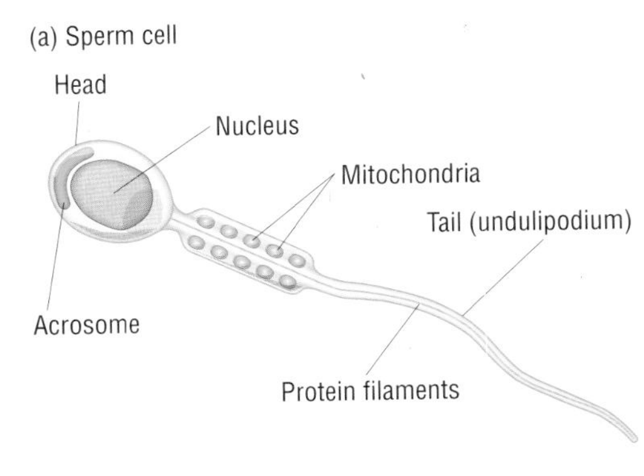

Sperm cells are specialised in a number of ways:
In organelle content:
(i) energy for movement of the undulipodium is generated by the many mitochondria present within the cell;
(ii) the sperm head contains a specialised lysosome (called an acrosome) that releases enzymes onto the outside of the egg so that the sperm nucleus can penetrate the egg in order to fertilise it.
In shape:
(i) sperm cells are very small, long and thin to help in easing their movement;
(ii) the single long undulipodium (flagellum) helps to propel the cell up the uterine tract towards the egg.
In content:
The sperm cell nucleus contains half the number of chromosomes of an adult cell in order to fulfil its role as a gamete.

(b) Root hair cell

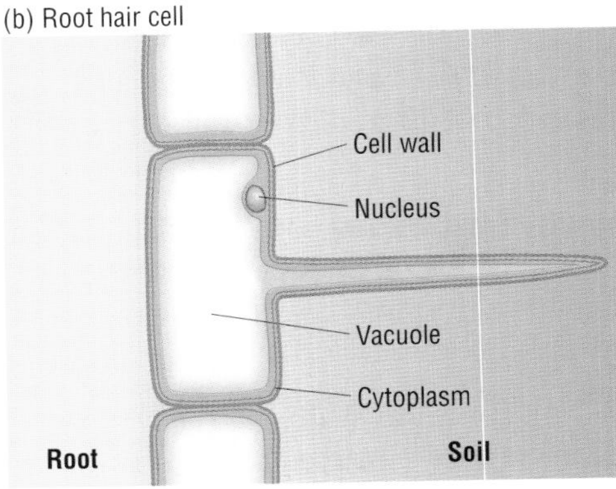

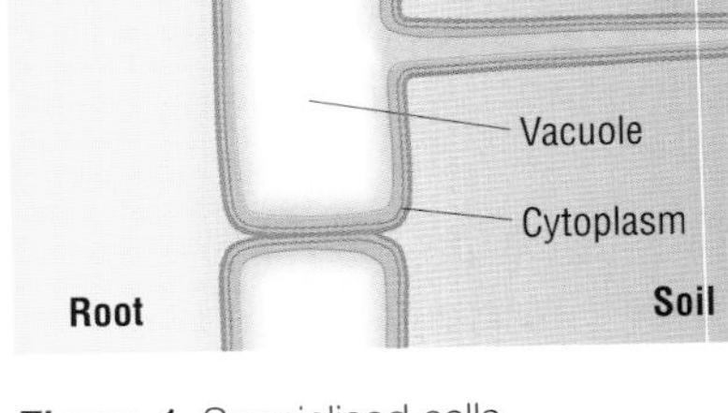

Root hair cells appear on the outside (epidermal layer) of young plant roots. The cells have a hair-like projection from their surface out into the soil. This greatly increases the surface area of root available to absorb water and minerals from the soil.

Figure 1 Specialised cells

Cells destined to become neutrophils keep their nucleus. Their cytoplasm appears granular because enormous numbers of **lysosomes** are produced. The role of neutrophils in the blood is to ingest invading microorganisms – so all those potent **enzymes** in the lysosomes enable the neutrophils to be specialised for killing microorganisms.

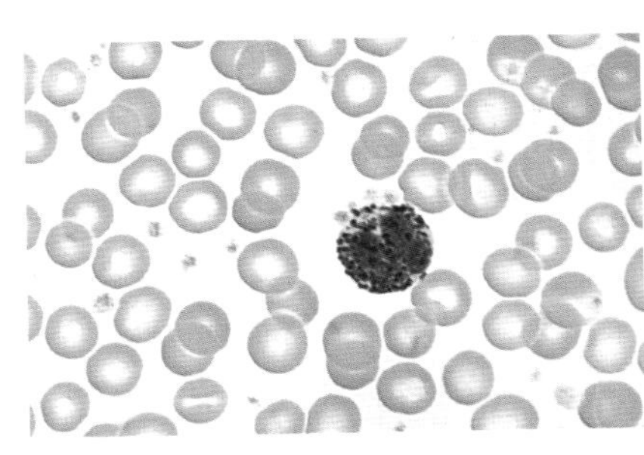

Figure 2 Blood smear showing erythrocytes, platelets and a single neutrophil stained differently (×600)

Organising the organism: tissues, organs and systems

In an organism consisting of many millions of cells, the activities and functions of all the cell types need to be organised to ensure the whole organism can survive. We describe the level of organisation seen in multicellular organisms under the following headings.

Tissues

A collection of cells that are similar to each other and perform a common function. They may be found attached to each other, but not always. Examples include xylem and phloem in plants; epithelial and nervous tissues in animals.

Organs

A collection of tissues working together to perform a particular function is called an organ. Examples include leaves of plants and the liver in animals.

Organ systems

An organ system is made up of a number of organs working together to perform an overall life function. Examples include the excretory system and the reproductive system.

As with many definitions in science, there are cell types or groups that do not easily fit into the categories we have defined. Unfortunately, organisms don't always behave in ways that make it easy for us to categorise them!

Questions

1 Explain why differentiation to produce erythrocytes involves a change in shape.
2 Red blood cells cannot divide as they have no nucleus. State two other processes that red blood cells cannot carry out.
3 Describe how the following are specialised for their roles: neutrophil, sperm cell, root hair cell.

Organising the organism

By the end of this spread, you should be able to . . .

* **Define the term differentiation with reference to the production of xylem vessels and phloem sieve tubes from cambium.**
* **Explain, with the aid of diagrams and photographs, how cells are organised into tissues such as squamous and ciliated epithelium, xylem and phloem.**
* **Discuss the importance of cooperation between cells, tissues, organs and organ systems.**

Key definition

A collection of similar cells that carry out a specific function is called a **tissue**. A collection of tissues that work together to carry out a specific function is called an **organ**.

Transport tissues – xylem and phloem

Plants need to move water and minerals from the soil, through their roots and stems, up into the leaves. They also need to move the products of photosynthesis from the leaves to other parts of the plant to use for growth or to store for later use. Xylem and phloem (Figure 1) come from dividing **meristem cells** (see spread 1.1.15) such as **cambium**. Meristem cells undergo **differentiation** to form the different kinds of cells in the transport tissues.

Xylem tissue consists of xylem vessels with **parenchyma** cells and fibres. Meristem cells produce small cells that elongate. Their walls become reinforced and waterproofed by deposits of **lignin**. This kills the cell contents. The ends of the cells break down so that they become continuous, long tubes with a wide **lumen**. Xylem tissue is well suited for transporting water and minerals up the plant. It also helps to support the plant.

Examiner tip

If you are asked to explain how the structure of a cell or organ is suited to carry out its function, you need to state the feature *and* explain how it enables the cell or organ to do its job.

Phloem tissue consists of **sieve tubes** and **companion cells**. The meristem tissue produces cells that elongate and line up end-to-end to form a long tube. Their ends do not break down completely, but form sieve plates between the cells. The sieve plates allow the movement of materials up or down the tubes. Next to each sieve tube is a companion cell. Companion cells are very metabolically active. Their activities play an important role in moving the products of photosynthesis up and down the plant in the sieve tubes.

Forming a lining – epithelial tissues in animals

Animal tissues are grouped into four main categories:

- epithelial tissue – layers and linings
- connective tissues – hold structures together and provide support (examples include cartilage, bone and blood)
- muscle tissue – cells specialised to contract and move parts of the body
- nervous tissue – cells that can convert stimuli to electrical impulses and conduct those impulses.

Within these categories are smaller groups of tissues, for example in the epithelial tissues.

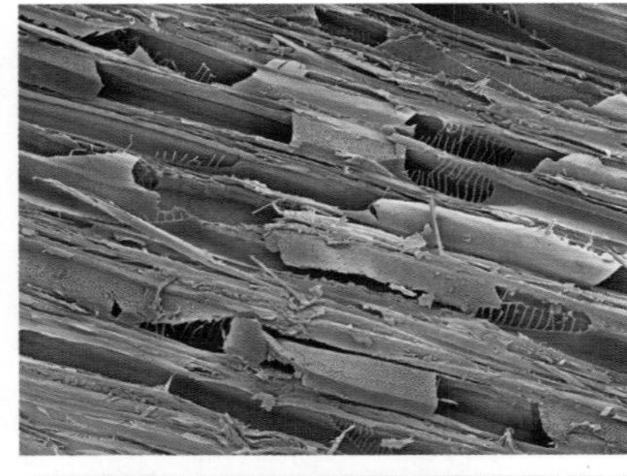

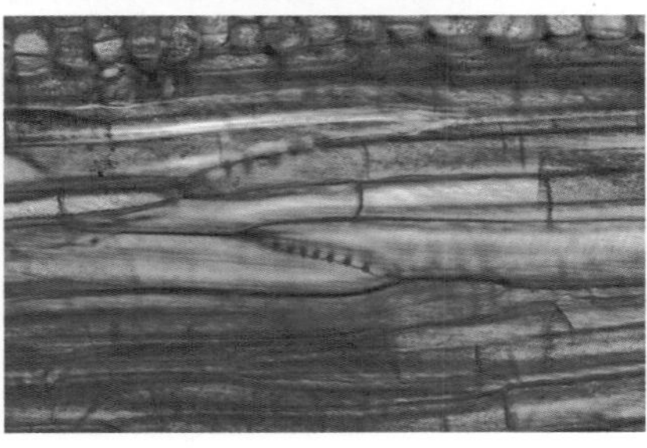

Figure 1 Xylem (top) and phloem (bottom) tissue, longitudinal sections (×33, ×60)

Squamous epithelial tissue is made up of cells that are flattened, so they are very thin. The cells together form a thin, smooth, flat surface. This makes them ideal for lining the insides of tubes such as blood vessels, where fluids can pass easily over them. Squamous epithelial tissue also forms thin walls, such as the walls of the alveoli in the lungs. It provides a short diffusion pathway for the exchange of oxygen and carbon dioxide.

The squamous cells are held in place by the basement membrane. This is secreted by epithelial cells. It is made of **collagen** and **glycoproteins**. The basement membrane attaches epithelial cells to connective tissue.

Ciliated epithelial tissue is made up of column-shaped cells. This type of tissue is often found on the inner surface of the tubes, for example in the trachea, bronchi and bronchioles (the airways in the lungs) and in the uterus and oviducts. The part of the cell surface that is 'exposed' in the tube space (lumen) is covered with tiny projections called **cilia**. Some cells produce **mucus**. The cilia wave in a synchronised rhythm and move the mucus.

In the breathing tract, small particles and microorganisms are trapped in the mucus. The cilia waft it up to the back of the throat to be swallowed. One of the problems with smoking is that nicotine in the smoke paralyses cilia, so they can't sweep to move the mucus. Cilia also become damaged by tar in the smoke and are destroyed, so there are fewer of them. The mucus becomes trapped in the lungs and microbes can't be cleared out of the airways to be killed by the stomach acid.

The rhythmic waves generated by ciliated epithelium move egg cells from the ovary along the oviduct.

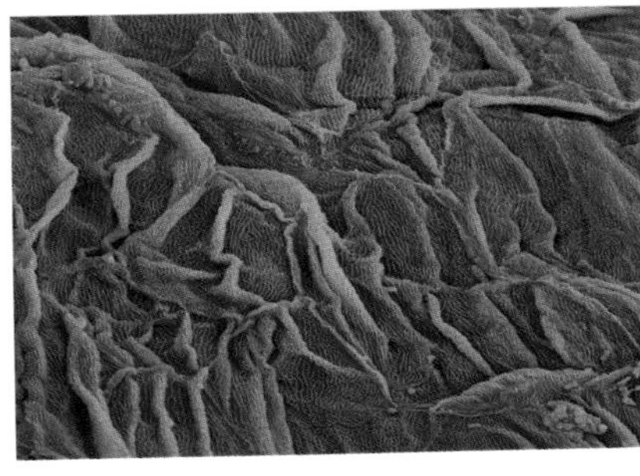

Figure 2 Ciliated (top) and squamous (bottom) epithelial tissue sections

Harvesting light – cooperation in action

Leaves are the major organs of photosynthesis in a plant. Their cells, tissues and overall shape are arranged to help maximise the rate of photosynthesis. The requirements for photosynthesis are:

- light
- a supply of water
- a supply of carbon dioxide
- the presence of chlorophyll.

As the products of photosynthesis build up, they need to be removed to where they are needed. The waste product, oxygen gas, must be excreted.

In order to balance these needs, the leaf is adapted in a number of ways.

- A transparent upper surface layer, the upper epidermis, lets light through.
- A palisade layer underneath consists of long, thin, tightly packed cells containing a lot of chloroplasts that contain chlorophyll.
- A loosely packed spongy mesophyll layer has many air spaces to allow circulation of gases.
- A lower epidermis layer has pores called stomata. These allow gases to be exchanged between the leaf and the outside air. The stomata each have two **guard cells** that can swell to open the pore. When the guard cells are not turgid, the stoma closes.
- A leaf vein system containing xylem and phloem tissues supports the leaf as well as carrying the transport tissues. These tissues transport water into the leaf, and the products of photosynthesis out of the leaf to other parts of the plant.

Opening and closing stomata – the role of guard cells

- Guard cells are specialised cells that appear in pairs on the lower epidermis. Unlike other lower epidermis cells, they contain chloroplasts, and their cell walls contain spiral thickenings of cellulose. When water is moved into these cells they become turgid, and because of the spirals in the walls of the inner edges, only the outer walls stretch. The two guard cells bulge at both ends so a pore opens between them. This pore is known as a **stoma** (plural stomata).

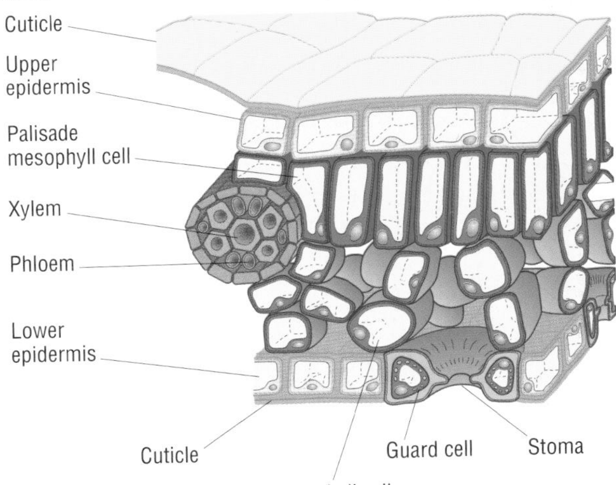

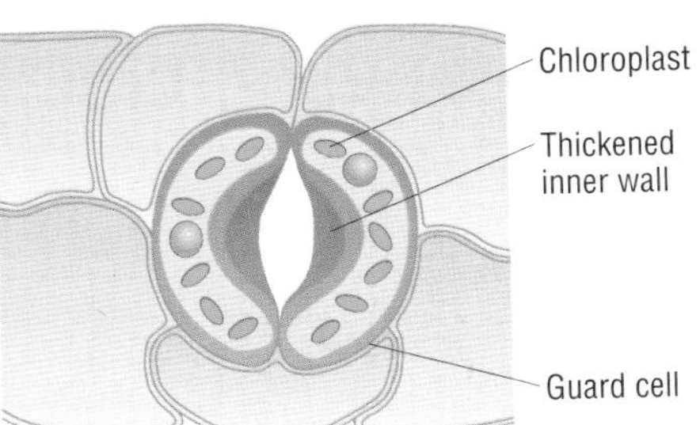

Figure 3 Cross-section through the leaf of a mesophytic dicotyledon, showing the layout of the cells, including a section through a vein. Below it is a diagram of a stoma surrounded by a pair of guard cells

Locomotion – an example of systems cooperation

How do you move from one place to another? The muscular and skeletal systems must work together in order for movement to take place. But this can happen only if the nervous system 'instructs' the muscles to coordinate their actions. As muscles and nerves work, they use energy. They require a supply of nutrients and oxygen from the circulatory system, which in turn receives these chemicals from the digestive and ventilation systems.

Questions

1. Why do plants need to move water to their leaves?
2. Explain how each of the tissue types that form the leaf organ contribute to maximising the rate of photosynthesis.
3. Suggest why it is important that the products of photosynthesis can be moved in both directions through the sieve tubes.

1.1 Cells summary

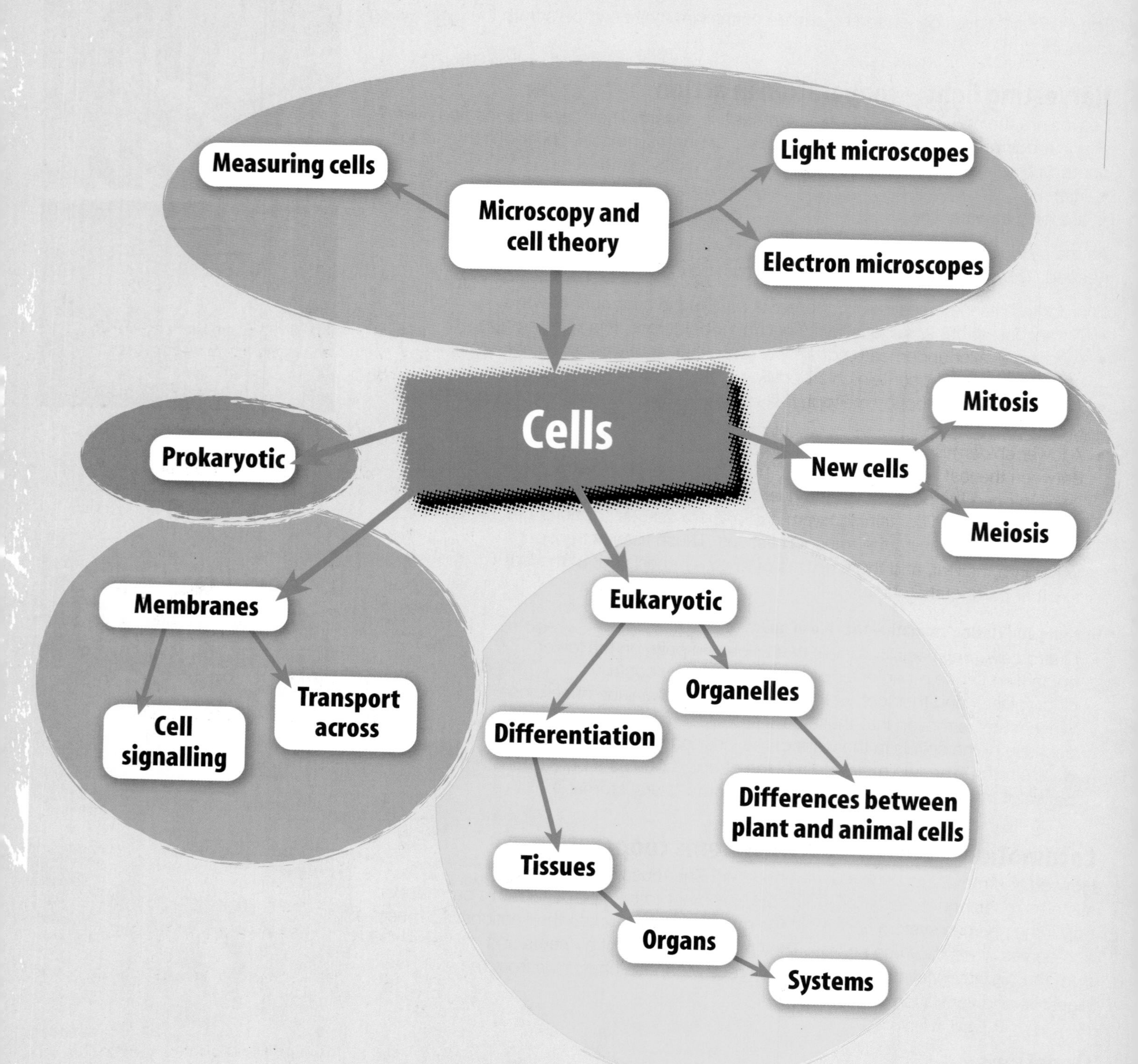

Practice questions

1. State two processes in which mitosis occurs in humans. [2]

2. Below is a list of parts of a cell, labelled **A–F**. Each of the list of statements, numbered **1–6**, refers to one of these parts of the cell.

A	nucleus	1	where some lipids, including steroids, are made
B	mitochondrion	2	controls entry of substances into the cell
C	plasma (cell surface) membrane	3	controls the activities of the cell
D	chloroplast	4	where polypeptides (proteins) are made
E	smooth endoplasmic reticulum	5	where photosynthesis takes place
F	ribosomes	6	where aerobic respiration takes place

Match a statement to each part of the cell. The first one has been done for you.

A ....3...... [5]

3. The diagram represents the structure of a plasma (cell surface) membrane.

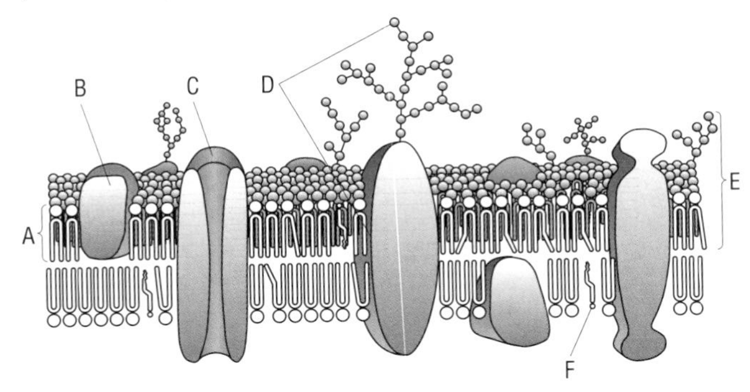

Figure 3.1

(a) Name the molecules **A–F**. [6]

(b) State the approximate width of the membrane. [1]

(c) Complete the table below to:

(i) show the direction in which water will move across the plasma (cell surface) membranes of cells in different conditions;

(ii) indicate whether or not the cell will burst.

Place a (✓) or a (✗) in each box of the table as appropriate.

	initial net movement of water		cell bursts
	in	**out**	
leaf mesophyll cell immersed in distilled water			
red blood cell immersed in concentrated salt solution			
red blood cell immersed in distilled water			

[6]

(d) With reference to Figure 3.1, outline the ways in which substances, other than water, cross plasma (cell surface) membranes.
(In this question, one mark is available for the quality of written communication.) [7]

4. Explain the role of mitotic cell division in plants. [4]

5. A diagram of a eukaryotic cell nucleus is shown.

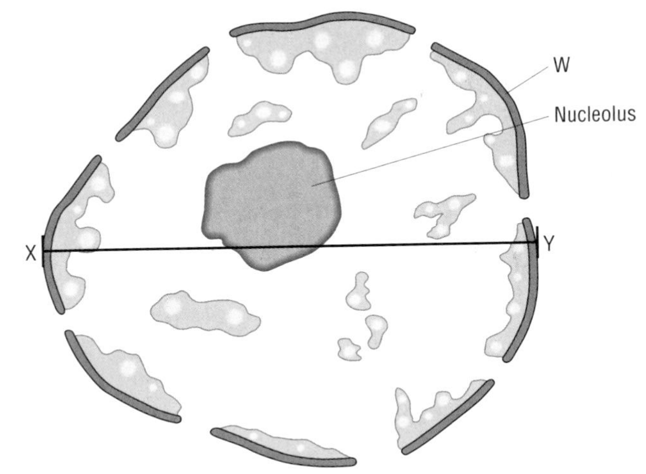

(a) Name the structure labelled **W**. [1]

(b) The actual diameter of the nucleus, measured along the line **XY**, is 7 μm.
Calculate the magnification of the nucleus. Show your working. [2]

6. Some single-celled organisms live in estuaries where the concentration of salt changes regularly.
Explain, **in terms of water potential**, the problem faced by these organisms. [2]

1.1 Examination questions

1 The table below compares the features of typical eukaryotic and prokaryotic cells.

(a) Complete the table by placing one of the following, as appropriate, in each empty box in the table:

a tick (✓)

a cross (✗)

the words 'sometimes present'

One of the two boxes in each row has been completed for you.

	eukaryotic cell	prokaryotic cell
cell wall	sometimes present	✓
nuclear envelope	✓	
Golgi apparatus		✗
undulipodium/ flagellum	sometimes present	
ribsomes		✓
carries out respiration	✓	
chloroplast	sometimes present	

[6]

(b) **(i)** Explain the meaning of the term *tissue*. [2]

(ii) State **one** example of a plant tissue. [1]

(iii) State **one** example of an animal tissue. [1]

[Total: 10]

(Jan01 2801)

2 Xylem is found in roots, stems and leaves of flowering plants. Xylem contains several types of cell, two of which are vessel elements and fibres. Vessel elements are arranged in columns with no end walls between them, forming long, hollow tubes. These tubes are called vessels and their function is to transport water and mineral salts. Fibres are narrow, very elongated, thick-walled cells that provide support.

(a) Use the information to explain why xylem is considered to be a tissue. [4]

Xylem tissue can be viewed using a light microscope or an electron microscope.

(b) With reference to **both** light and electron microscopy, explain and distinguish between the terms *magnification* and *resolution*. [4]

[Total: 8]

(Jan02 2801)

3 Figure 3.1 is a plan diagram of tissues in a transverse section of a dicotyledonous leaf.

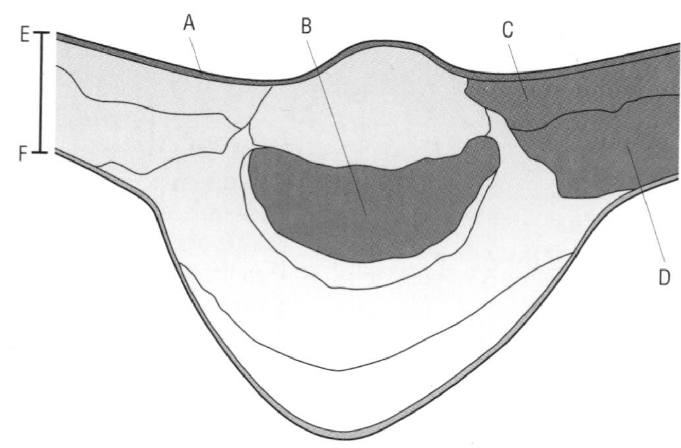

Figure 3.1

(a) Name tissues **A**–**D**. [4]

(b) The actual thickness of the leaf along the line **EF** is 0.6 mm.

Calculate the magnification of the diagram. Show your working.

Magnification = ×.................................. [2]

[Total: 6]

(Jan04 2801)

4 Figure 4.1 is a drawing of an animal cell as seen under an electron microscope.

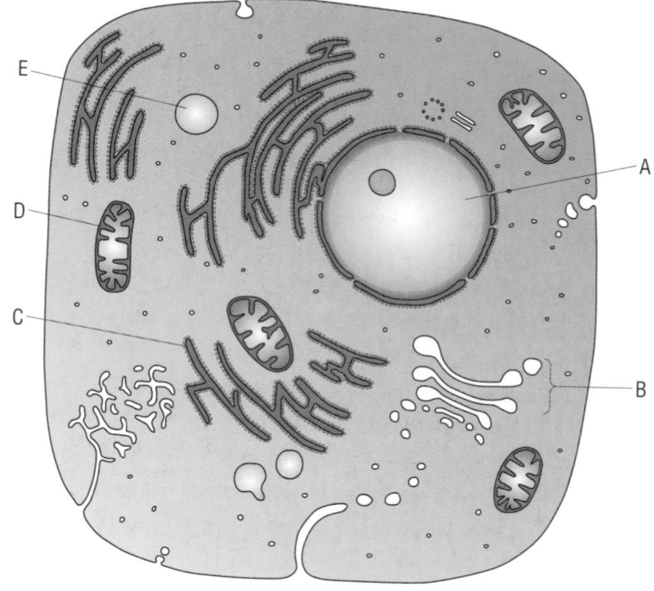

Figure 7.1

Complete the following table by:

- identifying the parts of the cell **A**–**E**
- naming the part of the cell responsible for the function stated.

Answers to examination questions will be found on the Exam Café CD.

The first one has been done for you.

function	part of cell	label
controls activities of the cell	nucleus	*A*
carries out aerobic respiration		
attaches to mRNA in protein synthesis		
produces secretory vesicles		
contains digestive enzymes		

[Total: 8]
(Jan05 2801)

5 **(a)** Figure 5.1 is a drawing of an organelle from a ciliated cell as seen with an electron microscope.

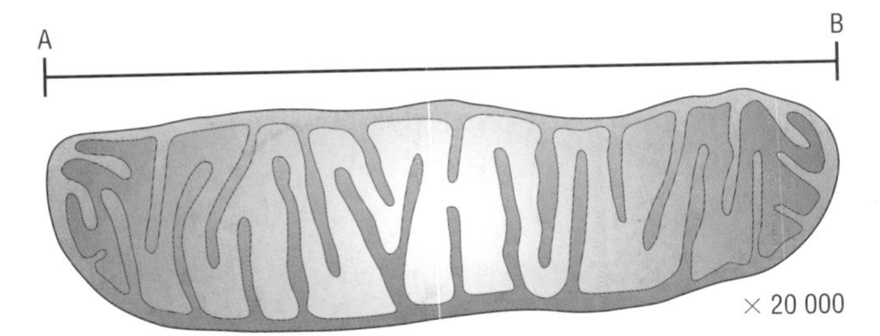

Figure 5.1

(i) Name the organelle shown in Figure 5.1. [1]
(ii) State the function of this organelle. [3]
(iii) State why ciliated cells contain relatively large numbers of these organelles. [1]
(iv) Calculate the actual length of the organelle as shown by the line **AB** in Figure 5.1.
Express your answer to the **nearest micrometre** (μm).
Show your working. [2]

(b) An image drawn to the same magnification as Figure 5.1 could be produced using a light microscope.
Explain why such an image would be of little use when studying cells. [2]

[Total: 8]
(Jun06 2801)

6 **(a)** Describe the role of mitosis. [3]
Figure 6.1 is a diagram that shows the stages of the mitotic cell cycle.

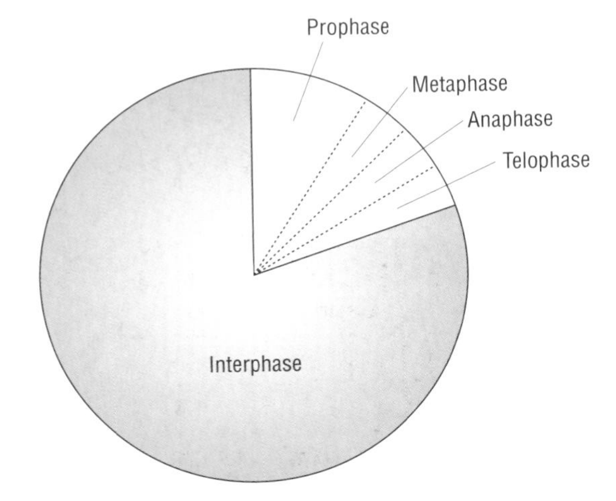

Figure 6.1

(b) (i) Which processes must occur in a cell during interphase before mitosis can take place? [3]
(ii) Draw an arrow on Figure 6.1 to indicate the sequence in which the stages occur during the mitotic cell cycle. [1]

(c) Name the stage of mitosis shown in Figure 6.1 in which each of the following events occurs.
(i) Chromosomes split at centromeres. [1]
(ii) Chromosomes become visible. [1]
(iii) Nuclear envelope re-forms. [1]
(iv) Chromatids move to opposite poles of the cell. [1]
(v) Chromosomes line up along the equator of the spindle. [1]

[Total: 12]
(Jun04 2801)

UNIT 1 Module 2
Exchange and transport

Introduction

All living things need to gain certain substances from the environment. For animals, these include:

- oxygen, which is used for respiration
- large organic nutrients, which will be digested
- smaller nutrients such as salts and vitamins
- water.

Plants need to gain:

- carbon dioxide for photosynthesis
- oxygen for respiration
- water
- minerals such as nitrate.

Living things also need to remove wastes such as carbon dioxide and urea.

Single-celled organisms can survive by exchanging these substances over their entire surface area. In larger and more active living things, special exchange surfaces are needed. (The gaseous exchange surface in humans is more than 50 times the area of the skin.) Larger and more active living things also need a way to transport these vital substances around the body so that all cells receive what they need. The xylem vessels of a tree, held open by rings of stiffer tissue as seen here, carry water and minerals up through the tree.

In this module you will learn about how these exchange surfaces work and how well adapted they are to their job. You will also learn about the ways in which large, multicellular plants and animals transport substances around their bodies.

Test yourself

1 What organ is used for exchange of gases in plants?
2 Where would you expect to find alveoli?
3 Where are nutrients absorbed in animals?
4 What is absorbed through the roots of a plant?
5 What creates the pressure that moves blood in the mammalian transport system?
6 What is transpiration?
7 What is so special about cacti?

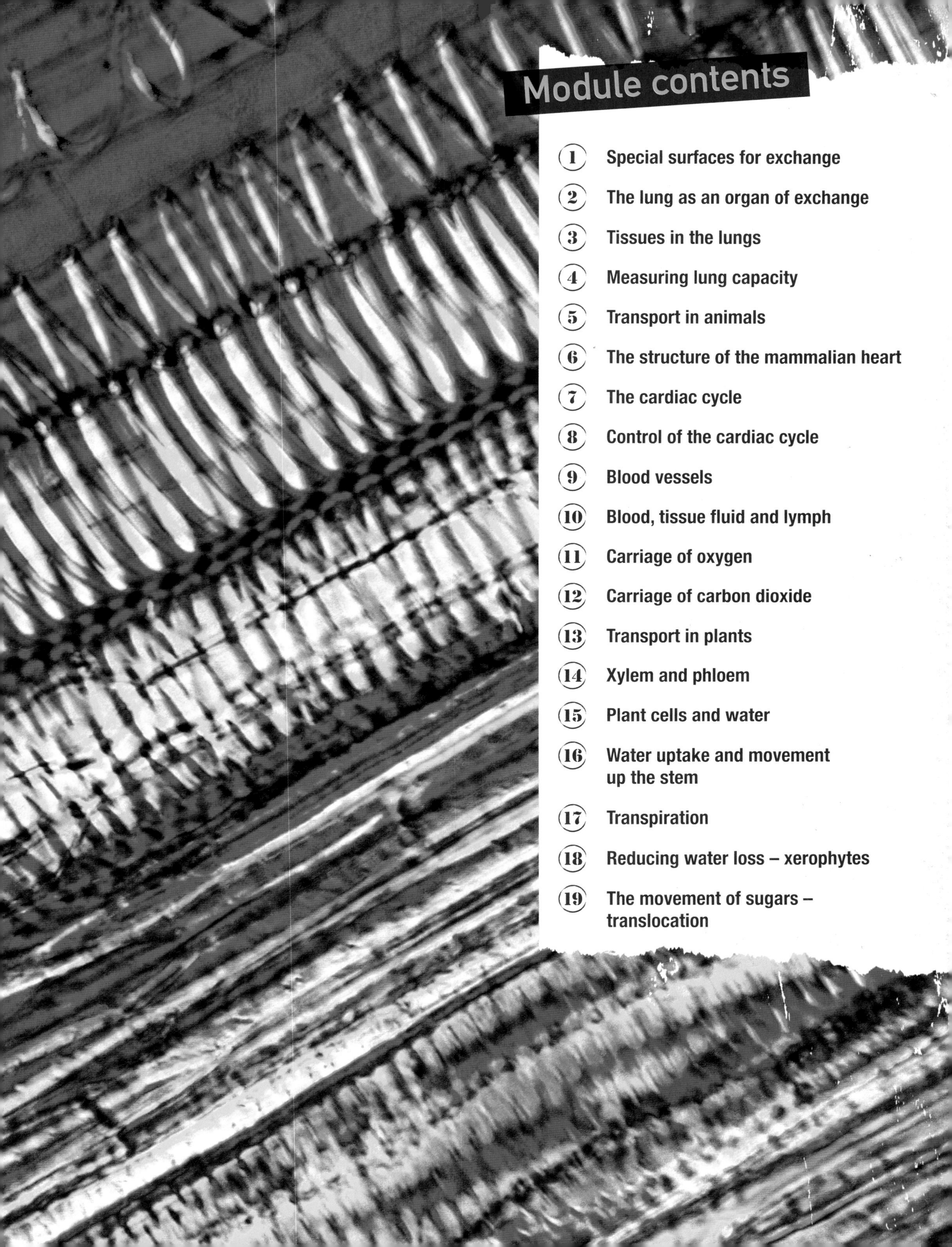

Module contents

1.2 (1) Special surfaces for exchange

By the end of this spread, you should be able to . . .

* **Explain, in terms of surface-area-to-volume ratio, why multicellular organisms need specialised exchange surfaces and single-celled organisms do not.**

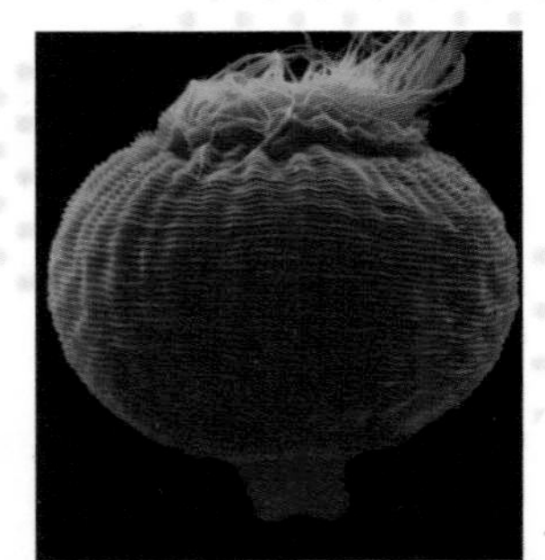

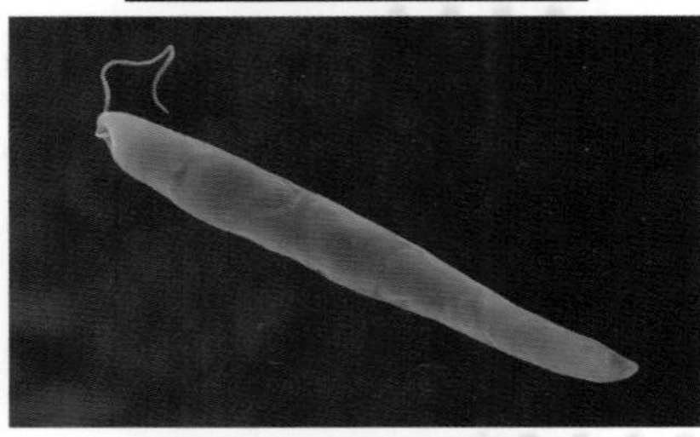

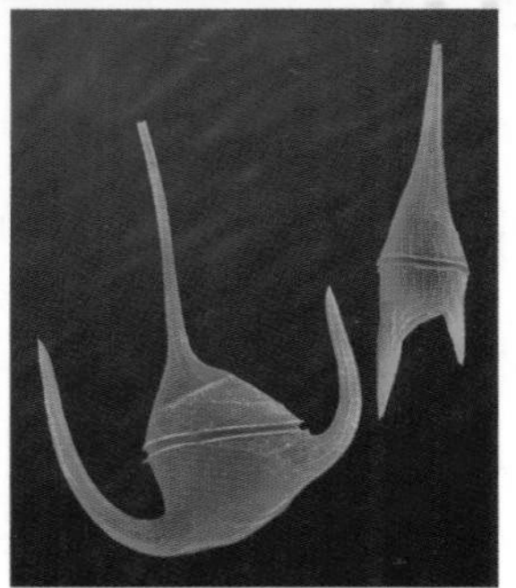

Figure 1 Some single-celled organisms, from top to bottom: *Vorticella*, *Euglena*, dinoflagellates (×1400, ×580, ×720)

Why organisms need special exchange surfaces

All living cells need certain substances to keep them alive. These substances include:

- oxygen for aerobic respiration
- glucose as a source of energy
- proteins for growth and repair
- fats to make membranes and to be a store of energy
- water
- minerals to maintain their water potential and to help enzyme action and other aspects of metabolism.

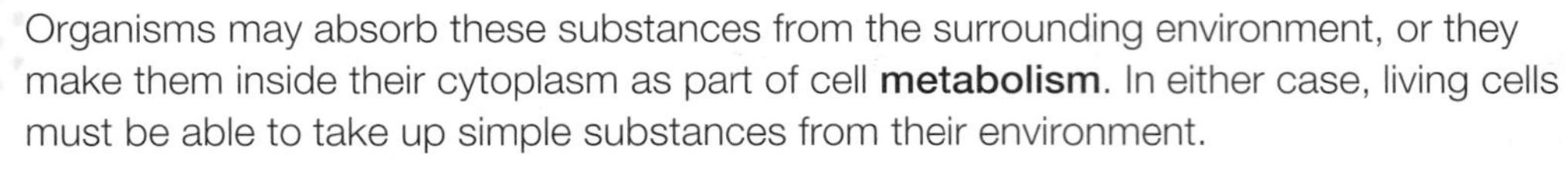

Organisms may absorb these substances from the surrounding environment, or they make them inside their cytoplasm as part of cell **metabolism**. In either case, living cells must be able to take up simple substances from their environment.

All living cells also need to remove waste products from the metabolic activities occurring in their cytoplasm. These include:

- carbon dioxide (in animals and microorganisms, and also from plant cells that are not actively carrying out **photosynthesis**)
- oxygen (from photosynthesis in some plant cells and some protoctists)
- other wastes such as ammonia or urea, which contain excess nitrogen.

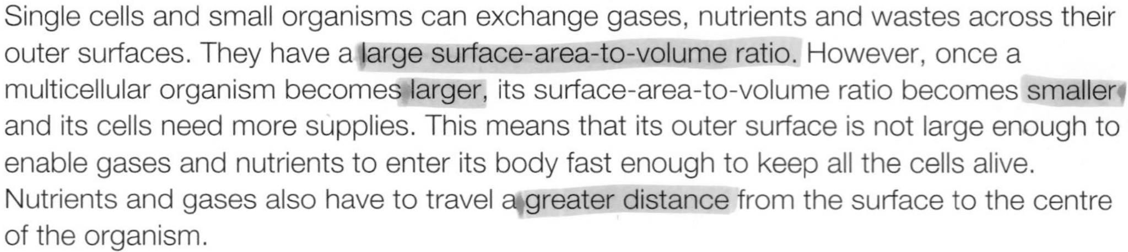

Single cells and small organisms can exchange gases, nutrients and wastes across their outer surfaces. They have a large surface-area-to-volume ratio. However, once a multicellular organism becomes larger, its surface-area-to-volume ratio becomes smaller and its cells need more supplies. This means that its outer surface is not large enough to enable gases and nutrients to enter its body fast enough to keep all the cells alive. Nutrients and gases also have to travel a greater distance from the surface to the centre of the organism.

Larger organisms need a larger area to exchange more substances. Often they combine this with a transport system to move substances around the body. (There is a more detailed explanation of surface-area-to-volume ratios in spread 1.2.5.)

What exchange surfaces look like

All good exchange surfaces have certain features in common:

- large surface area to provide more space for molecules to pass through – often achieved by folding the walls and membranes
- thin barrier to reduce the **diffusion** distance
- fresh supply of molecules on one side to keep the concentration high
- removal of required molecules on the other side to keep the concentration low.

The latter three features on this list are important to maintain a steep **diffusion gradient**.

Key definition

An **exchange surface** is a specialised area that is adapted to make it easier for molecules to cross from one side of the surface to the other.

Examiner tip

Use the term 'thin barrier' to avoid confusion between a 'thin wall of cells' and a 'thin cell wall'.

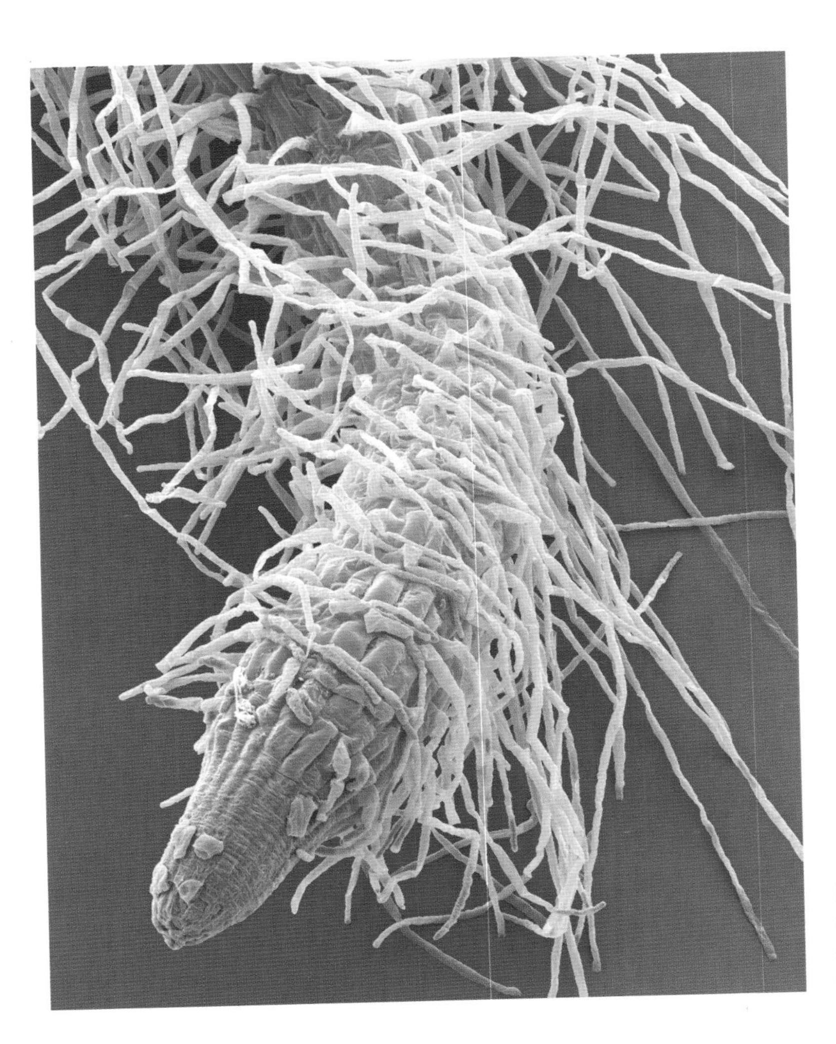

Figure 2 Scanning electron micrograph of root hairs at a root tip (×100)

Some exchange surfaces also use an **active transport** mechanism to increase exchange.

Some examples of specialised exchange surfaces

Exchange surfaces do not occur just at the surface of a large organism. They are also found in all the organs where substances are removed from the transport system, and where wastes are returned to the transport system. A good example is the walls of the alveoli in the lungs. You will find out more about the lungs later.

There are many other exchange surfaces in living organisms. These include:

- small intestine, where nutrients are absorbed
- liver, where levels of sugars in the blood are adjusted
- root hairs of plants, where water and minerals are absorbed
- hyphae of fungi, where nutrients are absorbed.

Questions

1 Explain why large, active organisms need special surfaces for exchange.
2 Describe and explain the features that make an exchange surface efficient.
3 How is a plant root adapted for taking in water and mineral ions?

1.2 ② The lung as an organ of exchange

By the end of this spread, you should be able to . . .

* **Describe the features of an efficient exchange surface with reference to diffusion of oxygen and carbon dioxide across an alveolus.**
* **Describe the features of the mammalian lung that adapt it to efficient gas exchange.**
* **Outline the mechanism of breathing (inspiration and expiration) in mammals, with reference to the function of the rib cage, intercostal muscles and diaphragm.**

Key definition

Gaseous exchange is the movement of gases by diffusion between an organism and its environment across a barrier such as the alveolus wall.

Examiner tip

Alveoli and capillaries both have thin walls. Their walls are one cell thick.

Don't say that alveoli have 'thin cell walls' – this is nonsense as they are animal cells!

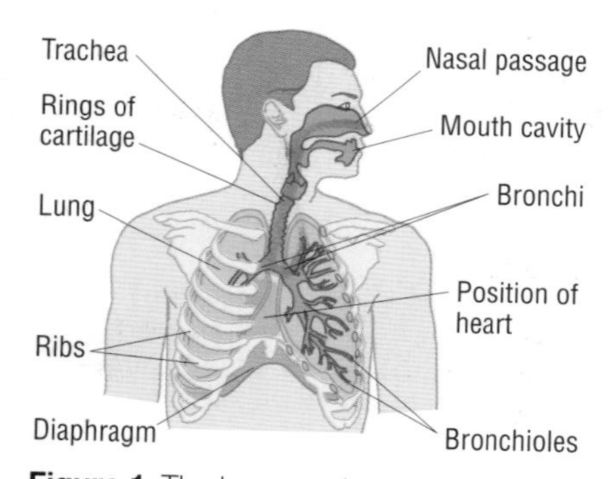

Figure 1 The lungs and associated structures

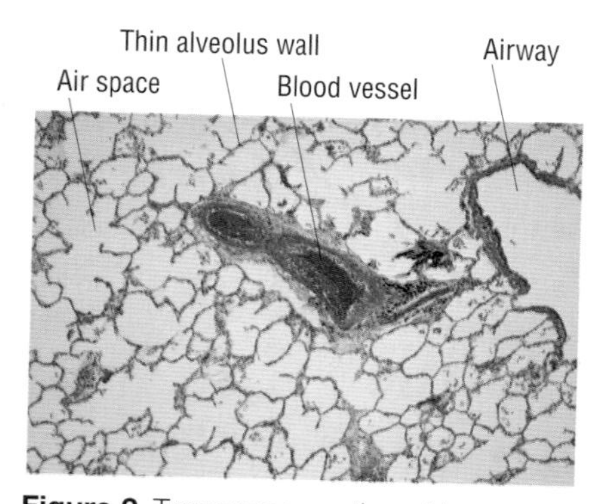

Figure 2 Transverse section of lung tissue showing alveoli and other structures (×50)

The lungs

The lungs are a large pair of inflatable structures lying in the chest cavity. Air can pass into the lungs through the nose and along the **trachea** (windpipe), **bronchi** and **bronchioles**. Each part of this airway is adapted to its function of allowing the passage of air. Finally the air reaches tiny, air-filled sacs called **alveoli**. The walls of the alveoli are the surface where the exchange of gases takes place.

The lungs are protected by the ribs. Movement of the ribs together with the action of the **diaphragm** (a layer of muscular tissue beneath the lungs) help to produce breathing movements (**ventilation**).

Gaseous exchange in the lungs

Gases pass both ways through the thin walls of the alveoli. Oxygen passes from the air in the alveoli to the blood in the capillaries. Carbon dioxide passes from the blood to the air in the alveoli.

How the lungs are adapted for exchange

Earlier we listed the features of a good exchange surface (spread 1.2.1). Now we shall see how these apply to the lungs.

Large surface area

The large surface area provides more space for molecules to pass through. The individual alveoli are very small – about 100–300 μm across. But they are so numerous that the total surface area of the lungs is much larger than that of our skin. It has been calculated that the total surface area of the lung exchange surface is about $70\,m^2$, or about half the size of a tennis court.

A barrier permeable to oxygen and carbon dioxide

The **plasma membranes** that surround the thin cytoplasm of the cells form the barrier to exchange. These readily allow the diffusion of oxygen and carbon dioxide.

Thin barrier to reduce diffusion distance

There are a number of adaptations to reduce the distance the gases have to diffuse:

- the alveolus wall is one cell thick
- the capillary wall is one cell thick
- both walls consist of squamous cells – this means flattened or very thin cells
- the capillaries are in close contact with the alveolus walls
- the capillaries are so narrow that the red blood cells are squeezed against the capillary wall, making them closer to the air in the alveoli and reducing the rate at which they flow past in the blood
- the total barrier to diffusion is only two flattened cells thick and is less than 1 μm thick.

A thin layer of moisture lines the alveoli. This moisture passes through the cell membranes from the cytoplasm of the alveolus cells. As we breathe out, it evaporates and is lost. The lungs must produce a substance called a **surfactant** to reduce the cohesive forces between the water molecules. Without the surfactant, the alveolus would collapse due to the cohesive forces between the water molecules lining the air sac.

Maintaining the diffusion gradient

For diffusion to be rapid, a steep diffusion gradient is needed. This means having a high concentration of molecules on the supply side of the exchange surface and a low concentration on the demand side. To maintain a steep diffusion gradient, a fresh supply of molecules on one side is needed to keep the concentration there high, and a way of removing molecules from the other side is needed to keep the concentration there low.

This is achieved by the action of the blood transport system and the ventilation (breathing) movements.

The blood brings carbon dioxide from the tissues to the lungs. This ensures that the concentration of carbon dioxide in the blood is higher than that in the air of the alveoli. It also carries oxygen away from the lungs. This ensures that the concentration of oxygen in the blood is kept lower than the concentration in the air inside the alveoli. The heart pumps the blood along the pulmonary artery to the lungs. In the lungs, the artery divides up to form finer and finer vessels. These eventually carry blood into tiny capillaries that are only just wide enough for a red blood cell to squeeze through. These capillaries lie over the surface of the alveoli.

The breathing movements of the lungs ventilate the lungs. They replace the used air with fresh air. This brings more oxygen into the lungs and ensures that the concentration of oxygen in the air of the alveolus remains higher than the concentration in the blood. Ventilation also removes air containing carbon dioxide from the alveoli. This ensures that the concentration of carbon dioxide in the alveoli remains lower than that in the blood.

This constant supply of gas to one side of the exchange surface and its removal from the other side ensures that diffusion, and therefore exchange, can continue.

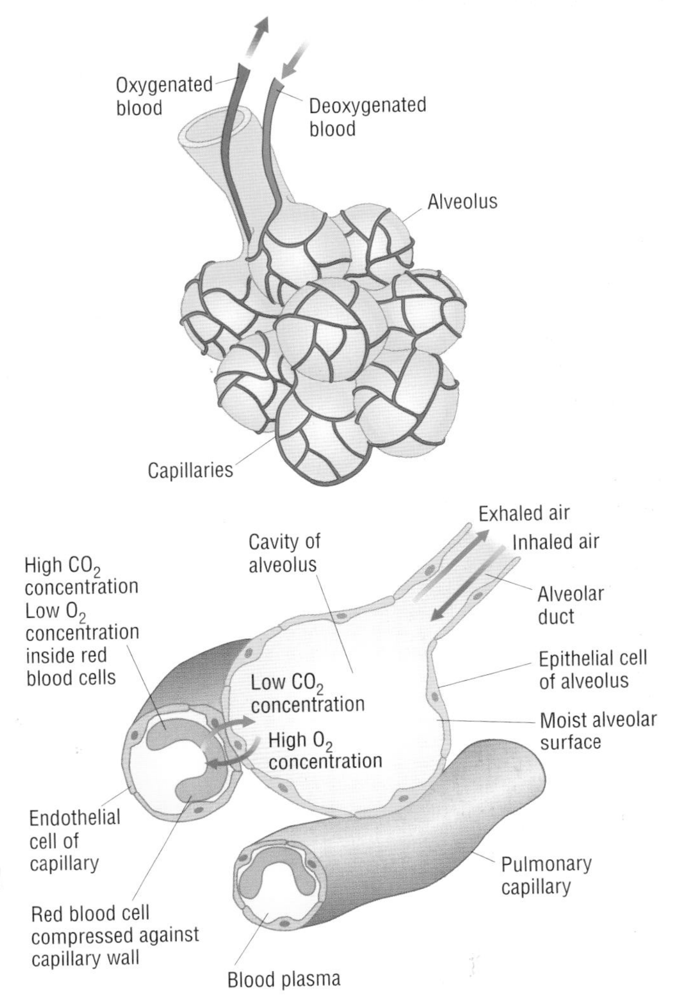

Figure 3 Capillary network over the surface of alveoli and details of gaseous exchange

Inhaling (inspiration)	Exhaling (expiration)
Diaphragm contracts to become flatter and pushes digestive organs down	Diaphragm relaxes and is pushed up by displaced organs underneath
External intercostal muscles contract to raise ribs	External intercostal muscles relax and ribs fall
Volume of chest cavity increases	Volume of chest cavity decreases
Pressure in chest cavity drops below atmospheric pressure	Pressure in lungs increases and rises above atmospheric pressure
Air moves into lungs	Air moves out of lungs

Table 1 Breathing movements in humans

Questions

1. State three ways in which the structure of the lungs allows efficient gas exchange.
2. Explain why the barrier to diffusion must be as thin as possible.
3. Describe how a steep diffusion gradient is achieved in the lungs.

Tissues in the lungs

By the end of this spread, you should be able to . . .

* **Describe the distribution of cartilage, ciliated epithelium, goblet cells and smooth muscle and elastic fibres in the trachea, bronchi and bronchioles and alveoli of the mammalian gaseous exchange system.**
* **Describe the functions of cartilage, cilia, goblet cells, smooth muscle and elastic fibres in the mammalian gaseous exchange system.**

The lungs

The trachea, bronchi and bronchioles are airways that allow passage of air into the lungs and out again. To be effective, these airways must meet certain requirements:

- the larger airways must be large enough to allow sufficient air to flow without obstruction
- they must also divide into smaller airways to deliver air to all the **alveoli**
- the airways must be strong enough to prevent them collapsing when the air pressure inside is low (this low pressure occurs during inhalation)
- they must be flexible, to allow movement
- they must also be able to stretch and recoil.

Figure 1 Bronchus in transverse section showing ciliated epithelium (×40)

The trachea and bronchi

The trachea and bronchi have a similar structure. They differ only in size – the bronchi are narrower than the trachea. They have relatively thick walls that have several layers of tissue.

- Much of the wall consists of **cartilage**.
- The cartilage is in the form of incomplete rings or C-rings in the trachea, but is less regular in the bronchi.
- On the inside surface of the cartilage is a layer of glandular tissue, connective tissue, **elastic fibres**, **smooth muscle** and blood vessels. This is often called the 'loose tissue'.
- The inner lining is an epithelium layer that has two types of cell. Most of the cells have **cilia**. This is called **ciliated epithelium**. Among the ciliated cells are **goblet cells**.

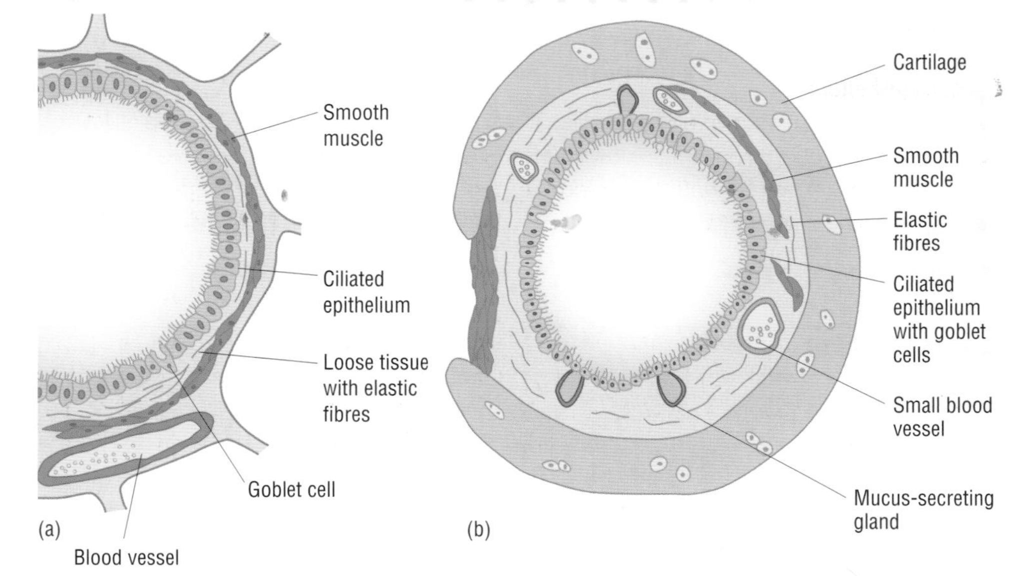

Figure 2 a Bronchiole (×400) and **b** trachea in transverse section (×2)

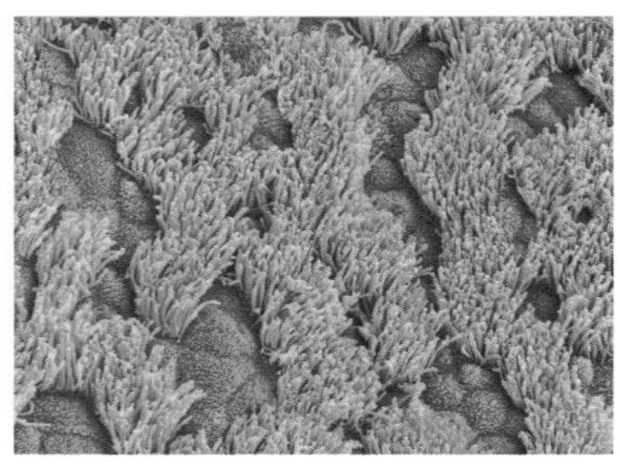

Figure 3 Ciliated epithelium (×1500)

The bronchioles

The bronchioles are much narrower than the bronchi. The larger bronchioles may have some cartilage, but smaller ones have no cartilage. The wall is made mostly of smooth muscle and elastic fibres. The smallest bronchioles have clusters of alveoli (air sacs) at their ends.

What is the role of each tissue?

Cartilage

The cartilage plays a structural role. It supports the trachea and bronchi, holding them open. This prevents collapse when the air pressure inside is low during inhalation. The cartilage does not form a complete ring so that there is some flexibility. This allows you to move your neck without constricting the airways. It also allows for the oesophagus (the tube that carries food down to the stomach) to expand during swallowing.

Smooth muscle

The smooth muscle can contract. When the smooth muscle contracts, it will constrict the airway. This makes the **lumen** of the airway narrower. The effect of the smooth muscle is most obvious in the bronchioles. Constricting the lumen can restrict the flow of air to and from the alveoli.

Controlling the flow of air to the alveoli may be important if there are harmful substances in the air. The contraction of the smooth muscle and control of airflow is not a voluntary act. Some people have an allergic reaction to certain substances in the air and their bronchioles constrict, making it difficult to breathe. This is one of the causes of asthma.

Elastic fibres

When the smooth muscle contracts, it reduces the diameter of the lumen of the airway. The smooth muscle cannot reverse this effect. When the airway constricts, it deforms the elastic fibres in the loose tissue. As the smooth muscle relaxes, the elastic fibres recoil to their original size and shape. This helps to dilate (widen) the airway.

Goblet cells and glandular tissue

The goblet cells and glandular tissue under the epithelium secrete **mucus**. The role of the mucus is to trap tiny particles from the air. These particles may include pollen and bacteria. Trapping the bacteria so that they can be removed will reduce the risk of infection.

Ciliated epithelium

The epithelium consists of ciliated cells. These cells have numerous tiny, hair-like structures projecting from their membrane. These are the **cilia**. Cilia move in a synchronised pattern to waft the mucus up the airway to the back of the throat. Once there, the mucus is swallowed, and the acidity in the stomach will kill any bacteria.

Examiner tips

Don't confuse *contraction* and *constriction* – the muscle contracts and the airway is constricted.

Also, don't confuse *contraction* with elastic *recoil*. The muscles contract and the elastic fibres recoil.

Questions

1. Explain the importance of the cartilage found in the trachea and bronchi.
2. Describe the action of the cilia and mucus in helping to reduce the risk of infection.
3. The action of smooth muscle and elastic fibres in the bronchioles can be described as antagonistic. Explain the meaning of this term, using the smooth muscle and elastic fibres to illustrate your answer.

1.2 (4) Measuring lung capacity

By the end of this spread, you should be able to . . .

* Explain the meanings of the terms tidal volume and vital capacity.
* Describe how a spirometer can be used to measure vital capacity, tidal volume, breathing rate and oxygen uptake.
* Analyse and interpret data from a spirometer.

Key definition

Inspiration (or inhalation) and **expiration** (or exhalation) are the terms used to describe breathing in and breathing out.

Examiner tip

Don't confuse *breathing* with **respiration**. Respiration happens in cells and is the release of energy from glucose (or other respiratory substrates). Breathing refers to the movements of the ribs and diaphragm that cause air to enter and leave (ventilate) the lungs.

Breathing

Think about how you breathe when you are resting. Air moves in and out of your lungs about 12 times per minute as your **diaphragm** and **intercostal muscles** contract and relax. Each breath refreshes some of the air in your lungs and carries away some of the carbon dioxide generated by your body. If you exercise, or if you are frightened, you breathe more deeply and more quickly. This gets more oxygen-rich air into your lungs and removes more carbon dioxide-rich air out of your lungs.

Different elements of lung volume

We can measure the different volumes of air moved in and out during breathing.

- *Tidal volume* is the volume of air moved in and out of the lungs with each breath when you are at rest. It is approximately $0.5\,dm^3$ and provides the body with enough oxygen for its resting needs while removing enough carbon dioxide to maintain a safe level.
- *Vital capacity* is the largest volume of air that can be moved into and out of the lungs in any one breath. It is approximately $5\,dm^3$ but varies between men and women. It also varies with the person's size and age. Regular exercise increases vital capacity.
- *Residual volume* is the volume of air that always remains in the lungs, even after the biggest possible exhalation. It is about $1.5\,dm^3$.
- *Dead space* is the air in the bronchioles, bronchi and trachea. There is no gas exchange between this air and the blood.
- *Inspiratory reserve volume* is how much more air can be breathed in (inspired) over and above the normal tidal volume when you take in a big breath. You call on this reserve when exercising.
- *Expiratory reserve volume* is how much more air can be breathed out (expired) over and above the amount that is breathed in a tidal volume breath.

The different elements of lung volume can be shown on a trace diagram (see Figure 1).

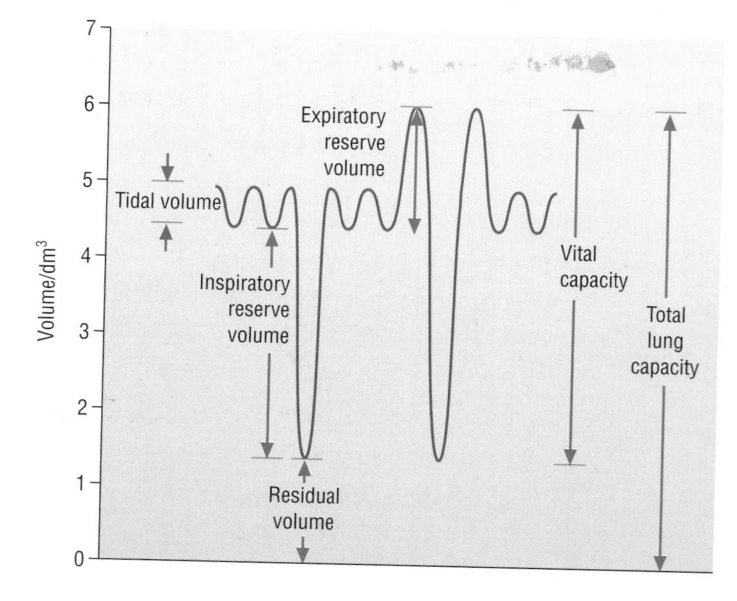

Figure 1 Spirometer trace showing tidal volume, residual volume, inspiratory reserve volume, expiratory reserve volume, vital capacity and total lung capacity

The spirometer and lung volume

A spirometer consists of a chamber filled with oxygen that floats on a tank of water. A person breathes from a disposable mouthpiece attached to a tube connected to the chamber of (medical-grade) oxygen. Breathing in takes oxygen from the chamber, which then sinks down. Breathing out pushes air into the chamber, which then floats up.

The movements of the chamber are recorded using a datalogger so that a spirometer trace can be produced.

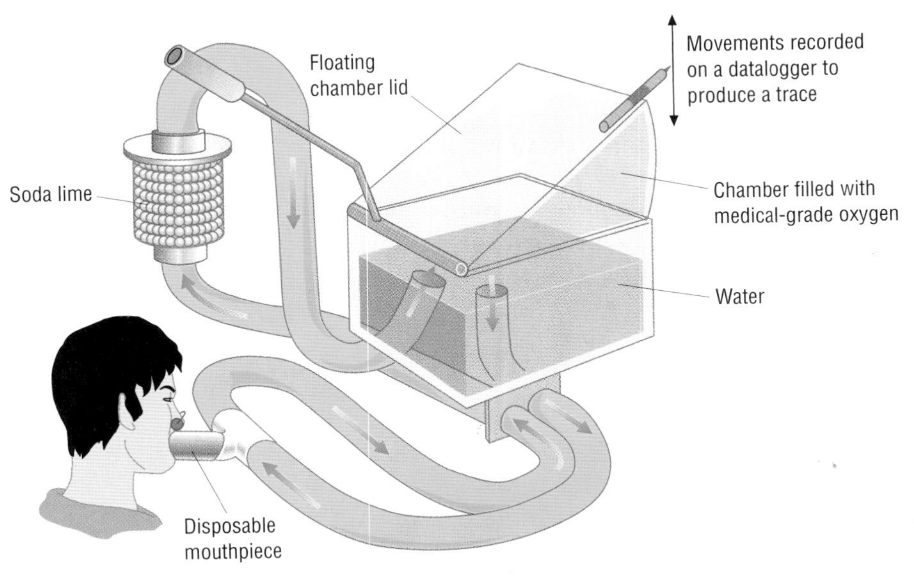

Figure 2 Using a spirometer

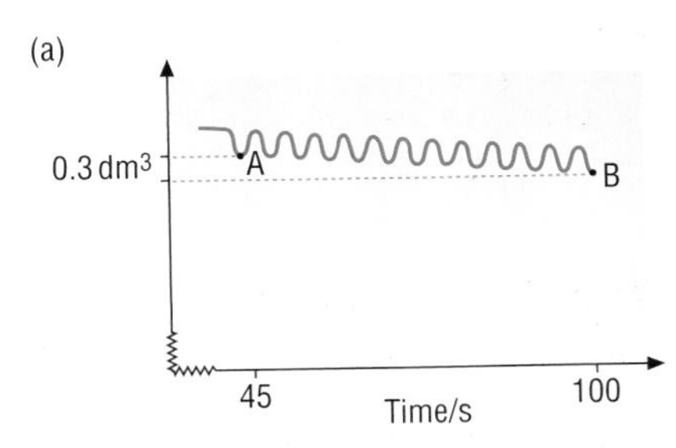

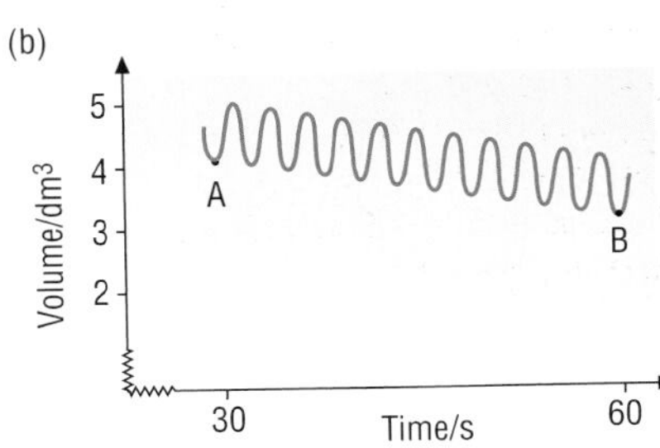

Figure 3 Spirometer traces of a subject **a** at rest and **b** during exercise

The person breathing in and out of the spirometer can be asked to breathe normally at rest, take deep breaths, or to do some exercise so that different patterns of breathing can be shown.

Measuring oxygen uptake

If someone breathes in and out of a spirometer for a period of time, the level of carbon dioxide will increase dangerously.

To avoid this, soda lime is used to absorb the carbon dioxide that is exhaled. This means that the total volume of gas in the spirometer will go down. Because the volume of carbon dioxide breathed out is the same as the volume of oxygen breathed in, as the carbon is removed this total reduction is equal to the volume of oxygen used up by the person breathing in and out. This allows us to make calculations of oxygen use under different conditions.

Worked example

From Figure 3a, we can calculate oxygen uptake per minute for the subject at rest.

The *y*-axis shows a reduction in chamber total volume by 0.3 dm^3 between points A and B.

The time taken for this reduction is 100–45 s.

So in 55 s, 0.3 dm^3 of oxygen is used up. This gives:

$$\frac{0.3}{55}\ dm^3\,s^{-1}$$

$$= \frac{0.3 \times 60}{55}\ dm^3\,min^{-1}$$

$$= 0.33\ dm^3\,min^{-1}$$

Questions

1. What process is responsible for oxygen moving into the blood from the alveoli?
2. Measure the tidal volume and breaths per minute shown on each trace in Figure 3.
3. Calculate the oxygen uptake in dm^3 per minute shown in Figure 3b.
4. Suggest how a spirometer trace from a trained athlete would differ from a spirometer trace from an untrained individual.
5. Suggest why the trace in Figure 3b is not started immediately as the subject begins to exercise.

Transport in animals

By the end of this spread, you should be able to . . .

* **Explain the need for transport systems in multicellular animals in terms of size, activity and surface-area-to-volume ratio.**
* **Explain the meaning of the terms single and double circulatory systems with reference to the circulatory systems of fish and mammals.**

Key definition

Transport is the movement of oxygen, nutrients, hormones, waste and heat around the body.

Large animal transport systems

All living animal cells need a supply of oxygen and nutrients to survive. They also need to remove waste products so that these do not build up and become toxic. Very small animals do not need a separate transport system, because all their cells are surrounded by (or very close to) the environment in which they live. Diffusion will supply enough oxygen and nutrients to keep the cell alive.

Once an animal has a complex anatomy with more than two layers of cells, diffusion alone will be too slow.

There are three main factors that affect the need for a transport system:

- size
- surface-area-to-volume ratio
- level of activity.

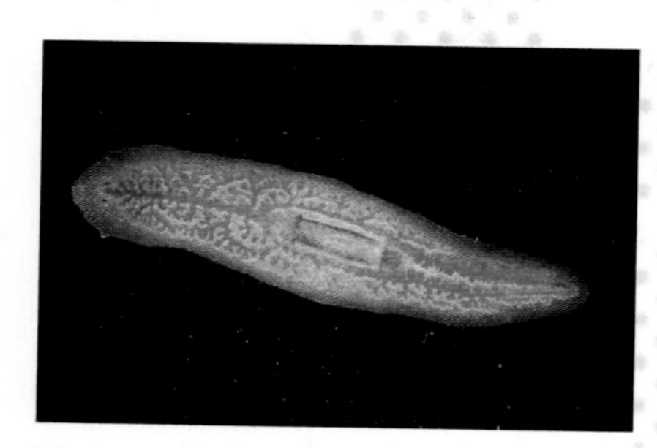

Figure 1 Planarian flatworn. It can grow to a length of nearly 10 cm, but the body is flat and thin so oxygen can diffuse through its surface and reach all the cells. The parts stained red are branches of the intestine

Size

Once an animal has several layers of cells, any oxygen or nutrients diffusing in from the outside will be used up by the outer layers of cells. The oxygen and nutrients will not reach the cells deeper within the body.

Surface-area-to-volume ratio

Small animals have a large surface area compared with their volume. This is known as their surface-area-to-volume ratio. This ratio is affected by an animal's shape. A flatworm has a very thin, flat body, which gives it a large surface-area-to-volume ratio. But such a body form limits the overall size that the animal can reach.

To allow animals to grow to a large size, they need a range of tissues and structural support to give the body strength. Their volume increases as their body gets thicker. But the surface area does not increase as much (Figure 2). So the surface-area-to-volume ratio of a large animal is relatively small. In larger animals, the surface area is not large enough to supply all the oxygen and nutrients needed by the internal cells.

	Small cube	Medium cube	Large cube
Length of side/cm	1	5	10
Surface area (SA)/cm^2	6	150	600
Volume (V)/cm^3	1	125	1000
SA/V ratio	6	1.2	0.6

Figure 2 The surface-area-to-volume ratio changes with increasing size

Level of activity

Animals need energy from food so that they can move around. Releasing energy from food by **respiration** requires oxygen. If an animal is very active, its cells need good supplies of nutrients and oxygen to supply the energy for movement. Those animals, such as mammals, that keep themselves warm need even more energy.

Features of a good transport system

An effective transport system will include:

- a fluid or medium to carry nutrients and oxygen around the body – this is the blood
- a pump to create pressure that will push the fluid around the body – this is the heart
- exchange surfaces that enable oxygen and nutrients to enter the blood and to leave it again where they are needed.

An efficient transport system will also include:

- tubes or vessels to carry the blood
- two circuits – one to pick up oxygen and another to deliver oxygen to the tissues.

Single and double circulatory systems

Fish have a single circulatory system. The blood flows from the heart to the gills and then on to the body before returning to the heart.

heart → gills → body →heart

Mammals have a circulation that involves two separate circuits. This is known as a double circulatory system. One circuit carries blood to the lungs to pick up oxygen. This is the **pulmonary circulation**. The other circuit carries the oxygen and nutrients around the body to the tissues. This is the **systemic circulation**.

The mammalian heart is adapted to form two pumps – one for each circulation. Blood flows through the heart twice for each circulation of the body.

heart → body → heart → lungs → heart

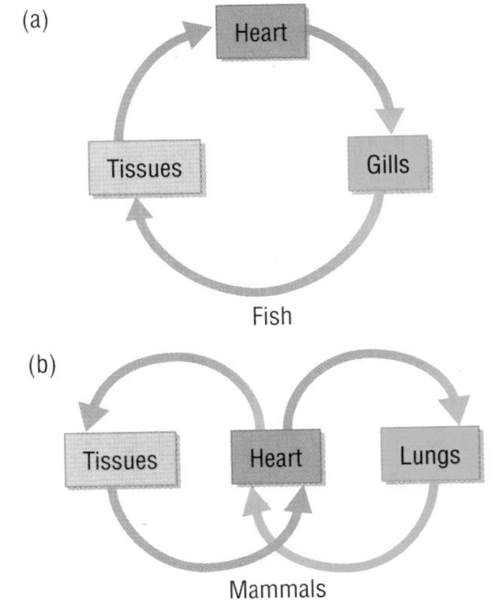

Figure 3 **a** Single and **b** double circulations

Advantages of a double circulation

An efficient circulatory system will deliver oxygen and nutrients quickly to the parts of the body where they are needed. The blood can be made to flow more quickly by increasing the blood pressure created by the heart.

By comparing the transport systems of fish and mammals, we shall see the advantages of a double circulatory system.

In the fish single circulatory system:

- the blood pressure is reduced as blood passes through the tiny capillaries of the gills
- this means it will not flow very quickly to the rest of the body
- this limits the rate at which oxygen and nutrients are delivered to respiring tissues.

Fish are not as active as mammals and they do not maintain their body temperature, so they need less energy. Their single circulatory system delivers oxygen and nutrients quickly enough for their needs, so for them it is efficient.

In the mammal double circulatory system:

- the heart can increase the pressure of the blood after it has passed through the lungs, so blood flows more quickly to the body tissues
- the systemic circulation can carry blood at a higher pressure than the pulmonary circulation
- the blood pressure must not be too high in the pulmonary circulation, otherwise it may damage the delicate capillaries in the lungs.

Mammals are active animals and maintain their body temperature. Both the energy for activity and the heat needed to keep the body warm require energy from food. The energy is released from food in the process of respiration. To release a lot of energy, the cells need good supplies of both nutrients and oxygen.

Questions

1. List the features of a good transport system.
2. Explain why large, active animals need a transport system.
3. With reference to fish and mammals, compare the single and double circulatory systems.
4. Explain why a double circulatory system is more efficient than a single circulatory system.

The structure of the mammalian heart

By the end of this spread, you should be able to . . .

- **Describe the external and internal structure of the mammalian heart.**
- **Explain the differences in thickness of the walls of the different chambers of the heart in terms of their functions.**

Key definition

The **heart** is a muscular pump that creates pressure (blood pressure) to propel the blood through the arteries and around the body.

Examiner tip

If you are asked how the pressure in the arteries is produced, you need to explain that it is due to the *contraction of the left ventricle walls.*

The heart

The mammalian heart is a muscular double pump. It is divided into two sides. The right side pumps **deoxygenated** blood to the lungs to be **oxygenated**. The left side pumps oxygenated blood to the rest of the body. On both sides the heart squeezes the blood, putting it under pressure. This pressure forces the blood along the arteries.

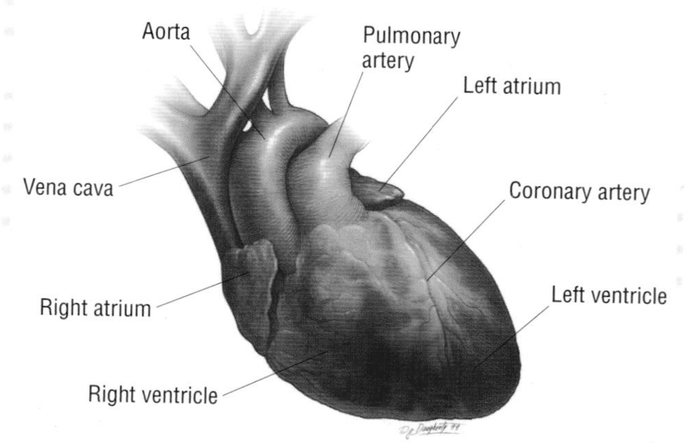

Figure 1 External features of the heart ×0.3

External features of the heart

The heart sits slightly off-centre to the left of the chest cavity. It lies with the atria (singular: **atrium**) in the middle of the cavity. The main part of the heart consists of dark red muscle, which feels very firm. This is the muscle surrounding the two main pumping chambers – the **ventricles**. Above the ventricles are two thin-walled chambers – the atria. These are much smaller than the ventricles and easy to overlook. (When you buy a heart from a butcher, the atria have often been removed.)

The **coronary arteries** lie over the surface of the heart. They carry oxygenated blood to the heart muscle itself. As the heart is a hard-working organ, these arteries are very important. If they become constricted, it can have severe consequences for the health of the heart and of the animal. Restricted blood flow to the heart muscle reduces the delivery of oxygen and nutrients such as fatty acids. This may cause **angina** or a heart attack (myocardial infarction).

At the top of the heart are a number of tubes. These are the veins that carry blood into the heart and the arteries that carry blood out of the heart.

Internal features of the heart

The heart is divided into four chambers. The two upper chambers are atria. These receive blood from the major veins. Deoxygenated blood from the body flows from the **vena cava** into the right atrium. Oxygenated blood from the lungs flows from the **pulmonary vein** into the left atrium.

From the atria, blood flows down through the **atrioventricular valves** into the ventricles. These valves are thin flaps of tissue arranged in a cup shape. When the ventricles contract, the valves fill with blood and remain closed. This ensures that the blood flows upwards into the major arteries and not back into the atria. Inside the ventricles are string-like **tendinous cords**. These attach the valves to the walls of the ventricle and prevent the flimsy valves from turning inside out, which would allow blood to flow up into the atria.

A wall of muscle called the **septum** separates the ventricles from each other. This ensures that the oxygenated blood in the left side of the heart and the deoxygenated blood in the right side are kept separate.

Deoxygenated blood leaving the right ventricle flows into the pulmonary artery leading to the lungs. Oxygenated blood leaving the left ventricle flows into the aorta. This carries blood to a number of arteries that supply all parts of the body. At the base of the major arteries, where they exit the heart, are valves called **semilunar valves**, which prevent blood returning to the heart as the ventricles relax.

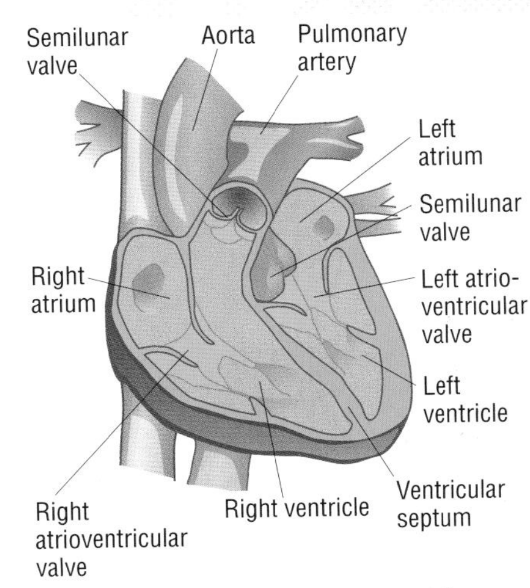

Figure 2 The internal structure of the heart

Blood pressure

The muscle of each chamber contracts to create increased pressure in the blood. The higher the pressure created in the heart, the further it will push the blood.

Atria

The muscle of the atria is very thin. This is because these chambers do not need to create much pressure. Their function is to push the blood into the ventricles.

Right ventricle

The walls of the right ventricle are thicker than the walls of the atria. This enables the right ventricle to pump blood out of the heart.

But the walls of the right ventricle are much thinner than those of the left ventricle. The right ventricle pumps deoxygenated blood to the lungs. The lungs are in the chest cavity beside the heart, so the blood does not need to travel very far. Also, the lungs contain a lot of very fine capillaries that are in close contact with the walls of the alveoli. The alveoli walls are very thin and there is very little or no tissue fluid. So the capillaries are not supported and could easily burst. The pressure of the blood must be kept down to prevent the capillaries in the lungs bursting.

Left ventricle

The walls of the left ventricle can be two or three times thicker than those of the right ventricle. The blood from the left ventricle is pumped out through the aorta and needs sufficient pressure to overcome the resistance of the systemic circulation.

Units of pressure

The SI unit of pressure is the pascal, but because blood pressure used to be measured with a tube containing mercury we still use mmHg as a pressure measurement.

Questions

1. Describe the function of the atrioventricular valves.
2. Describe the role of the tendinous cords, which can be seen in the ventricles.
3. Explain why the wall of the left ventricle needs to be much thicker than the wall of the right ventricle.
4. Explain why it might be harmful if the right ventricle creates too much pressure.

1.2 ⑦ The cardiac cycle

By the end of this spread, you should be able to . . .

* **Describe the cardiac cycle with reference to the action of the valves in the heart.**

Key definition

The **cardiac cycle** is the sequence of events in one heartbeat.

Examiner tip

Use the term **atrioventricular valve.** It refers to the same valves as the words bicuspid (on the left) and the tricuspid (on the right), but you don't have to recall which side is which.

Sequence of contraction

It is important that the chambers of the heart all contract in a coordinated fashion. If the chambers contract out of sequence, this will lead to inefficient pumping. The sequence of events involved in one heartbeat is called the **cardiac cycle**.

As the series of events is a cycle, it has no clear beginning or end. To make it easier, we can describe the cycle in a number of phases.

Filling phase

While both the atria (singular: **atrium**) and the **ventricles** are relaxing, the internal volume increases and blood flows into the heart from the major veins. The blood flows into the atria, then through the open **atrioventricular valves** and into the ventricles. This phase of the cycle is called **diastole**.

Atrial contraction

The heart beat starts when the atria contract. Both right and left atria contract together. The small increase in pressure created by this contraction helps to push blood into the ventricles. This stretches the walls of the ventricles and ensures thay are full of blood. Contraction of the atria is called atrial **systole**. Once the ventricles are full they begin to contract. Blood fills the atrioventricular valve flaps causing them to snap shut. This prevents blood returning to the atria.

Ventricular contraction

Now there is a short period when all four heart valves are closed. The walls of the ventricles contract. This is called ventricular systole. This raises the pressure in the ventricles very quickly. The contraction starts at the apex (base) of the heart so this pushes the blood upwards towards the arteries. The semilunar valves open and blood is pushed out of the heart. The contraction only lasts for a short time. Then the ventricle walls relax allowing the heart to start filling again.

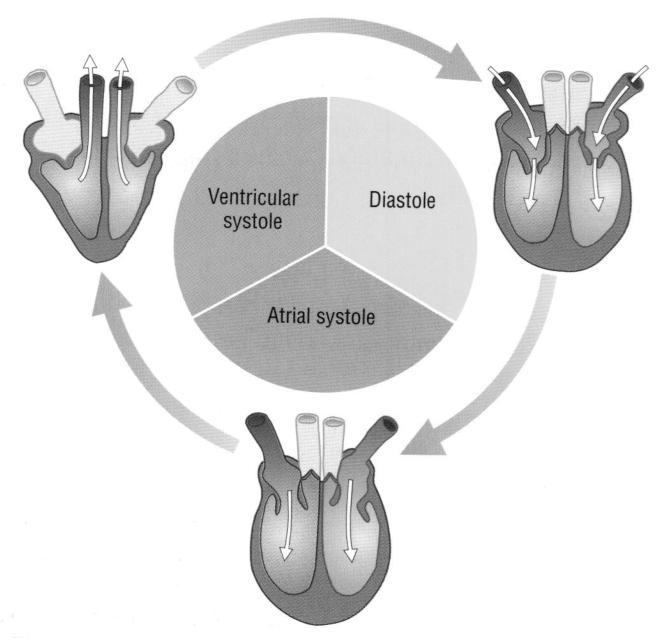

Figure 1 The cardiac cycle

How valves work

The valves ensure that blood flows in the correct direction. They are opened and closed by changes in the blood pressure in the various chambers of the heart.

Atrioventricular valves

When the ventricular walls relax and recoil (go back into shape) after contracting, the pressure in the ventricles drops below the pressure in the atria. This causes the atrioventricular valves to open. Blood entering the heart flows straight through the atria and into the ventricles. The pressure in the atria and the ventricles slowly rises as they fill with blood. The valves remain open while the atria contract.

As the ventricles begin to contract, the pressure of the blood in the ventricles rises. When the pressure rises above that in the atria, the blood starts to move upwards. This movement fills the valve pockets and keeps them closed. This prevents the blood flowing back into the atria.

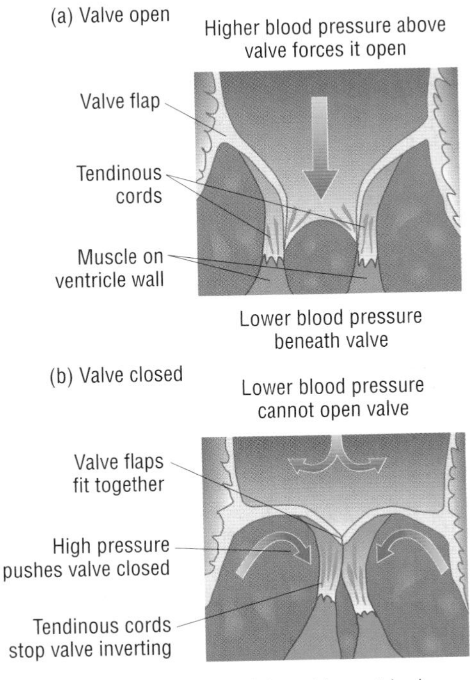

Figure 2 Action of the atrioventricular valves

Semilunar valves

When the ventricles start to contract, the pressure in the major arteries is higher than the pressure in the ventricles. This means that the semilunar valves are closed. As the ventricles contract, the pressure inside rises very quickly because the blood cannot escape. Once the pressure in the ventricles rises above the pressure in the aorta and pulmonary arteries, the semilunar valves are pushed open. The blood is under very high pressure, so it is forced out of the ventricles in a powerful spurt.

Once the ventricle walls have finished contracting, the heart muscle starts to relax. Elastic tissue in the walls of the ventricles recoils to stretch the muscle out again and return the ventricle to its original size. This causes the pressure in the ventricle to drop quickly. As it drops below the pressure in the major arteries, the semilunar valves are pushed closed by blood starting to flow back towards the ventricles and collecting in the pockets of the valves. This prevents blood returning to the ventricles.

The sound of the heart

The familiar lub-dup sound made by the heart is actually made by the valves closing.

- The first sound, lub, is made by the atrioventricular valves closing as the ventricles start to contract.
- The second sound, dup, is made by the semilunar valves closing as the ventricles start to relax.

The atrioventricular valves snap shut, so this noise is louder than the closing of the semilunar valves, which shut because blood is accumulating in their pockets.

Questions

1 Describe the cardiac cycle with reference to the action of the valves in the heart.
2 Explain what causes the atrioventricular valves to close.

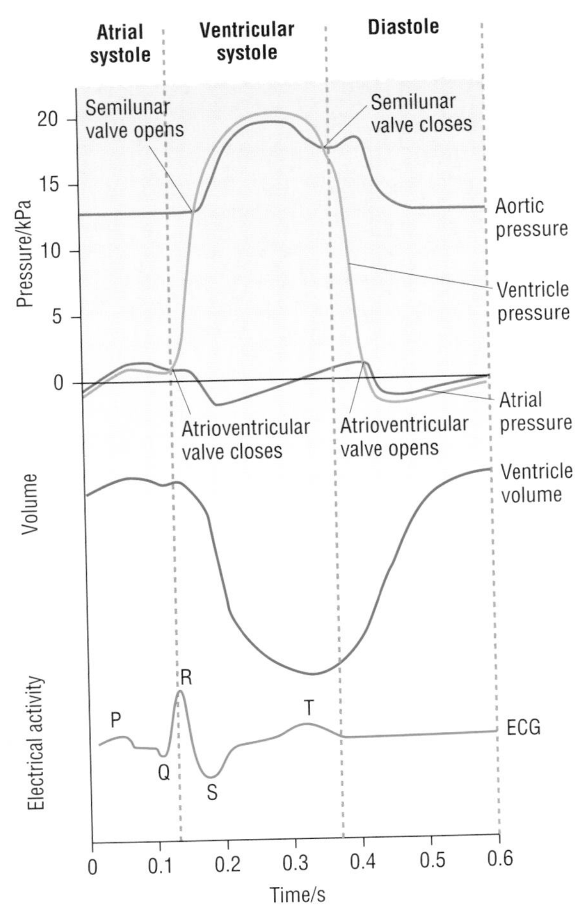

Figure 3 Pressures changes in the aorta, left ventricle and left atrium during one heart beat; volume changes in left ventricle and ECG (electrical activity in the heart muscle) trace over the same period

1.2 (8) Control of the cardiac cycle

By the end of this spread, you should be able to . . .

* **Describe how heart action is coordinated with reference to the sinoatrial node, the atrioventricular node and the Purkyne tissue.**
* **Interpret and explain electrocardiogram (ECG) traces with reference to normal and abnormal heart activity.**

Key definitions

The **sinoatrial node** (SAN) is the heart's pacemaker. It is a small patch of tissue that sends out waves of electrical excitation at regular intervals to initiate contractions.

Purkyne tissue is specially adapted muscle fibres that conduct the wave of excitation from the AVN down the **septum** to the **ventricles**.

Examiner tip

If you are asked to calculate a heart rate, don't forget the units – beats min^{-1}.

Never say that the *heartbeat* increases – this is not precise enough. Always say whether the *heart rate* increases, the *force of contraction* increases, or the *stroke volume* increases.

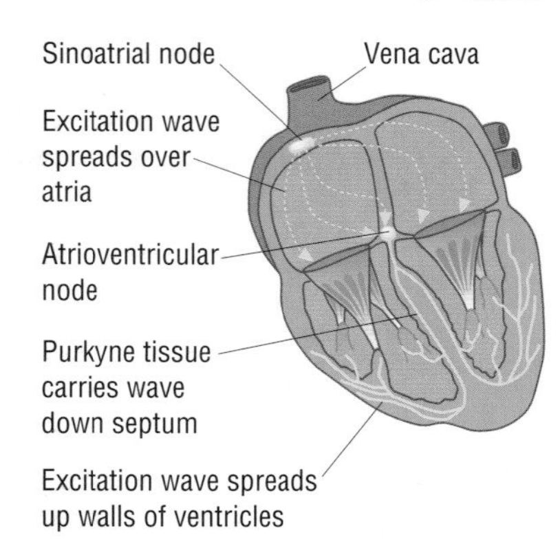

Figure 1 The pathway followed by the wave of excitation

The need for coordination

Heart muscle or **cardiac muscle** is unusual in that it can initiate its own contraction. Because of this property, the heart muscle is described as **myogenic**. The muscle will contract and relax rhythmically even if it is not connected to the body. The muscles from the **atria** and the **ventricles** each have their own natural frequency of contraction. The atrial muscle tends to contract at a higher frequency than the ventricular muscle.

This property of the muscle could cause inefficient pumping (a condition known as **fibrillation**) if the contractions of the chambers are not synchronised. So the heart needs a mechanism that can coordinate the contractions of all four chambers.

How the heartbeat starts

At the top of the right atrium, near the point where the **vena cava** empties blood into the atrium, is the **sinoatrial node** (SAN). This is a small patch of tissue that generates electrical activity. The SAN initiates a wave of excitation at regular intervals. In a human, this occurs approximately 55–80 times a minute. The SAN is also known as the pacemaker.

Contraction of the atria

The wave of excitation quickly spreads over the walls of both atria. It travels along the membranes of the muscle tissue. As the wave of excitation passes, it causes the cardiac muscle cells to contract. This is atrial **systole** (see spread 1.2.7).

At the base of the atria is a disc of tissue that cannot conduct the wave of excitation. So the excitation cannot spread directly to the ventricle walls. At the top of the inter-ventricular septum (the septum separating the two ventricles) is another node – the **atrioventricular node** (AVN). This is the only route through the disc of non-conducting tissue. The wave of excitation is delayed in the node. This allows time for the atria to finish contracting and for the blood to flow down into the ventricles before they begin to contract.

Contraction of the ventricles

After this delay, the wave of excitation is carried away from the AVN and down specialised conducting tissue. This is the **Purkyne tissue** and it runs down the inter-ventricular septum. At the base of the septum, the wave of excitation spreads out over the walls of the ventricles. As the excitation spreads upwards from the base (apex) of the ventricles, it causes the muscles to contract. This means that the ventricles contract from the base upwards, pushing blood up to the major arteries at the top of the heart.

Electrocardiograms

We can monitor the electrical activity of the heart using an **electrocardiogram** (ECG). This involves attaching a number of sensors to the skin. Some of the electrical activity generated by the heart spreads through the tissues next to the heart and onwards to the skin. The sensors on the skin pick up the electrical excitation created by the heart and convert this into a trace.

The trace of a healthy person has a particular shape. It consists of a series of waves that are labelled P, Q, R, S and T. Wave P shows the excitation of the atria. QRS indicates the excitation of the ventricles. T shows **diastole** (see spread 1.2.7).

The shape of the ECG trace can sometimes indicate when part of the heart muscle is not healthy. It can show if the heart is beating irregularly (arrhythmia), if it is in fibrillation (the beat is not coordinated), or if it has suffered a heart attack (myocardial infarction). It can also show if the heart has enlarged or if the Purkyne system is not conducting electrical activity properly.

The heart muscle cells **respire** fatty acids and must have a continuous supply of oxygen because fat can only be respired aerobically. So a blood clot in the coronary artery starves part of the heart muscle of oxygen and those cells die. This is a heart attack.

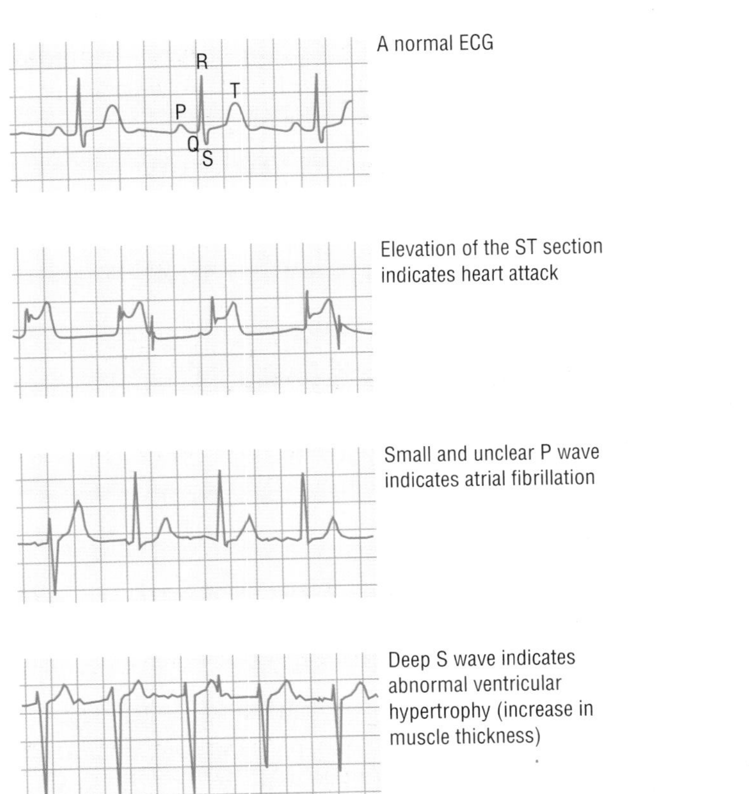

Figure 2 A normal ECG trace compared with others indicating an unhealthy heart

Questions

1. Explain why the sinoatrial node is called the pacemaker.
2. Explain why atrial fibrillation decreases the efficiency of the heart.
3. Explain why the ventricles contract from the apex upwards.
4. Describe the shape of an electrocardiogram trace.
5. Explain why the QRS complex has a larger peak than the P wave.

1.2 (9) Blood vessels

By the end of this spread, you should be able to . . .

- **Explain the meaning of the terms closed and open circulatory systems with reference to the circulatory systems of insects and fish.**
- **Describe the structures and functions of arteries, veins and capillaries.**

In order to be active, the muscles need a supply of both oxygen and nutrients, such as glucose, amino acids and fatty acids, and the rapid removal of carbon dioxide.

Open circulatory systems

In mammals, the blood never gets out into the body cavity – it is always confined to the blood vessels. But many animals, including all insects, have an open circulatory system. This means that the blood is not always held within blood vessels. Instead, the blood fluid circulates through the body cavity, so the tissues and cells of the animal are bathed directly in blood.

In some animals, the action of the body muscles during movement may help to circulate the blood. In others, such as insects, there is a muscular pumping organ much like a heart. This is a long, muscular tube that lies just under the dorsal (upper) surface of the insect. Blood from the body enters the heart through pores called ostia. The heart then pumps the blood towards the head by **peristalsis**. At the forward end of the heart (nearest the head), the blood simply pours out into the body cavity.

Some larger and more active insects, such as locusts, have open-ended tubes attached to the heart. These direct the blood towards active parts of the body, such as the leg and wing muscles.

Why don't all animals have an open system?

An open system works for insects because they are small. The blood does not have to travel far. Also, they do not rely on blood to transport oxygen and carbon dioxide. They use a separate transport system for this.

Larger organisms rely on the blood to transport oxygen and carbon dioxide. In an open circulatory system the blood remains at a low pressure and the flow is very slow. This would not be sufficient to supply the needs of the muscles in a large, active animal. It would also mean that many other parts of the body do not receive sufficient oxygen or nutrients.

Closed circulatory systems

In larger animals the blood always stays entirely inside vessels. A separate fluid called **tissue fluid** bathes the tissues and cells. This enables the heart to pump the blood at a higher pressure, so that it flows more quickly. This means that it can deliver oxygen and nutrients more quickly, and remove carbon dioxide and other wastes more quickly.

Fish have a closed single circulation (see spread 1.2.5). This means the blood remains in vessels that carry it on a single pathway around the body.

heart → arteries → gills → veins → body tissues → veins → heart

There must be exchange surfaces at the gills and at the body tissues, to allow materials to be exchanged between blood and tissue fluid.

Key definitions

In an **open circulatory system** the blood is not always in vessels.

In a **closed circulatory system** the blood always remains within vessels.

Examiner tips

The terms *constrict* and *contract* don't mean the same thing. The smooth muscle in the artery walls contracts (becomes shorter) and the artery then constricts (becomes narrower).

Don't confuse *capillary walls* and *cell walls*. It is not correct to say that animal cells have 'thin cell walls'. Animal cells do not have cell walls. The capillary walls are made of a single layer of cells.

Open circulation of a locust

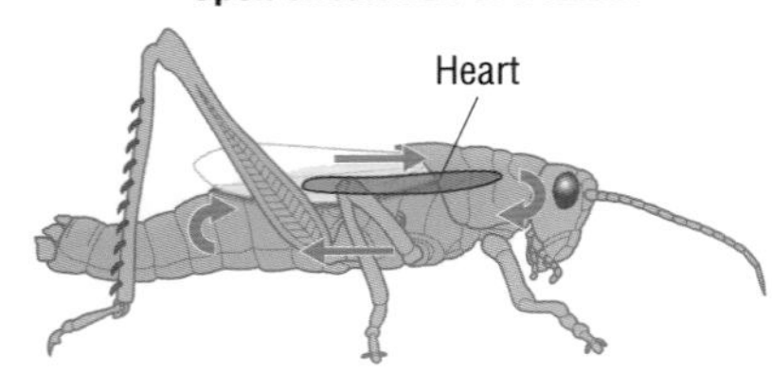

Closed circulation of a fish

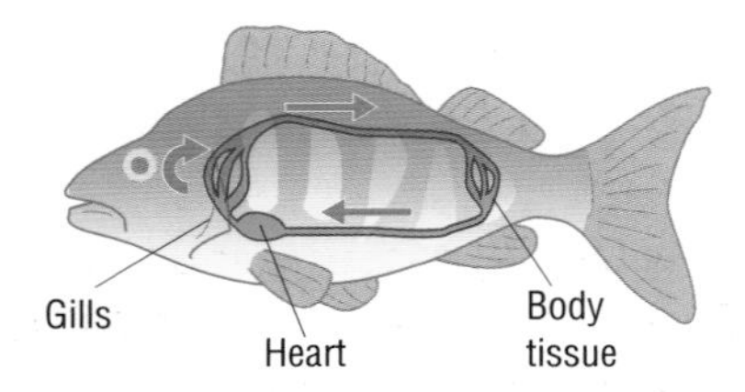

Figure 1 Circulation in an insect and a fish

Blood vessels

Blood flows through a series of vessels. Each is adapted to its particular role in relation to its distance from the heart. All types of blood vessels have an inner layer or lining, made of a single layer of cells called the **endothelium.** This is a thin layer that is particularly smooth to reduce friction with the flowing blood.

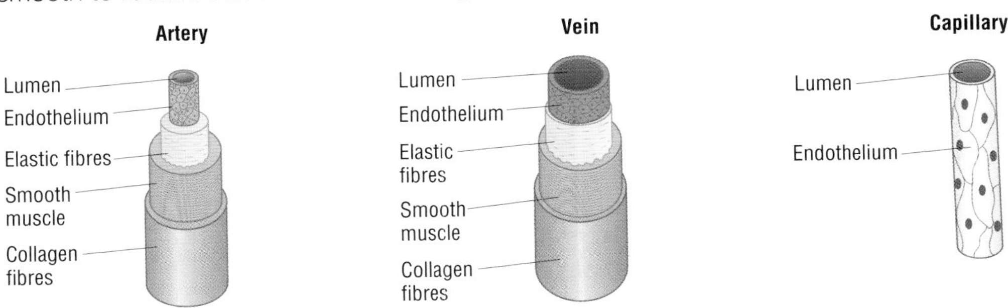

Figure 2 Artery, vein and capillary

Arteries

Arteries carry blood away from the heart. The blood is at high pressure, so the artery wall must be able to withstand that pressure.

- The lumen is relatively small to maintain high pressure.
- The wall is relatively thick and contains collagen, a fibrous protein, to give it strength to withstand high pressure.
- The wall has **elastic tissue** that allows the wall to stretch and then recoil when the heart pumps. This is felt as a pulse in areas where the arteries lie close to the skin. The recoil maintains the high pressure while the heart relaxes.
- The wall also contains **smooth muscle** that can contract and constrict the artery. The constriction narrows the **lumen** of the artery. (In arterioles this is used to limit blood flow to certain organs or tissues so that it can be directed to other tissues.)
- The endothelium is folded and can unfold when the artery stretches.

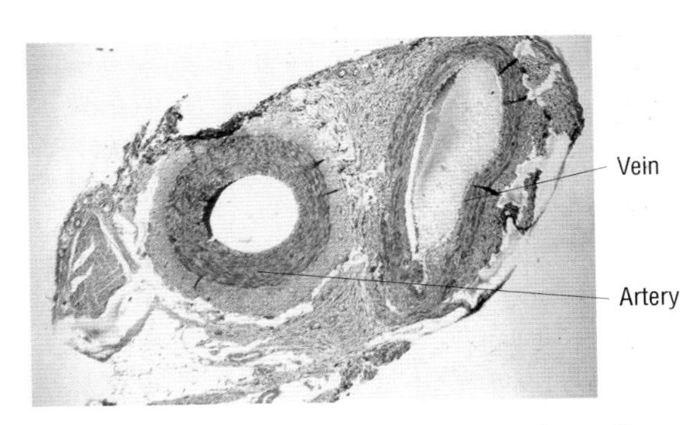

Figure 3 Light micrograph of a section through an artery and a vein (×36)

Veins

Veins carry blood back to the heart. The blood is at low pressure and the walls do not need to be thick.

- The lumen is relatively large to ease the flow of blood.
- The walls have thinner layers of collagen, smooth muscle and elastic tissue. They do not need to stretch and recoil, and are not actively constricted to reduce blood flow.
- The main feature of veins is that they contain valves to help the blood flow back to the heart and to prevent it flowing in the opposite direction. As the walls are thin, the vein can be flattened by the action of the surrounding skeletal muscle. Pressure is applied to the blood, forcing the blood to move along in a direction dictated by the valves.

Capillaries

Capillaries have very thin walls. They allow exchange of materials between the blood and cells of tissues via the tissue fluid.

- The walls consist of a single layer of flattened endothelial cells that reduces the diffusion distance for the materials being exchanged.
- The lumen is very narrow – its diameter is the same as that of a red blood cell (7 μm). This ensures that the red blood cells are squeezed as they pass along the capillaries. This helps them give up their oxygen because it presses them close to the capillary wall, reducing the diffusion path to the tissues.

Questions

1. Compare open and closed circulatory systems with reference to locusts and fish.
2. Discuss why an open circulatory system is not as efficient as a closed system.
3. Draw a table to compare the structure of arteries, capillaries and veins.
4. Explain how the structure of arteries, capillaries and veins enables them to carry out their functions.

Blood, tissue fluid and lymph

By the end of this spread, you should be able to . . .

* **Explain the differences between blood, tissue fluid and lymph.**
* **Describe how tissue fluid is formed from plasma.**

Key definitions

Blood is held in the heart and blood vessels.

Tissue fluid bathes the cells of individual tissues.

Lymph is held within the lymphatic system.

Examiner tip

Give examples of nutrients carried in blood (e.g. glucose) at least once in your answers – don't just write 'nutrients'. Similarly, give examples of waste products produced by cells (e.g. urea) rather than 'wastes'.

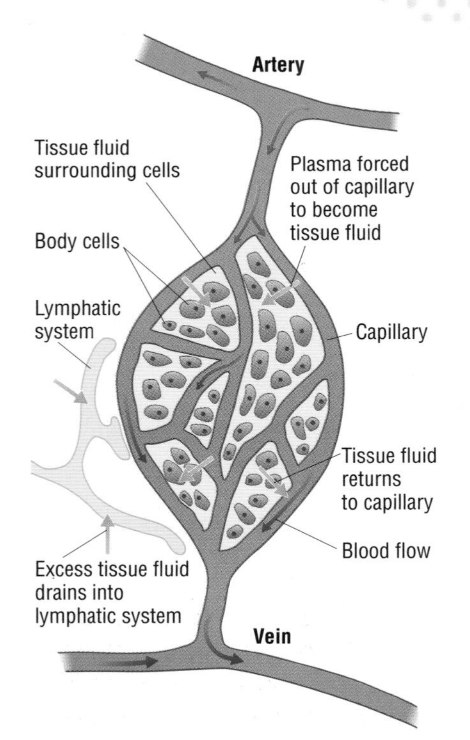

Figure 1 Artery linked to vein by a capillary bed

Blood and tissue fluid

Blood is the liquid held in our blood vessels. It consists of blood cells in a watery fluid called plasma. The plasma contains many dissolved substances, including oxygen, carbon dioxide, salts, glucose, fatty acids, amino acids, hormones and **plasma proteins**. The cells include the red blood cells (**erythrocytes**), various white blood cells (**leucocytes**) and fragments called **platelets**.

Tissue fluid is similar to blood, but it does not contain most of the cells found in blood, nor does it contain plasma proteins. The role of tissue fluid is to transport oxygen and nutrients from the blood to the cells, and to carry carbon dioxide and other wastes back to the blood.

How tissue fluid is formed

When an artery reaches the tissues it branches into smaller arterioles, then into a network of capillaries. These eventually link up with venules to carry blood back to the veins. So blood flowing into an organ or tissue is contained in the capillaries. At the arterial end of a capillary, the blood is under high pressure due to the contraction of the heart muscle. This is known as **hydrostatic pressure**. It will tend to push the blood fluid out of the capillaries. The fluid can leave through the tiny gaps in the capillary wall.

The fluid that leaves the blood consists of plasma with dissolved nutrients and oxygen. All the red blood cells, platelets and most of the white blood cells remain in the blood, as do the plasma proteins. These are too large to be pushed out through the gaps.

The fluid that leaves the capillary is known as the tissue fluid. This fluid surrounds the body cells, so exchange of gases and nutrients can occur across the cell surface membranes. This exchange occurs by **diffusion** and **facilitated diffusion**. Oxygen and nutrients enter the cells; carbon dioxide and other wastes leave the cells.

How does the fluid return to the blood?

The hydrostatic pressure of the blood is not the only force acting on the fluid. The tissue fluid itself has some hydrostatic pressure, which will tend to push the fluid back into the capillaries. Both the blood and the tissue fluid also contain solutes, giving them a negative **water potential** (see spread 1.1.12). The water potential of the tissue fluid is less negative than that of the blood. This means that water tends to move back into the blood from the tissue fluid by **osmosis**, down the water potential gradient.

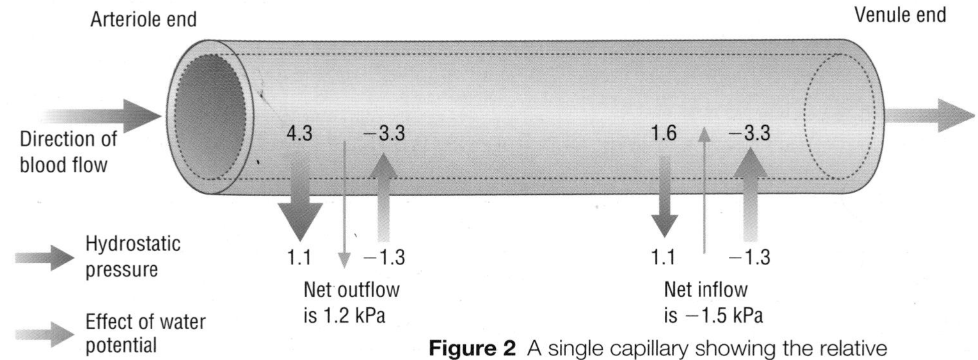

Figure 2 A single capillary showing the relative hydrostatic and osmotic (water potential) forces (in kPa)

At the venous (vein) end of the capillary, the blood has lost its hydrostatic pressure. The combined effect of the hydrostatic pressure in the tissue fluid and the osmotic force of the plasma proteins is sufficient to move fluid back into the capillary. It carries with it any dissolved waste substances, such as carbon dioxide, that have left the cells.

Worked example

(all figures in kPa)

In tissue fluid:
HP = 1.1
SP = −1.3

At arterial end:
HP in capillary = 4.3
SP in capillary = −3.3
Effective HP = 4.3 − 1.1 = 3.2
Effective SP = −3.3 − (−1.3) = −2
Effective blood pressure = 3.2 − 2 = 1.2
Fluid is pushed out of capillary.

At venous end:
HP in capillary = 1.6
SP in capillary = −3.3
Effective HP = 1.6 − 1.1 = 0.5
Effective SP = −3.3 − (−1.3) = −2
Effective blood pressure = 0.5 − 2 = −1.5
Fluid is pushed back into capillary.

HP: hydrostatic pressure
SP: solute potential, which is numerically equal to water potential

Formation of lymph

Not all the tissue fluid returns to the blood capillaries. Some is drained away into the lymphatic system. The lymphatic system consists of a number of vessels (tubes) that are similar to capillaries. They start in the tissues and drain the excess fluid into larger vessels, which eventually rejoin the blood system in the chest cavity.

Lymph fluid is similar to tissue fluid and contains the same solutes. There will be less oxygen and fewer nutrients, as these have been absorbed by the body cells. There will also be more carbon dioxide and wastes that have been released from the body cells. Lymph also has more fatty material that has been absorbed from the intestines.

The main difference between tissue fluid and lymph is that the lymph contains many **lymphocytes**. These are produced in the lymph nodes. The lymph nodes are swellings found at intervals along the lymphatic system. They filter any bacteria and foreign material from the lymph fluid. The lymphocytes can then engulf and destroy these bacteria and foreign particles. This is part of the immune system that protects the body from infection.

Feature	Blood	Tissue fluid	Lymph
Cells	Erythrocytes, leucocytes and platelets	Some phagocytic white blood cells	Lymphocytes
Proteins	Hormones and plasma proteins	Some hormones, proteins secreted by body cells	Some proteins
Fats	Some transported as lipoproteins	None	More than in blood (absorbed from lacteals in intestine)
Glucose	80–120 mg per 100 cm^3	Less (absorbed by body cells)	Less
Amino acids	More	Less (absorbed by body cells)	Less
Oxygen	More	Less (absorbed by body cells)	Less
Carbon dioxide	Little	More (released by body cells)	More

Table 1 Differences between blood, tissue fluid and lymph

Questions

1. To compare blood, tissue fluid and lymph, construct a table to show the location, direction of flow and what causes the flow.
2. Explain why blood contains many proteins that are not found in the tissue fluid or lymph.
3. What produces the hydrostatic pressure in the blood?
4. Describe how fluid can pass through the capillary wall from the plasma to the tissue fluid.

Carriage of oxygen

By the end of this spread, you should be able to . . .

* **Describe the role of haemoglobin in carrying oxygen.**
* **Explain the significance of the different affinities for oxygen of fetal haemoglobin and adult haemoglobin.**

Examiner tip

Haemoglobin becomes *oxygenated*, it does not become *oxidised*.

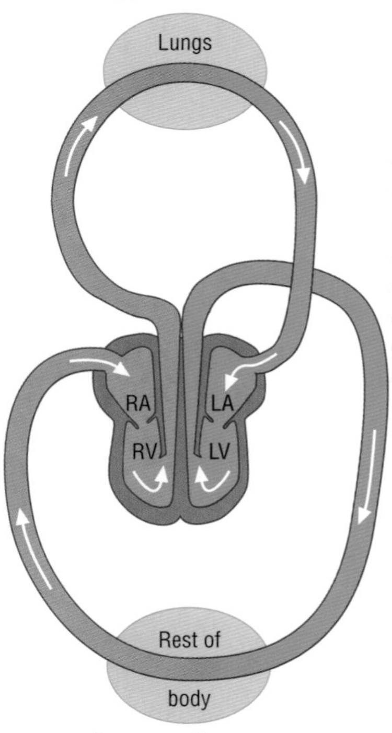

Figure 1 Where oxygen is combined with haemoglobin and where it is released

Haemoglobin

Oxygen is transported in the **erythrocytes** (red blood cells). These cells contain the protein **haemoglobin**. When haemoglobin takes up oxygen, it becomes **oxyhaemoglobin**.

haemoglobin + oxygen → oxyhaemoglobin

Haemoglobin is a complex protein with four subunits. Each subunit consists of a polypeptide (protein) chain and a **haem** (non-protein) group. The haem group contains a single iron atom in the form of Fe^{2+}. This iron ion can attract and hold an oxygen molecule. The haem group is said to have an **affinity** (attraction) for oxygen. As each haem group can hold one oxygen molecule, each haemoglobin molecule can carry four oxygen molecules.

Taking up oxygen

Oxygen is absorbed into the blood in the lungs. Oxygen molecules diffusing into the blood plasma enter the red blood cells. Here they are taken up by the haemoglobin. This takes the oxygen molecules out of solution and so maintains a steep diffusion gradient. This diffusion gradient allows more oxygen to enter the cells.

Releasing oxygen

In the body tissues, cells need oxygen for aerobic **respiration**. Therefore the oxyhaemoglobin must be able to release the oxygen. This is called **dissociation**.

Haemoglobin and oxygen transport

The ability of haemoglobin to take up and release oxygen depends on the amount of oxygen in the surrounding tissues. The amount of oxygen is measured by the relative pressure that it contributes to a mixture of gases. This is called the **partial pressure** or pO_2. It is also called the **oxygen tension** and is measured in units of pressure (kPa).

With a normal liquid, you might expect the amount of oxygen absorbed into the liquid to be directly proportional to the oxygen tension in the surrounding air. A graph of percentage saturation plotted against oxygen tension would be a straight line. This is not the case with blood containing haemoglobin.

Haemoglobin can take up oxygen in a way that produces an S-shaped curve. This is called the **oxyhaemoglobin dissociation curve**. At low oxygen tension, the haemoglobin does not readily take up oxygen molecules. This is because the haem groups that attract the oxygen are in the centre of the haemoglobin molecule. This makes it difficult for the oxygen molecule to reach the haem group and associate with it. This difficulty in combining with the first oxygen molecule accounts for the low saturation level of haemoglobin at low oxygen tensions.

As the oxygen tension rises, the diffusion gradient into the haemoglobin molecule increases. Eventually one oxygen molecule diffuses into the haemoglobin molecule and associates with one of the haem groups. This causes a slight change in the shape of the

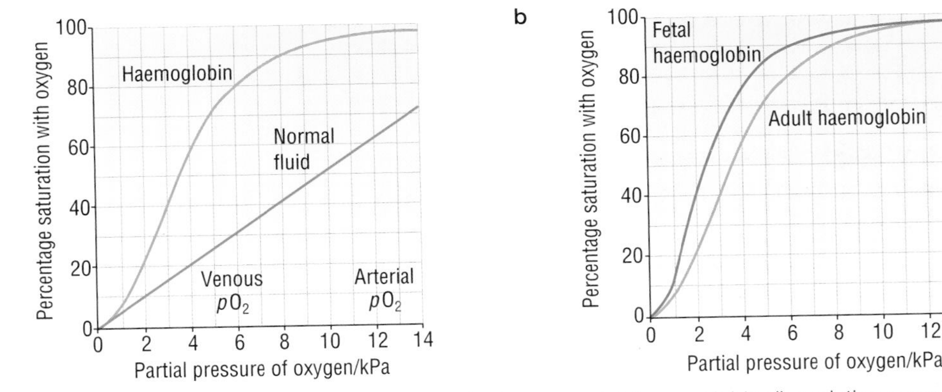

Figure 2a Oxyhaemoglobin dissociation curve; **b** fetal versus adult haemoglobin dissociation curve

haemoglobin molecule known as a *conformational change*. It allows more oxygen molecules to diffuse into the haemoglobin molecule and associate with the other haem groups relatively easily. This accounts for the steepness of the curve as the oxygen tension rises.

Once the haemoglobin molecule contains three oxygen molecules, it becomes more difficult for the fourth molecule to diffuse in and associate with the last available haem group. This means it is difficult to achieve 100% saturation of all the haemoglobin molecules, even when the oxygen tension is very high. So the curve levels off as saturation approaches 100%, despite an increasing oxygen tension.

Mammalian haemoglobin is well adapted to transporting oxygen to the tissues of a mammal. The oxygen tension found in the lungs is sufficient to produce almost 100% saturation. The oxygen tension in respiring body tissues is sufficiently low to cause oxygen to dissociate readily from the oxyhaemoglobin.

Fetal haemoglobin

The haemoglobin of a mammalian fetus has a higher affinity for oxygen than that of adult haemoglobin. Fetal haemoglobin must be able to 'pick up' oxygen from an environment that makes adult haemoglobin release oxygen. In the placenta, the fetal haemoglobin must absorb oxygen from the fluid in the mother's blood. This reduces the oxygen tension within the blood fluid, which in turn makes the maternal haemoglobin release oxygen. So the oxyhaemoglobin dissociation curve for fetal haemoglobin is to the left of the curve for adult haemoglobin.

Questions

1. State where the red blood cells 'pick up' oxygen and where they are likely to release it again.
2. What part of the haemoglobin molecule binds to oxygen?
3. Explain the meaning of the term 'affinity'.
4. Describe and explain the shape of the adult oxyhaemoglobin dissociation curve.
5. Explain why fetal haemoglobin must have a higher affinity for oxygen than adult haemoglobin.

Carriage of carbon dioxide

By the end of this spread, you should be able to . . .

* **Describe the role of haemoglobin in carrying carbon dioxide.**
* **Describe and explain the significance of the dissociation curves of adult oxyhaemoglobin at different carbon dioxide levels (the Bohr effect).**

Key definition

The Bohr effect (Bohr shift) refers to a change in the shape of the oxyhaemoglobin curve when carbon dioxide is present – this causes the oxyhaemoglobin to release oxygen more readily.

Examiner tip

When asked how carbon dioxide is transported, remember that there are three ways. Also remember that the hydrogencarbonate ions diffuse back into the plasma – so most carbon dioxide is carried in the plasma, not in the red blood cells.

How is carbon dioxide transported?

Carbon dioxide is released from **respiring** tissues. It must be removed from these tissues and transported to the lungs. Carbon dioxide in the blood is transported in three ways:

- about 5% is dissolved directly in the plasma
- about 10% is combined directly with **haemoglobin** to form a compound called **carbaminohaemoglobin**
- About 85% is transported in the form of hydrogencarbonate ions (HCO_3^-).

How are hydrogencarbonate ions formed?

As carbon dioxide diffuses into the blood, some of it enters the red blood cells. It combines with water to form a weak acid called carbonic acid. This is catalysed by the enzyme carbonic anhydrase.

$$CO_2 + H_2O \rightarrow H_2CO_3$$

This carbonic acid dissociates to release hydrogen ions (H^+) and hydrogencarbonate ions (HCO_3^-).

$$H_2CO_3 \rightarrow HCO_3^- + H^+$$

The hydrogencarbonate ions diffuse out of the red blood cell into the plasma. The charge inside the red blood cell is maintained by the movement of chloride ions (Cl^-) from the plasma into the red blood cell. This is called the **chloride shift**.

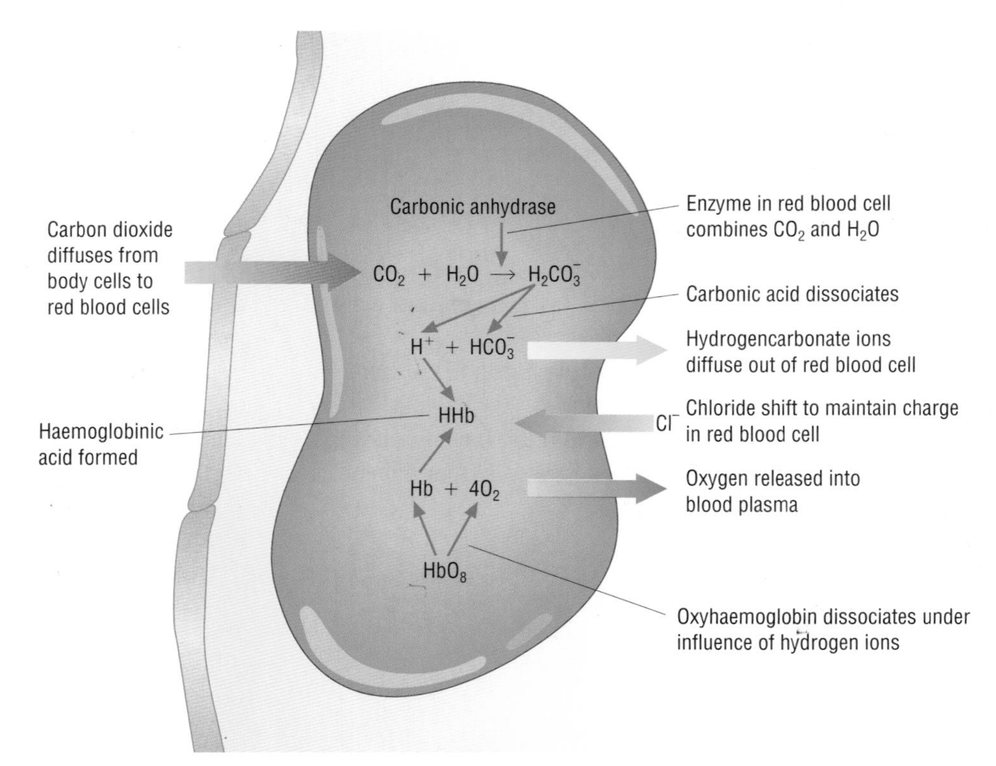

Figure 1 How carbon dioxide is converted to hydrogencarbonate ions

The hydrogen ions could cause the contents of the red blood cell to become very acidic. To prevent this, the hydrogen ions are taken up by haemoglobin to produce **haemoglobinic acid**. The haemoglobin is acting as a **buffer** (a compound that can maintain a constant pH).

Releasing oxygen

As the blood enters respiring tissues, the haemoglobin is carrying oxygen in the form of **oxyhaemoglobin**. The **oxygen tension** of the respiring tissues is lower than that in the lungs because oxygen has been used in respiration. As a result, the oxyhaemoglobin begins to dissociate and releases oxygen to the tissues.

Releasing more oxygen – the Bohr effect

The hydrogen ions released from the **dissociation** of carbonic acid compete for the space taken up by oxygen on the haemoglobin molecule. So when carbon dioxide is present, the hydrogen ions displace the oxygen on the haemoglobin. As a result, the oxyhaemoglobin releases more oxygen to the tissues.

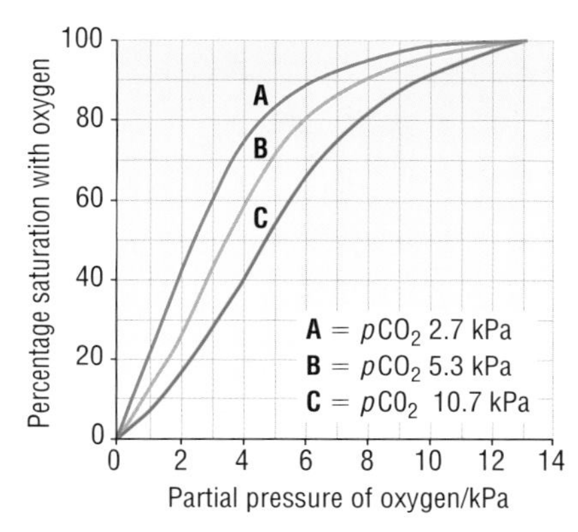

Figure 2 Dissociation curves showing the Bohr effect

Where tissues (such as contracting muscles) are respiring more, there will be more carbon dioxide. As a result there will be more hydrogen ions produced in the red blood cells. This makes the oxyhaemoglobin release more oxygen. This is the **Bohr effect**. At any particular oxygen tension, the oxyhaemoglobin releases more oxygen when more carbon dioxide is present. So when more carbon dioxide is present, haemoglobin is less saturated with oxygen. This makes the oxyhaemoglobin **dissociation curve** shift downwards and to the right (the Bohr shift).

The Bohr effect results in oxygen being more readily released where more carbon dioxide is produced from respiration. This is just what the muscles need for aerobic respiration to continue!

Questions

1 List the ways carbon dioxide can be transported in the blood.
2 Describe how carbon dioxide is converted to hydrogencarbonate ions.
3 Explain the need for the chloride shift.
4 Explain how the presence of carbon dioxide can reduce the affinity of haemoglobin for oxygen.
5 Describe how haemoglobin can supply more oxygen to actively respiring tissues than to those that have a lower level of respiration.

1.2 (13) Transport in plants

By the end of this spread, you should be able to . . .

* **Explain the need for transport systems in multicellular plants in terms of size and surface-area-to-volume ratio.**
* **Describe the distribution of xylem and phloem tissues in roots, stems and leaves of dicotyledonous plants.**

Key definitions

Xylem transports water up the plant.

Phloem transports sugars and other assimilates up and down the plant.

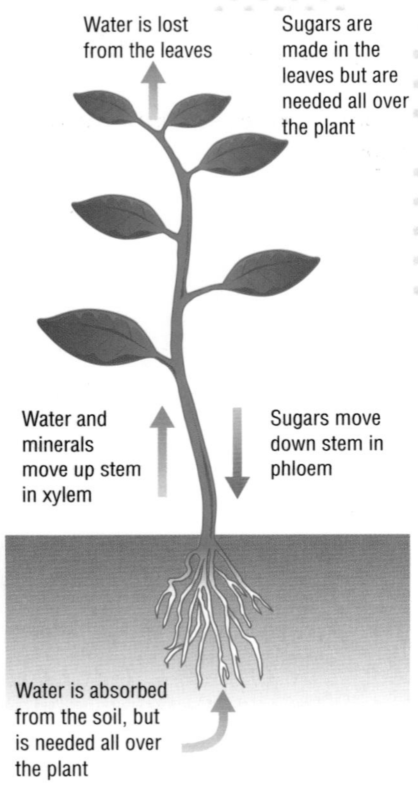

Figure 1 Overall water relations

Why do plants need a transport system?

All living things need to take substances from their environment and return wastes to their environment. Every cell of a multicellular plant needs a regular supply of water and nutrients. In a large multicellular plant the epithelial (surface) cells, which are close to the supply, could gain all they need by simple diffusion. But there are many cells inside the plant that are further from the supply. These cells would not receive enough water or nutrients to survive.

The particular problem in plants is that the roots can obtain water fairly easily, but they cannot absorb sugars from the soil. The leaves can produce sugars, but cannot obtain water from the air.

What substances need to be moved?

The transport system in plants moves water in special tissue called **vascular tissue**:

- water and soluble minerals travel *upwards* in **xylem** tissue
- sugars travel *up or down* in **phloem** tissue.

Both tissues are highly specialised to carry out their transport function. But, unlike in animals, there is no pump and respiratory gases are not carried by these systems.

The vascular tissues

The vascular tissue is distributed throughout the plant. The xylem and phloem are found together in **vascular bundles**. These bundles also often contain other types of tissue that give the bundle some strength and help to support the plant.

Xylem and phloem in the young root

The vascular bundle is found at the centre of a young root. There is a large central core of xylem, often in the shape of an X. The phloem is found in between the arms of the X-shaped xylem. This arrangement provides strength to withstand the pulling forces to which roots are exposed.

Around the vascular bundle is a special sheath of cells called the **endodermis**. The endodermis has a key role in getting water into the xylem vessels. Just inside the endodermis is a layer of **meristem cells** (cells that remain able to divide) called the **pericycle**.

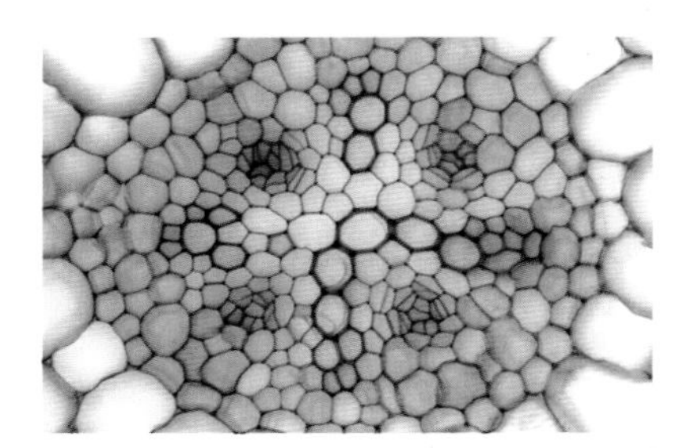

Figure 2 Photomicrograph of the vascular cylinder in the centre of a root of the lesser celandine (*Ranunculus ficaria*). Magnification ×100. The xylem cells, stained blue-black, show a cross-shaped region. The phloem regions, stained red, are within the corners of the xylem

Xylem and phloem in the stem

The vascular bundles are found near the outer edge of the stem. In non-woody plants the bundles are separate and discrete. In woody plants the bundles are separate in young stems, but become continuous in older stems. This means there is a complete ring of vascular tissue just under the bark of a tree. This arrangement provides strength and flexibility to withstand the bending forces to which stems and branches are exposed.

The xylem is found towards the inside of each vascular bundle. The phloem is found towards the outside of the bundle. In between the xylem and phloem is a layer of **cambium**. The cambium is a layer of meristem cells that divide to produce new xylem and phloem.

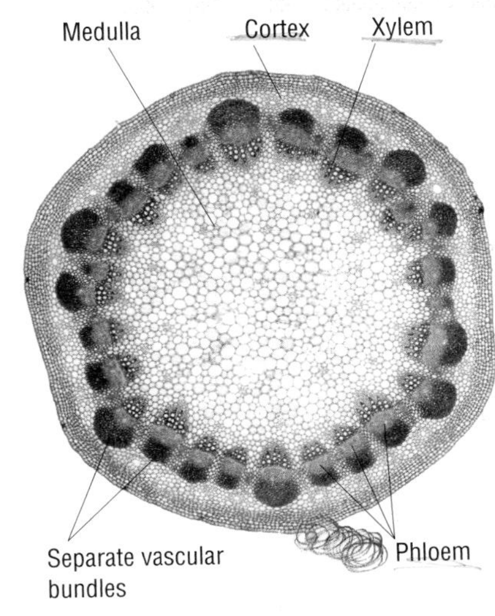

Figure 3 Transverse section of a stem (×40)

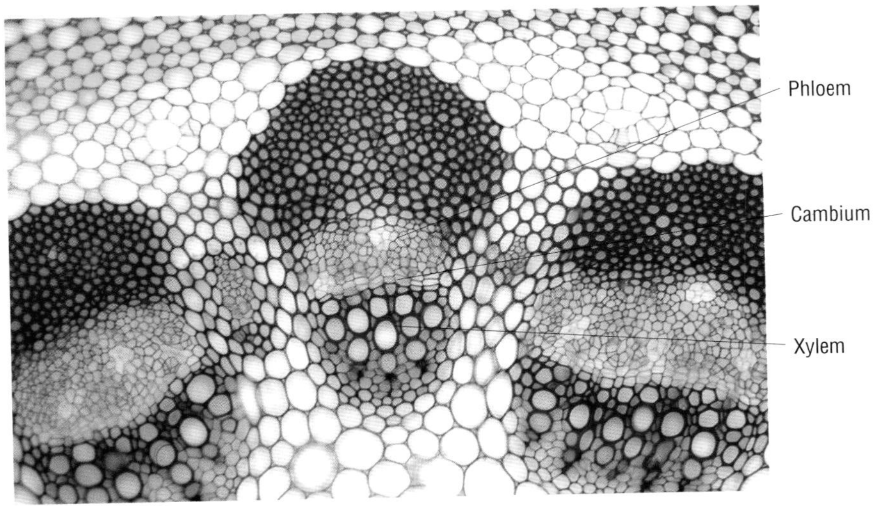

Figure 4 Vascular bundle of a sunflower stem, transverse section (×100)

Xylem and phloem in the leaf

The vascular bundles form the midrib and veins of a leaf. There are two major groups of flowering plants, dicotyledons and monocotyledons – these names are based on the number of first leaves (seed leaves, or cotyledons) they have. These two groups have different patterns of veins. A dicotyledon leaf has a branching network of veins that get smaller as they spread away from the midrib. Within each vein, the xylem can be seen on top of the phloem.

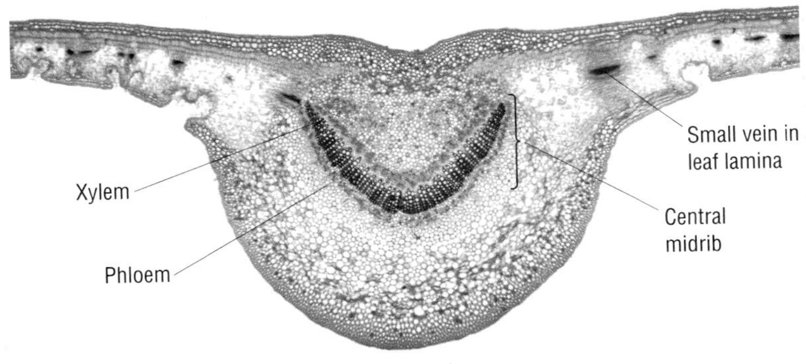

Figure 5 Transverse section of a leaf (×10)

Questions

1 What is transported in the xylem, and in which direction does it go?

2 Name the tissue that transports sugars, and describe its distribution in the stem and roots.

Xylem and phloem

By the end of this spread, you should be able to . . .

* **Describe the structure and function of xylem vessels, sieve tube elements and companion cells.**

Key definitions

Phloem is a plant transport tissue that carries the products of photosynthesis (e.g. sugars) to the rest of the plant. It consists of sieve tube elements and companion cells.

Xylem is a plant transport tissue that carries water from the roots to the rest of the plant. It consists of hollow columns of dead cells lined end-to-end and reinforced with lignin. It provides important support for the plant.

Structure of xylem

Xylem is used to transport water and minerals from the roots up to the leaves and other parts of the plant. Xylem tissue consists of tubes to carry the water and dissolved minerals, fibres to help support the plant and living **parenchyma** cells.

Xylem vessels

In dicotyledonous plants, the most obvious features of xylem are the **xylem vessel** elements. These are long cells with thick walls that have been impregnated by **lignin**. As the xylem develops, the lignin waterproofs the walls of the cells. As a result the cells die, and their end walls and contents decay. This leaves a long column of dead cells with no contents – a tube with no end walls – a xylem vessel. The lignin strengthens the vessel walls and prevents the vessel from collapsing. This keeps the vessels open even at times when water may be in short supply.

The lignin thickening forms patterns in the cell wall. These may be spiral, annular (rings) or reticulate (a network of broken rings). This prevents the vessel from being too rigid and allows flexibility of the stem or branch.

In some places this lignification is not complete. It leaves pores in the wall of the vessel, which are called **pits** or **bordered pits**. These allow water to leave one vessel and pass into another adjacent vessel, or pass into the living parts of the plant.

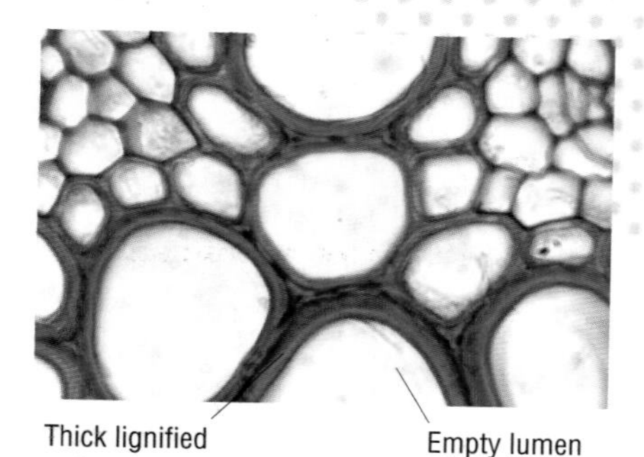

Figure 1 Xylem (transverse section ×400)

Adaptations of xylem to its function

Xylem tissue can carry water and minerals from roots to the very top of the plant because:

- it is made from dead cells aligned end-to-end to form a continuous column
- the tubes are narrow so the water column does not break easily and capillary action can be effective
- pits in the lignified walls allow water to move sideways from one vessel to another
- lignin deposited in the walls in spiral, annular or reticulate patterns allows xylem to stretch as the plant grows and enables the stem or branch to bend.

The flow of water is not impeded because:

- there are no end walls
- there are no cell contents
- there is no nucleus or cytoplasm
- lignin thickening prevents the walls from collapsing.

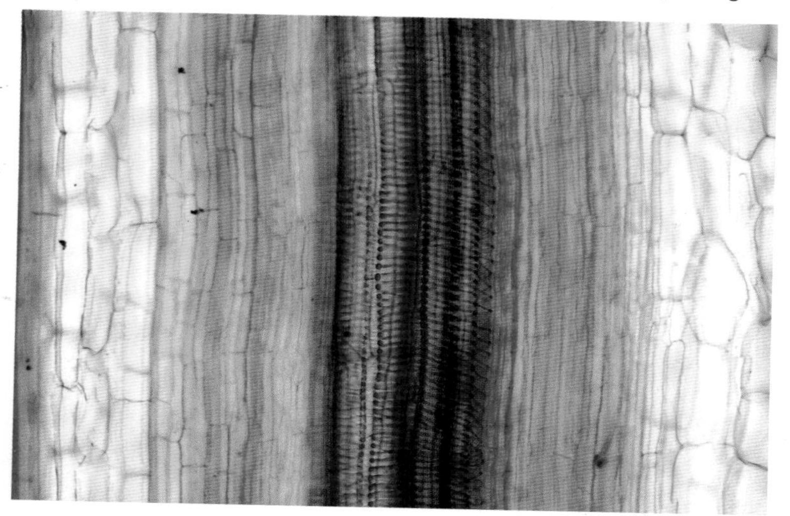

Figure 2 Spiral thickening of xylem (longitudinal section)

How does water move between cells?

When plant cells are touching each other, water molecules can pass from one cell to another. The water molecules will move from the cell with the higher water potential (less negative) to the cell with the lower water potential (more negative).

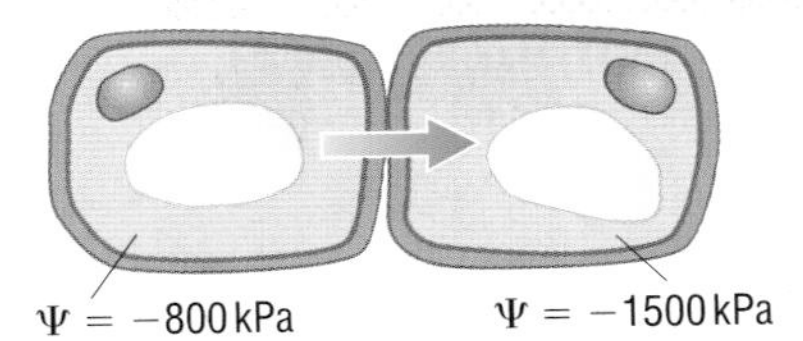

Water moves from the cell with the higher water potential (−800 kPa) to the cell with the lower (more negative) water potential (−1500 kPa)

Figure 3 Cells with different water potentials

What route can water take between cells?

There are three possible pathways that water molecules can take between cells:

- the **apoplast pathway**
- the **symplast pathway**
- the **vacuolar pathway**.

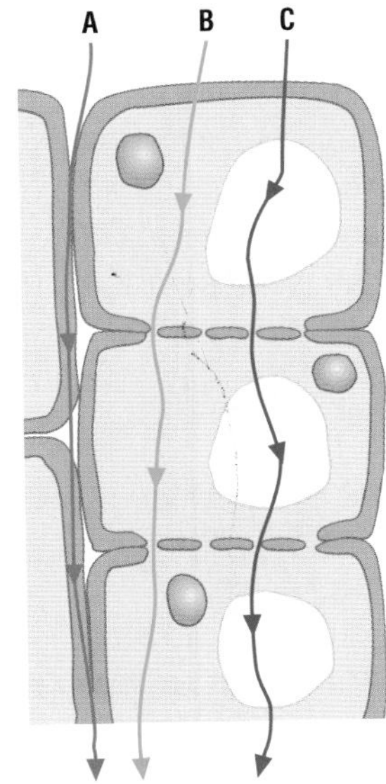

Figure 4 Apoplast (A), symplast (B) and vacuolar (C) pathways

The apoplast pathway (A)

The cellulose cell walls have many water-filled spaces between the cellulose molecules. Water can move through these spaces and between the cells. In this pathway – the apoplast pathway – the water does not pass through any plasma membranes. This means that dissolved mineral ions and salts can be carried with the water.

The symplast pathway (B)

Water enters the cell cytoplasm through the plasma membrane. It can then pass through the plasmodesmata (singular: **plasmodesma**) from one cell to the next. The plasmodesmata are gaps in the cell wall that contain a thin strand of cytoplasm. Therefore the cytoplasm of adjacent cells is linked. Once inside the cytoplasm, water can move through the continuous cytoplasm from cell to cell.

Key definition

A plasmodesma is a fine strand of cytoplasm that links the contents of adjacent cells.

The vacuolar pathway (C)

This is similar to the symplast pathway, but the water is not confined to the cytoplasm of the cells. It is able to enter and pass through the vacuoles as well.

Questions

1. Explain the following terms: turgid, plasmolysis.
2. A pair of adjacent cells, A and B, have water potentials of –1000 and –1200 kPa, respectively. Which cell will gain water from the other?
3. What is in the space between the cell wall and the plasma membrane of a plasmolysed cell?

Water uptake and movement up the stem

By the end of this spread, you should be able to . . .

* **Describe, with the aid of diagrams, the pathway by which water is transported from the root cortex to the air surrounding the leaves, with reference to the Casparian strip, apoplast pathway, symplast pathway, xylem and stomata.**
* **Explain the mechanism by which water is transported from the root cortex to the air surrounding the leaves, with reference to adhesion, cohesion and the transpiration stream.**

Key definitions

Cohesion is the attraction of water molecules for one another.

Adhesion is the attraction of water molecules to the walls of the xylem.

The **symplast pathway** moves water through the cell cytoplasm.

The **apoplast pathway** moves the water in the cell walls and between the cells.

Water uptake from the soil

Plant roots are surrounded by soil particles. The outermost layer of cells (the epidermis) contains **root hair cells** that increase the surface area of the root. These cells absorb minerals from the soil by **active transport** using **ATP** for energy. The minerals reduce the **water potential** of the cell cytoplasm (see spread 1.2.15). This makes the water potential in the cell lower than that in the soil. Water is taken up across the **plasma membrane** by **osmosis** as the molecules move down the water potential gradient.

Movement across the root

The movement of water across the root is driven by an active process that occurs at the endodermis. The endodermis is a layer of cells surrounding the **xylem**. It is also known as the starch sheath as it contains granules of starch – a sign that energy is being used. The endodermis consists of special cells that have a waterproof strip in some of their walls. This strip is called the **Casparian strip**. The Casparian strip blocks the apoplast pathway (see spread 1.2.15), forcing water into the symplast pathway.

The endodermis cells move minerals by active transport from the **cortex** into the xylem. This decreases the water potential in the xylem. As a result, water moves from the cortex through the endodermal cells to the xylem by osmosis.

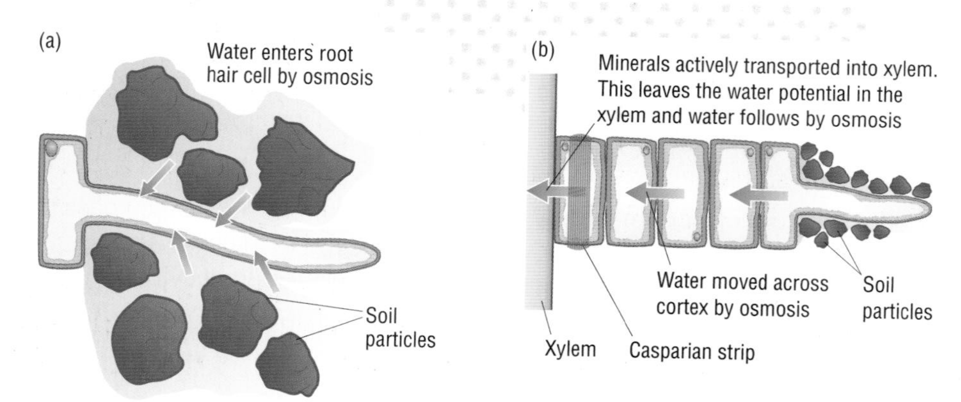

Figure 1 **a** A root hair cell taking up water; **b** movement across the cortex

This reduces the water potential in the cells just outside the endodermis. This, combined with water entering the root hair cells, creates a water potential gradient across the whole cortex. Therefore water is moved along the symplast pathway from the root hair cells, across the cortex and into the xylem.

At the same time, water can move through the apoplast pathway across the cortex. This water moves into the cells to join the symplast pathway just before passing through the endodermis.

What is the role of the Casparian strip?

- The Casparian strip blocks the apoplast pathway between the cortex and the xylem.
- This ensures that water and dissolved nitrate ions have to pass into the cell cytoplasm through cell membranes.
- There are transporter proteins in the cell membranes.
- Nitrate can be actively transported, from the cytoplasm of the cortex cells, into the xylem.
- This lowers the water potential in the xylem so water from cortex cells follows into the xylem by osmosis.
- Once the water has entered the xylem it cannot pass back into the cortex as the apoplast pathway of the endodermal cells is blocked.

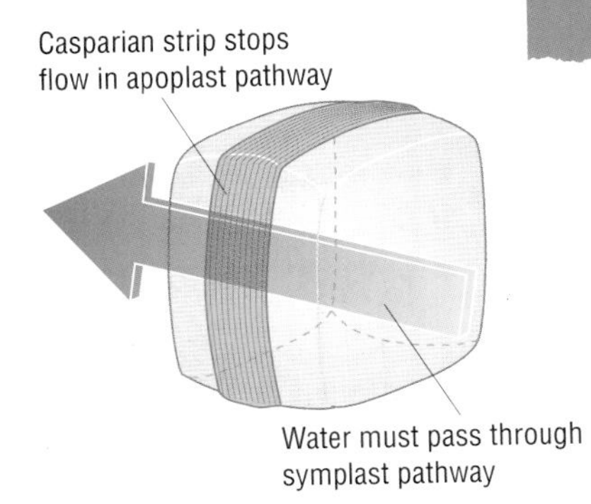

Figure 2 The Casparian strip

How does water move up the stem?

There are three processes that help to move water up the stem:

- root pressure
- transpiration pull
- capillary action.

Root pressure

The action of the endodermis moving minerals into the xylem by active transport drives water into the xylem by osmosis. This forces water into the xylem and pushes the water up the xylem. Root pressure can push water a few metres up a stem, but cannot account for water getting to the top of tall trees.

Transpiration pull

The loss of water by evaporation from the leaves must be replaced by water coming up from the xylem. Water molecules are attracted to each other by forces of **cohesion**. You can learn more about cohesion in spread 2.1.12. These cohesion forces are strong enough to hold the molecules together in a long chain or column. As molecules are lost at the top of the column, the whole column is pulled up as one chain. This creates the transpiration stream. The pull from above can create tension in the column of water. This is why the xylem vessels must be strengthened by **lignin**. The lignin prevents the vessel from collapsing under tension. Because this mechanism involves cohesion between the water molecules and tension in the column of water, it is called the **cohesion–tension theory**. It relies on the plant maintaining an unbroken column of water all the way up the xylem. If the water column is broken in one xylem vessel, the water column can still be maintained through another vessel via the **pits**.

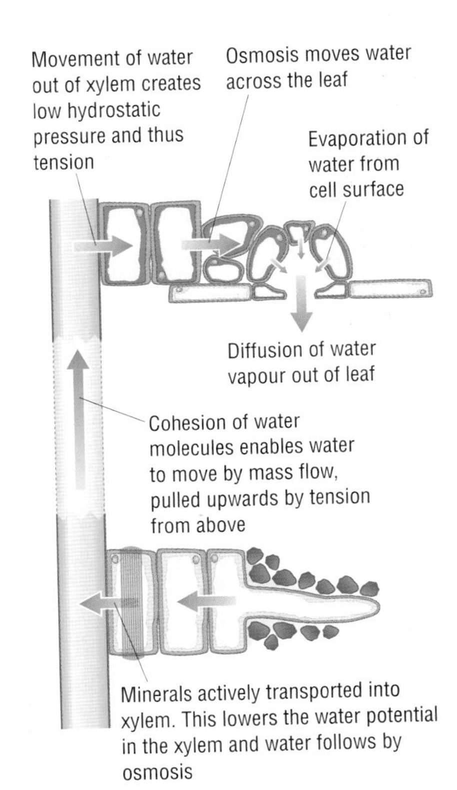

Figure 3 The route taken by water

Capillary action

The same forces that hold water molecules together also attract the water molecules to the sides of the xylem vessel. This is called **adhesion**. Because the xylem vessels are very narrow, these forces of attraction can pull the water up the sides of the vessel.

How water leaves the leaf

Most of the water that leaves the leaf exits through the **stomata**, tiny pores in the epidermis. Only a tiny amount leaves through the waxy cuticle. Water evaporates from the cells lining the cavity immediately below the **guard cells**. This lowers the water potential in these cells, causing water to enter them by osmosis from neighbouring cells. Water then enters these neighbouring cells from cells deeper in the leaf, and so on until eventually water leaves the xylem and enters into the innermost leaf cells. This is explained in more detail in spread 1.2.17.

Questions

1. Describe the role of the Casparian strip.
2. Explain how the cohesion–tension theory can account for water rising up a tall tree.

1.2 (17) Transpiration

By the end of this spread, you should be able to . . .

* **Define the term transpiration.**
* **Describe the factors that affect the transpiration rate.**
* **Describe how a potometer is used to estimate transpiration rates.**

Key definition

Transpiration is the loss of water by evaporation from the aerial parts of a plant.

Examiner tip

Water must evaporate to form vapour before it is lost. The water vapour will diffuse down a **water vapour potential** gradient.

Never use the term 'water concentration gradient'.

What is transpiration?

Transpiration is the loss of water vapour from the upper parts of the plant – particularly the leaves.

Water enters the leaves in the **xylem** and passes to the mesophyll cells by **osmosis**. The water evaporates from the surface of the mesophyll cells to form water vapour. The spongy mesophyll cells have large air spaces between them that help the water vapour to diffuse through the leaf tissue. As water vapour collects in these air spaces, the **water vapour potential** rises. Once the water vapour potential inside the leaf is higher than outside, water molecules will **diffuse** out of the leaf. Open stomata provide an easy route for the water vapour to leave the leaf. The stomata are open during the day to allow gaseous exchange for photosynthesis.

Transpiration involves three processes:

- osmosis from the xylem to mesophyll cells
- evaporation from the surface of the mesophyll cells into the intercellular spaces
- diffusion of water vapour from the intercellular spaces out through the stomata.

What is the transpiration stream?

As water leaves the xylem in the leaf, it must be replaced from below. Water moves up the xylem from the roots to replace the water lost.

This movement of water up the stem is useful to the plant in a number of ways:

- water is required in the leaves for photosynthesis
- water is required to enable cells to grow and elongate
- water keeps the cells turgid
- the flow of water can carry useful minerals up the plant
- evaporation of water can keep the plant cool.

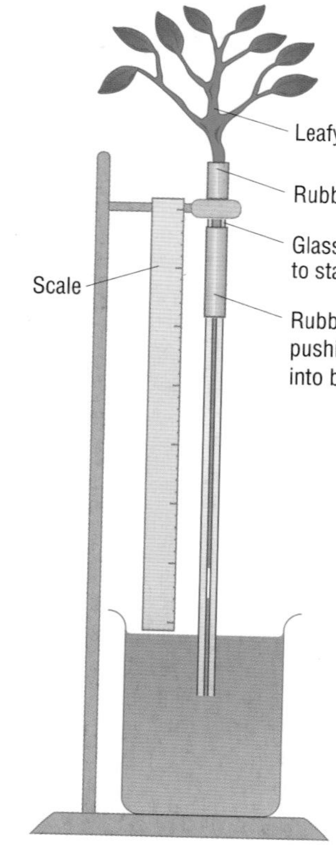

Figure 1 A simple potometer

How can we measure the rate of transpiration?

A piece of apparatus called a **potometer** can be used to estimate the rate of water loss. It is not an exact measure, as it actually measures the rate of water uptake by a cut shoot. However, as about 99% of water taken up is lost in transpiration, it does give a reasonable estimate of water loss.

Set up the potometer as shown in the diagram. It is important to make sure there are no air bubbles inside the apparatus. Water lost by the leaf is replaced from the water in the capillary tube. The movement of the meniscus at the end of the water column can be measured.

To study the effect of different environmental conditions on the rate of transpiration, you can place the whole apparatus in different situations.

What factors alter the rate of transpiration?

The rate of water loss can be affected by many things. Some of these are to do with the plant, others are environmental. Anything that increases the gradient between the water vapour potential inside the leaf and that outside the leaf will increase water loss.

Feature that affects the rate of water loss	How it affects water loss
Number of leaves	A plant with more leaves has a larger surface area over which water vapour can be lost.
Number, size and position of stomata	If the leaves have many large stomata, then water vapour is lost more quickly. If the stomata are on the lower surface, water vapour loss is slower.
Presence of cuticle	A waxy cuticle reduces evaporation from the leaf surface.
Light	In light, the stomata open to allow gaseous exchange for photosynthesis.
Temperature	A higher temperature will increase the rate of water loss in three ways. It will: • increase the rate of evaporation from the cell surfaces so that the water vapour potential in the leaf rises • increase the rate of diffusion through the stomata because the water molecules have more kinetic energy • decrease the relative water vapour potential in the air, allowing more rapid diffusion of molecules out of the leaf.
Relative humidity	Higher relative humidity in the air will decrease the rate of water loss. This is because there will be a smaller water vapour potential gradient between the air spaces in the leaf and the air outside.
Air movement or wind	Air moving outside the leaf will carry away water vapour that has just diffused out of the leaf. This will maintain a high water vapour potential gradient.
Water availability	If there is little water in the soil, then the plant cannot replace the water that is lost. Water loss in plants is reduced when stomata are closed or when the plants shed leaves in winter.

Table 1 Factors that affect water loss by transpiration

What happens if the plant loses too much water?

If water loss by transpiration is greater than water uptake from the soil, the plant cells will lose turgidity. Non-woody plants will wilt and eventually die. The leaves of woody plants will also wilt and the plant will eventually die.

Questions

1. Describe how water is lost from the leaf.
2. Describe how a potometer can be used to measure the rate of water uptake. Describe the precautions you should take to ensure the results will be valid and reliable.
3. List the environmental factors that can affect the rate of water loss.
4. Explain, in terms of water vapour potential, how each factor that you have named in question 3 can affect the rate of water loss.

Reducing water loss – xerophytes

By the end of this spread, you should be able to . . .

* **Explain why transpiration is a consequence of gaseous exchange.**
* **Describe, with the aid of diagrams, how the leaves of some xerophytes are adapted to reduce water loss by transpiration.**

Key definition

A **xerophyte** is a plant that is adapted to reduce water loss so that it can survive in very dry conditions.

Unavoidable losses

The loss of water by **transpiration** is unavoidable. This is because plants exchange gases with the atmosphere via their **stomata**. During the day, plants take up a lot of carbon dioxide. This is used in **photosynthesis**. They must also remove oxygen, which is a by-product of photosynthesis. So the stomata must be open during the day. While the stomata are open, there is an easy route for water to be lost.

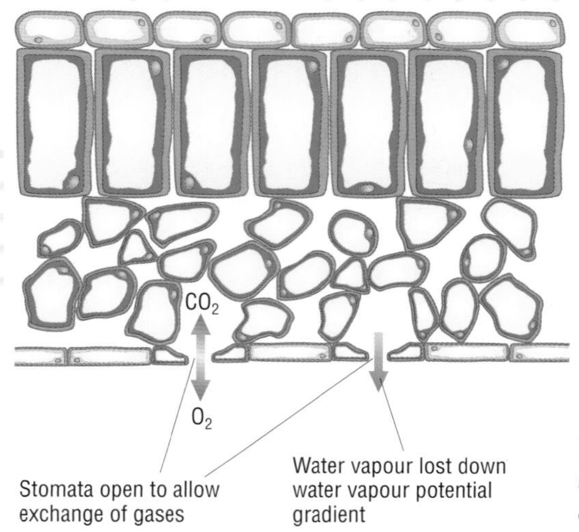

Figure 1 Why stomata need to stay open, and the consequences of them being open

Most plants can reduce these losses by structural and behavioural adaptations:

- a waxy cuticle on the leaf will reduce water loss due to evaporation through the epidermis
- the stomata are often found on the undersurface of leaves, not on the top surface – this reduces the evaporation due to direct heating from the sun
- most stomata are closed at night, when there is no light for photosynthesis
- deciduous plants lose their leaves in winter, when the ground may be frozen (making water less available) and when temperatures may be too low for photosynthesis.

What happens if a plant loses too much water?

If water loss is greater than water uptake, then a plant may suffer water stress. If this occurs, the cells may lose their **turgidity** and may even undergo **plasmolysis**. This will cause non-woody parts of the plant to wilt. Eventually, the plant will die.

Living in arid conditions

Some plants are particularly well adapted to living in very dry or arid conditions. These plants are known as **xerophytes**. They have a number of adaptations to reduce water loss from their leaves.

- Smaller leaves, particularly leaves shaped like needles. This reduces the total surface area of the leaves. The total leaf surface area is also reduced, so that less water is lost by transpiration. A typical example is the pine tree.

- Densely packed spongy mesophyll. This reduces the cell surface area that is exposed to the air inside the leaves. Less water will evaporate into the leaf air spaces, reducing the rate of water loss.
- A thicker waxy cuticle on the leaves reduces evaporation further. For example, holly leaves have very thick waxy cuticles.
- Closing the stomata when water availability is low will reduce water loss and so reduce the need to take up water.
- Hairs on the surface of the leaf trap a layer of air close to the surface. This air can become saturated with moisture and will reduce the diffusion of water vapour out through the stomata. This is because the gradient of the **water vapour potential** between the inside of the leaf and the outside has been reduced.
- **Pits** containing stomata at their base also trap air that can become saturated with water vapour. This will reduce the gradient in the water vapour potential between inside and outside the leaf, so reducing loss by diffusion.
- Rolling the leaves so that the lower epidermis is not exposed to the atmosphere can trap air that becomes saturated. This is another way to reduce or even eliminate the water vapour potential gradient.
- Some plants have a low water potential inside their leaf cells. This is achieved by maintaining a high salt concentration in the cells. The low water potential reduces the evaporation of water from the cell surfaces as the water potential gradient between the cells and the leaf air spaces is reduced.

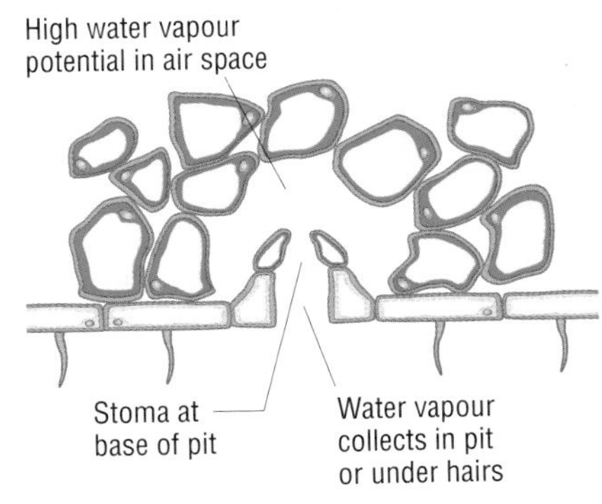

Figure 2 How water loss is reduced

Marram grass – a special case

Marram grass (*Ammophila*) specialises in living on sand dunes. The conditions are particularly harsh because any water in the sand drains away quickly, the sand may be salty and the leaves are often exposed to very windy conditions.

Marram grass shows many of the xerophytic features described above.

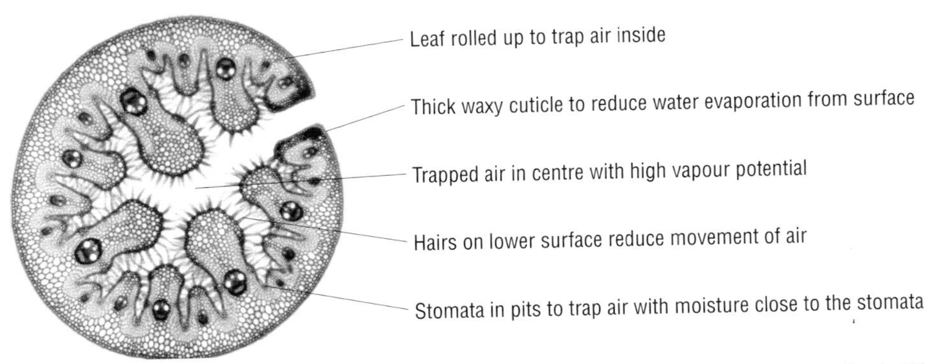

Figure 3 A cross section of a marram grass leaf showing adaptations to conserve water (×10)

Questions

1. Why must the stomata remain open?
2. Explain, in terms of water vapour potential, why water vapour is lost to the atmosphere when the stomata are open.
3. List the features of a xerophyte that enable it to survive in dry conditions. Explain how each feature helps to reduce water loss.
4. Describe how marram grass is adapted to living on sand dunes.

1.2 (19) The movement of sugars – translocation

By the end of this spread, you should be able to . . .

* **Describe the mechanism of transport in phloem involving active loading at the source and removal at the sink, and the evidence for and against this mechanism.**
* **Explain translocation as an energy-requiring process transporting assimilates (especially sucrose) between the sources and sinks.**

Key definitions

Translocation is the transport of assimilates throughout the plant, in the phloem tissue.

A **source** releases sucrose into the phloem.

A **sink** removes sucrose from the phloem.

Translocation

The movement of assimilates (sugars and other chemicals made by plant cells) is called **translocation**. Sugars are transported in the **phloem** in the form of sucrose. A part of the plant that releases sucrose into the phloem is called a **source**. A part of the plant that removes sucrose from the phloem is called a **sink**.

How does sucrose enter the phloem?

Sucrose is loaded into the phloem by an active process. **ATP** is used by the **companion cells** to actively transport hydrogen ions (protons) out of their cytoplasm and into the surrounding tissue. This sets up a **diffusion gradient** and the hydrogen ions diffuse back into the companion cells. This diffusion occurs through special **cotransporter proteins**. These proteins enable the hydrogen ions to bring sucrose molecules into the companion cells. As the concentration of sucrose molecules builds up inside the companion cells, they diffuse into the **sieve tube elements** through the numerous **plasmodesmata**.

Examiner tip

Water molecules do not flow along a gradient. They *diffuse* down a *water potential gradient.*

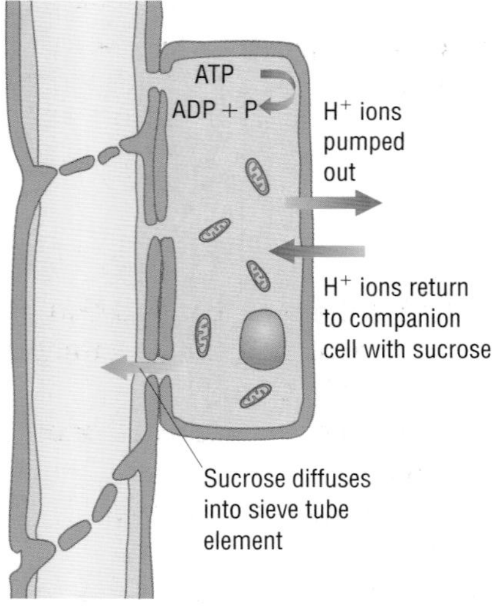

Figure 1 Active loading at the source

Movement of sucrose along the phloem

At the source

Sucrose entering the sieve tube element reduces the **water potential** inside the sieve tube. As a result, water molecules move into the sieve tube element by **osmosis** from surrounding tissues. This increases the **hydrostatic pressure** in the sieve tube at the source.

At the sink

Sucrose is used in the cells surrounding the phloem. The sucrose may be converted to starch for storage, or may be used in a metabolic process such as **respiration**. This reduces the sucrose concentration in these cells. Sucrose molecules move by **diffusion** or **active transport** from the sieve tube element into the surrounding cells. This increases the water potential in the sieve tube element, so water molecules move into the surrounding cells by osmosis. This reduces the hydrostatic pressure in the phloem at the sink.

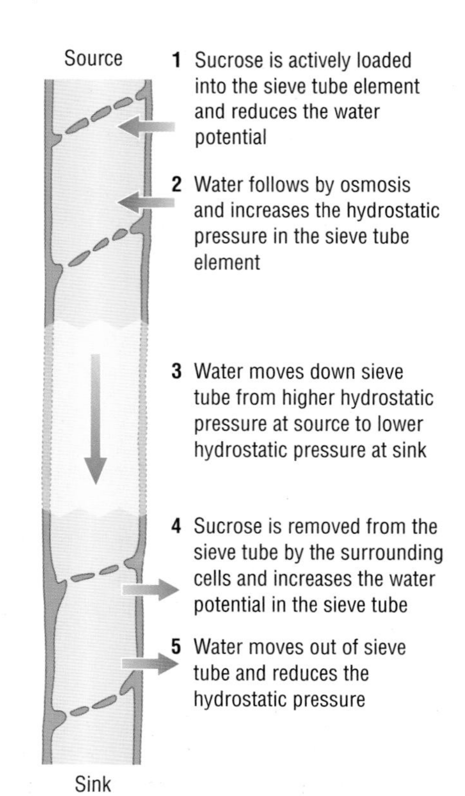

Figure 2 Mass flow in the phloem

Along the phloem

Water entering the phloem at the source, moving down the hydrostatic pressure gradient and leaving the phloem at the sink, produces a flow of water along the phloem. This flow carries sucrose and other assimilates along the phloem. This is called mass flow. It can occur in either direction – up or down the plant – depending on where sugars are needed.

Mass flow may occur up or down the plant in the same phloem tube at different times. It may be moving assimilates up the plant in some tubes and down the plant in other tubes at the same time.

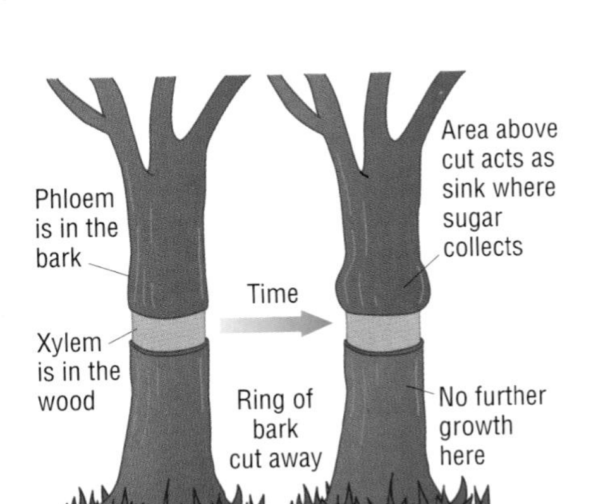

Figure 3 Ringing a tree causes sugars to collect above the ring

Where the sources and sinks are

Perhaps the most obvious source is a leaf. Sugars made during photosynthesis are converted to sucrose and loaded into the phloem. This occurs during late spring, summer and early autumn, while the leaves are green. In early spring, when the leaves are growing, they need energy. This energy is supplied from stores in other parts of the plant, and the leaves act as a sink.

Other sources include the roots, where stored carbohydrates are released into the phloem. This occurs particularly in spring, when other parts of the plant need energy for growth. During summer and autumn the roots store sugars as starch. Thus the roots can also act as a source at some times of year and as a sink at other times.

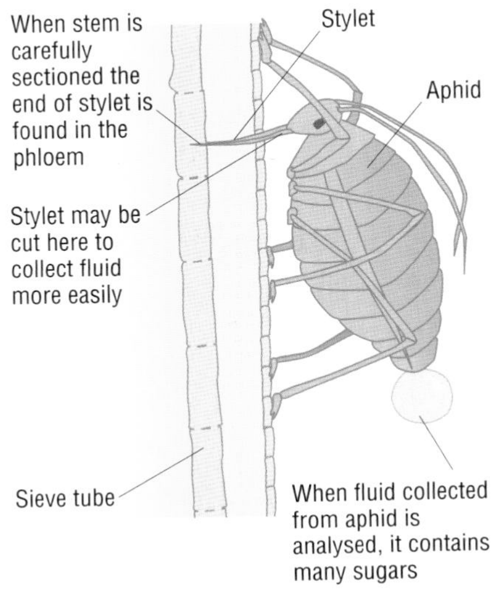

Figure 4 An aphid feeds from the phloem

Evidence for this mechanism of translocation

How we know the phloem is used

- If a plant is supplied with radioactively labelled carbon dioxide (which will be used in photosynthesis), the labelled carbon soon appears in the phloem.
- Ringing a tree to remove the phloem results in sugars collecting above the ring (Figure 3).
- An aphid feeding on a plant stem can be used to show that the mouthparts are taking food from the phloem (Figure 4).

How we know it needs metabolic energy (ATP)

- The companion cells have many **mitochondria**.
- Translocation can be stopped by using a metabolic poison that inhibits the formation of ATP.
- The rate of flow of sugars in the phloem is so high that energy must be needed to drive the flow. It has been calculated that sugars move up to 10 000 times more quickly by mass flow than they could through diffusion alone.

How we know it uses this mechanism

- The pH of the companion cells is higher than that of surrounding cells.
- The concentration of sucrose is higher in the source than in the sink.

Is there evidence against this mechanism?

- Not all the solutes in the phloem sap move at the same rate.
- Sucrose is moved to all parts of the plant at the same rate, rather than going more quickly to areas with a low concentration.
- The role of sieve plates is unclear.

Questions

1. Define the term 'translocation' and explain why it is important in the life of plants.
2. Define the terms 'source' and 'sink'.
3. What is meant by 'active loading'?
4. Describe the role of hydrogen ions in active loading.
5. Compare the mechanism of translocation with that of transpiration.
6. Describe how an aphid can be used to show that sugars travel in the phloem.
7. Why is the pH of the companion cells higher than in the surrounding cells?
8. Suggest a possible function that could be performed by the sieve plates.

1.2 Exchange and transport summary

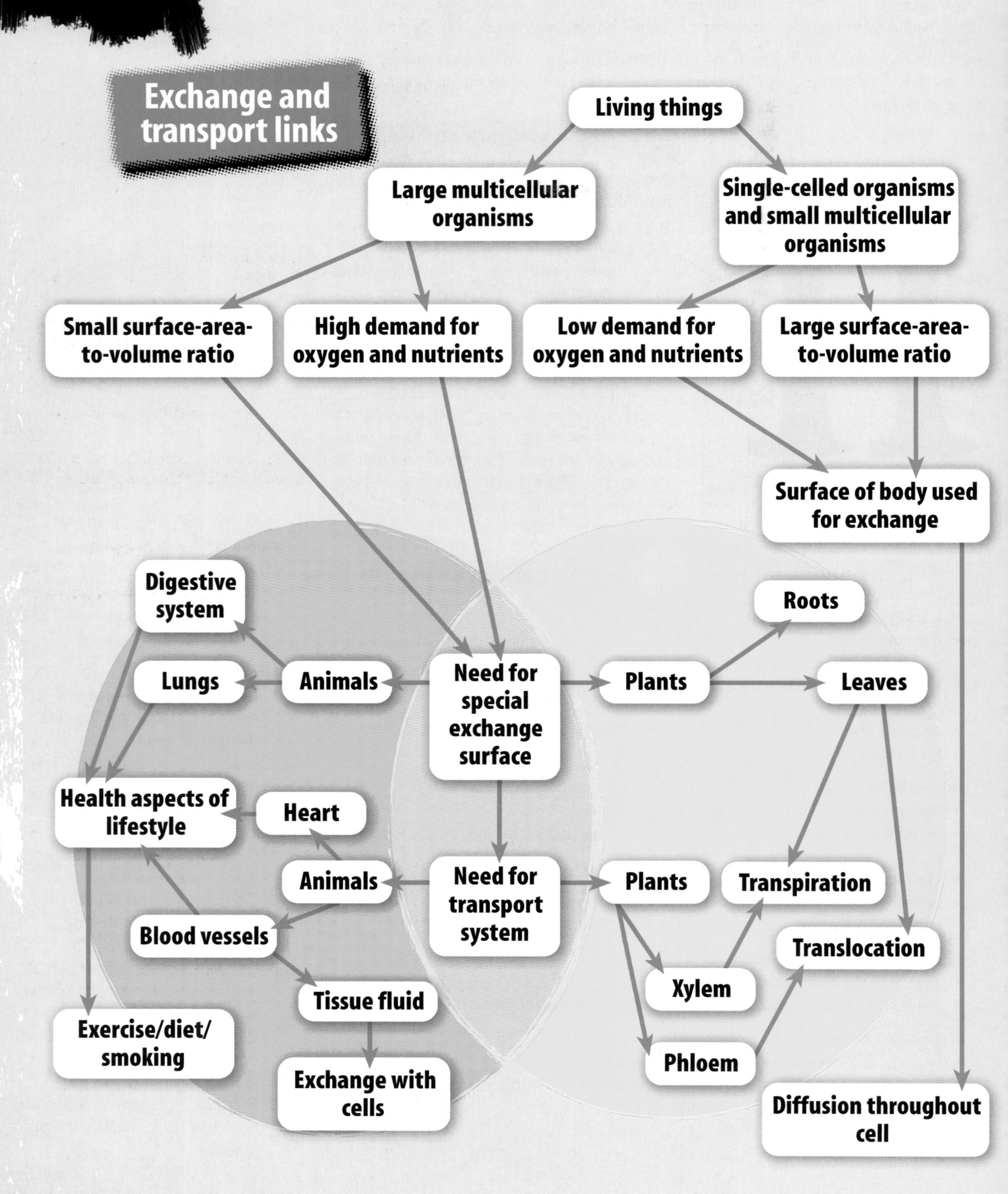

Practice questions

1 (a) State the names of the tissues labelled **A–D** below. [4]

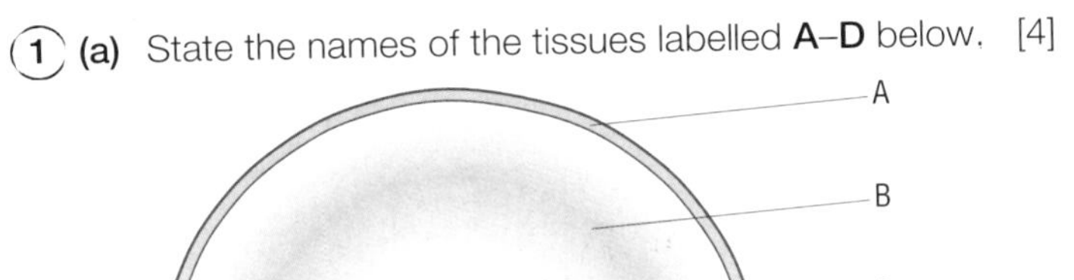

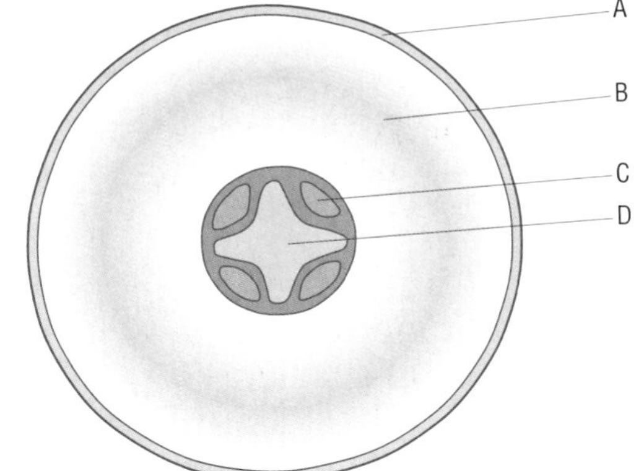

Figure 1

(b) Name the organ represented by Figure 1. [1]

2 (a) State the name of the type of muscle found in the heart. [1]

(b) Explain why the muscle surrounding the left ventricle is thicker than that surrounding the right ventricle. [2]

(c) Name the blood vessels that:
(i) carry blood away from the ventricles;
(ii) carry blood back towards the heart. [2]

3 (a) State two features of a good gaseous exchange system. [2]

(b) Describe the route taken by air as it is inhaled. [3]

(c) Name the air sacs in the lungs. [1]

4 (a) What immediate effect does exercise have on the heart rate? [1]

(b) What happens to the tidal volume of the lungs during exercise? [1]

(c) Name two other exchange surfaces. [2]

5 Complete the following paragraph by filling in the blank spaces.

Blood is .... in the lungs. The red pigment .... has a high affinity for oxygen. The pumping action of the .... creates pressure which pushes the blood around the body. In the tissues the partial pressure of .... is low. This causes the .... of the oxyhaemoglobin. In the tissues, the oxygen is used in the process of .... Most of the carbon dioxide produced in this process enters the...................... .... cells. Here it is converted to carbonic acid by the action of the enzyme carbonic anhydrase. The carbon dioxide is transported as .... back to the lungs. [8]

6 (a) Explain the role of elastic tissue in the alveoli. [2]

(b) Name the two types of cell found in the epithelium of the airways. [2]

(c) Describe the action of the cilia in bronchi. [2]

7 (a) In a plant, name two tissues that may act as a source. [2]

(b) Name the tissue that is used to transport sugars in a plant. [1]

8 (a) Explain why a large organism with a low surface-area-to-volume ratio needs a special surface for gaseous exchange. [3]

(b) Explain how a higher concentration of carbon dioxide in the blood can cause extra oxygen to dissociate from oxyhaemoglobin. [4]

9 (a) Describe how sucrose is loaded into the sieve tube element. [4]

(b) Explain, using the term water vapour potential gradient, why increasing the wind speed will increase the rate of transpiration. [3]

10 (a) Explain the importance of the Purkyne tissue in coordination of the cardiac cycle. [3]

(b) Describe and explain the role of the coronary arteries. [3]

1.2 Examination questions

1 Figure 1 is a drawing of a part of an alveolus together with an associated blood capillary.

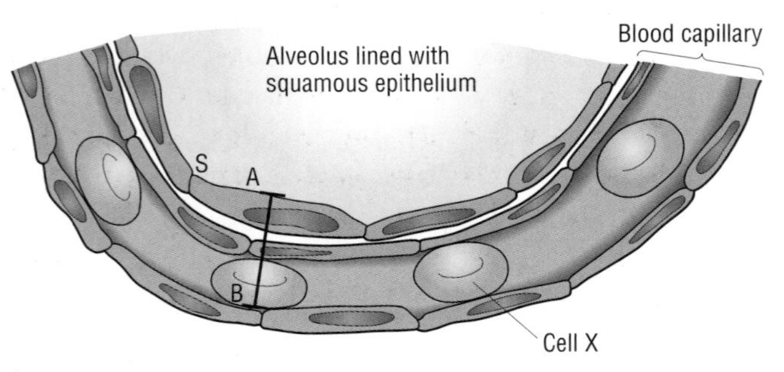

Figure 1

(a) (i) State a feature, visible in Figure 1, that shows that squamous epithelial cells are eukaryotic. [1]

(ii) State why squamous epithelium is described as a tissue. [1]

(iii) State **two** features of a gas exchange surface such as the lining of the alveolus. [2]

(b) The line **AB** in Figure 1 represents an actual distance of 1.5 μm.
Calculate the magnification of the drawing. Show your working. [2]

(c) Describe the mechanism by which oxygen gets from point **S** on Figure 1 to the red blood cell. [1]

(d) Once in the red blood cells, oxygen is picked up by haemoglobin. Explain how two features of red blood cells, other than the presence of haemoglobin, make them efficient in the collection of oxygen and its transport to the tissues. [4]

[Total: 11]

(OCR 2801 Jun06)

2 Figure 2 shows the effect of different levels of carbon dioxide on the dissociation of oxyhaemoglobin.

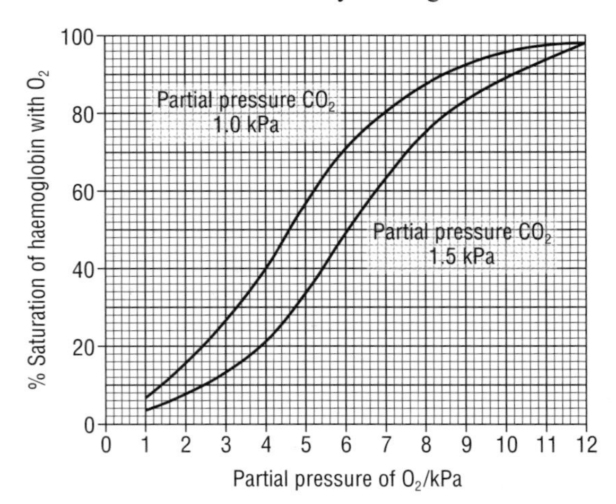

Figure 2

With reference to Figure 2:

(a) Name the effect shown. [1]

(b) Calculate the difference in oxygen percentage saturation between the partial pressures of carbon dioxide at a partial pressure of oxygen of 5 kPa. [1]

(c) Outline how this effect ensures more efficient delivery of oxygen to muscles when exercising. [3]

[Total: 5]

(OCR 2803–01 Jun01)

Answers to examination questions will be found on the Exam Café CD.

3 Transpiration is the loss of water by evaporation. Figure 3 shows a potometer, an apparatus used to estimate transpiration rates.

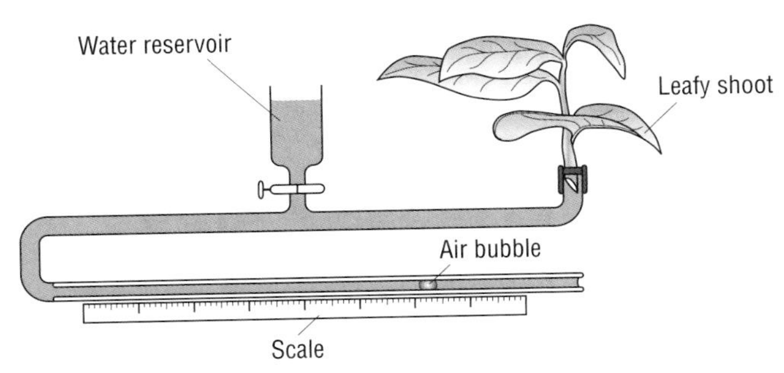

Figure 3

(a) (i) Describe how the apparatus should be set up so that valid measurements can be obtained. [4]

(ii) Transpiration itself is not measured by the apparatus. State precisely what is measured by using the apparatus. [1]

(b) A student investigated the rate of transpiration of a plant under two conditions, **A** and **B**, in the laboratory using a potometer. In both cases the temperature, humidity and duration were the same. A fan was placed next to the potometer and turned on for condition **A** but not for **B**.

The results are shown in Table 3.

reading	estimate of transpiration rate/ arbitrary units	
	condition A	condition B
1	45	107
2	39	99
3	41	106
4	46	101
5	38	103
mean	42	

Table 3

(i) Calculate the mean estimated transpiration rate for condition **B**. Express your answer to the nearest whole number and write it in Table 3. [1]

(ii) Explain why the mean estimated transpiration rate for condition **B** is greater than for condition **A**. [2]

(c) The student wanted to compare the mean estimated transpiration rate of two species of plant using the potometer in Figure 3.

Suggest what the student would need to do in order to get a valid comparison of the two species. [2]

[Total: 10]

(OCR 2803–03 Jun05)

UNIT 2

Module 1 Biological molecules

Introduction

Every living organism on the planet is built from a relatively few different chemical constituents. These come together to form all the carbohydrates, proteins, lipids and nucleic acids needed in order to make living things.

These molecules exist in many forms. In this module, you will learn how the simple chemical elements can form this vast array of biologically important molecules. Many biological molecules are huge, and can consist of many millions of atoms bonded together. Biological molecules are bonded by simple reactions, repeated over and over again, that can be described in simplified terms. You will also begin to understand how the molecules work together – in processes that not only shape and drive living processes, but also shape how life moves from generation to generation.

Enzymes are one type of biological molecule. They are all proteins, and are vital to living processes. The reactions of metabolism are far too slow if they are not catalysed and controlled by enzymes. Some enzymes work inside cells, some outside (like amylase, pictured here), but all speed up chemical reactions by several million times. You will learn about enzyme structure and function, and how various factors that affect enzyme action can be investigated and understood. Examples are used to show some of the key metabolic processes that depend on enzyme function. A number of disease states related to enzyme function are also described.

One very small, but very important, biological molecule is water. Water is not like any other liquid. It is the nature of water that allows trees to grow, blood to flow, and animals and plants to survive freezing temperatures. You will learn just how important to life this tiny molecule really is.

Test yourself

1. What is a polymer? What is a monomer?
2. What functions do the different biological molecules perform?
3. What sort of bonds hold atoms together to form molecules?
4. What do some organisms gain from regulating their body temperature?
5. Why are all enzymes proteins?
6. Why do eukaryotic organisms enclose many enzymes in lysosomes?
7. Why do enzymes need to be controlled?
8. Why is water so important to life?
9. Which type of molecule is responsible for the unique characteristics of organisms?

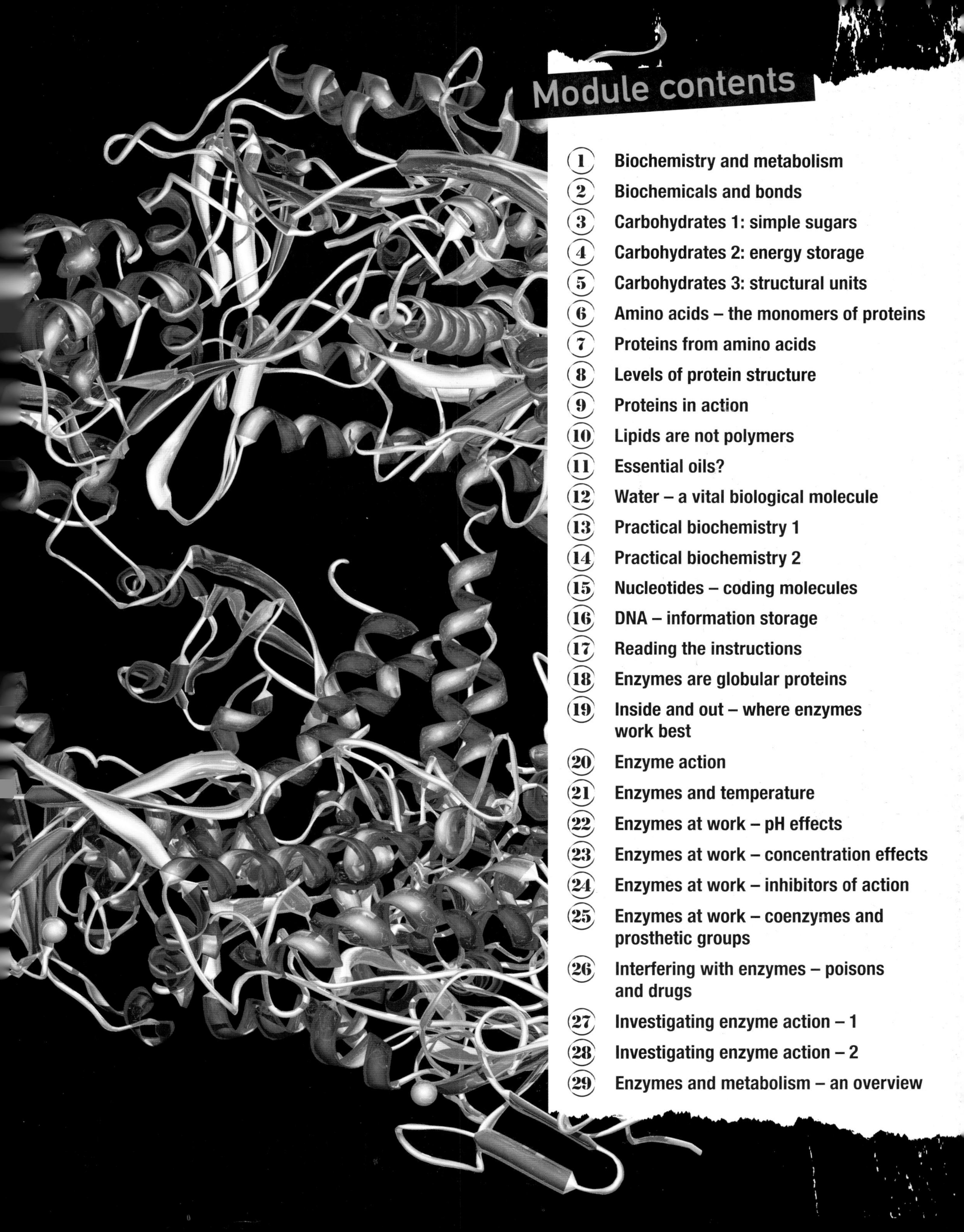

Module contents

① Biochemistry and metabolism

By the end of this spread, you should be able to . . .

- ✱ **State the functions of biological molecules in organisms.**
- ✱ **Define metabolism.**

You are what you eat

You are – literally – what you eat. The nutrients any organism takes in build and maintain a healthy body because:

- they become part of your body – often after being broken down into smaller pieces (digested), then rebuilt to form the different parts of the organism
- they are used to provide the energy needed to drive **metabolism** (living processes).

If substances taken into the body cannot be digested or take part in metabolic processes, they are removed from the body.

Organisms are adapted to remove these substances in a variety of ways. For example, the human intestine is adapted to work best when indigestible fibre is present in food. The fibre does not provide nutrients for building the body. It does not provide energy, but it is an essential component of the diet. This is because it eases the flow of materials through the gut by giving gut muscles a bulk to push against.

Fibre also helps to remove some waste products such as excess bile salts, which can be toxic. A diet low in fibre is a **risk factor** for a number of diseases including intestinal cancer and some other cancers.

Key definition

A **risk factor** is a factor that increases your chance of developing a particular disease.

The nutrients required

All organisms are made up of many thousands of different molecules. In order to survive, an organism must be able to either make or take in all the molecules required. These molecules fall into a small number of chemical groups, with differing roles within the organism (see Table 1).

Chemical group	Role
Carbohydrates	Energy storage and supply, structure (in some organisms)
Proteins	Structure, transport, **enzymes**, **antibodies**, most hormones
Lipids	Membranes, energy supply, thermal insulation, protective layers/padding, electrical insulation in neurones, some hormones
Vitamins and minerals	Form parts of some larger molecules and take part in some metabolic reactions, some act as **coenzymes** or enzyme activators
Nucleic acids	Information molecules, carry instructions for life
Water	Takes part in many reactions, support in plants, solvent/medium for most metabolic reactions, transport

Table 1 Chemical groups and their roles within organisms

Note that fibre is usually included in 'nutrients required' lists for humans – fibre is a type of carbohydrate.

Biological molecules

The key biological molecules are carbohydrates, lipids, proteins and nucleic acids. The structure of these molecules is closely related to their functions within living organisms.

The chemical elements found in the biological molecules are carbon, hydrogen, oxygen and nitrogen. These elements make up over 99% of all organisms. A few biological molecules also contain some phosphorus or sulfur. We will consider the minerals, including magnesium, iron and iodine, separately.

Water is sometimes described as a biological molecule because of its importance to life.

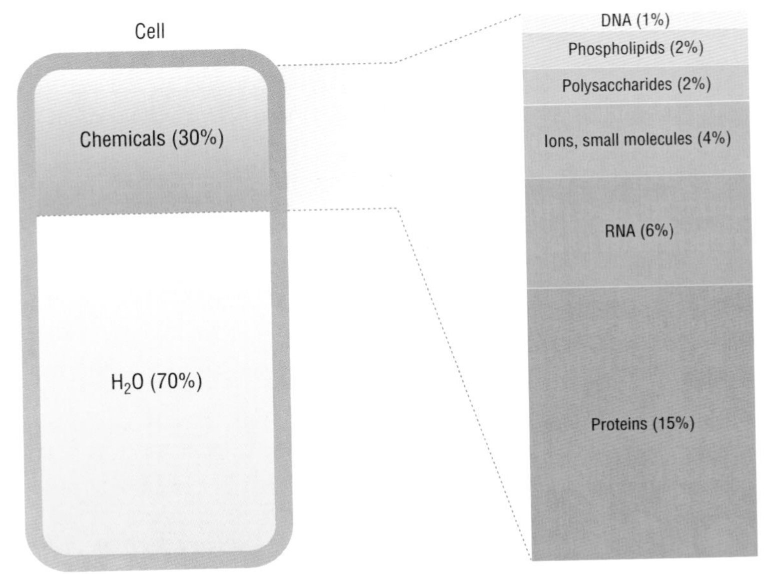

Figure 1 Biological elements and compounds in the human body

Key definition

Metabolism is the sum total of all the biochemical reactions taking place in the cells of an organism.

Examiner tip

Remember that all organisms have a ***metabolism*** and require ***nutrients*** in order to survive. Make sure you consider the metabolic needs of organisms other than animals in your revision.

Biochemistry and metabolism

Biochemistry refers to the chemical reactions involving biological molecules. In a hospital pathology laboratory, for instance, the biochemistry technicians are responsible for measuring the level of enzymes in the blood of a patient who has suffered a heart attack (myocardial infarction).

Metabolism refers to the sum total of all the chemical reactions that take place in an organism. These reactions may be involved with breaking larger molecules into smaller ones. Such metabolic reactions are called catabolic reactions – one example is digestion. Reactions that involve building smaller molecules into larger ones are called anabolic reactions – an example is muscle growth.

Carbon is very special

Organic chemistry is the study of chemical reactions that involve carbon. All the biological molecules we shall consider (with the exception of water) are carbon-based, so it is important to understand how carbon can form such a wide variety of molecules.

Carbon atoms can bond together to form long chains and rings. It is also possible to bond other atoms to the chains and rings to form many different molecules with different structures and properties. Because of this multiple-bonding feature, carbon is a kind of framework atom that can form the basis of all the biological molecules necessary for life.

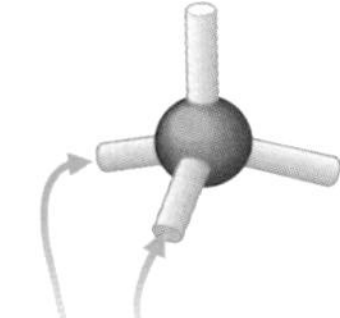

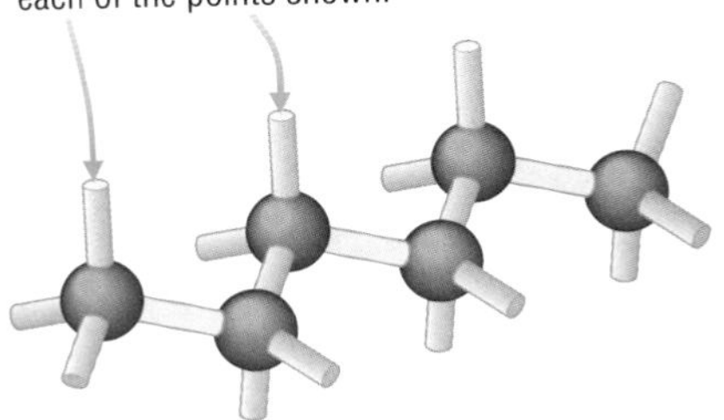

Figure 2 Carbon can make up to four possible bonds with other atoms – it has a valency of four

Questions

1 Why do some nutrients have to be broken down before they can be made into part of an organism?
2 Suggest a definition for 'metabolic rate' and describe how it might be measured.
3 What does the term 'balanced diet' mean?

2.1 (2) Biochemicals and bonds

By the end of this spread, you should be able to . . .

* **Name the monomers and polymers of carbohydrates, proteins and nucleic acids.**
* **Describe the general features of condensation and hydrolysis reactions.**

Key definition

Covalent bonds are formed when electrons are shared between atoms. These bonds are very strong. Covalently bonded atoms form new molecules.

Carbon chains and rings

Atoms are at their most stable when their outer energy levels contain a specific number of electrons. This number is usually eight. Because carbon atoms have four electrons in their outer orbitals, they can gain stability by sharing four electrons with other atoms. The sharing of electrons forms a strong bond between the atoms, known as a **covalent bond**. The bonded atoms form a molecule.

A molecule is very stable because each atom in the molecule is sharing in the electrons to give a stable, full outer energy level. So carbon is able to make four covalent bonds. The bonds made can be between atoms of the same type (other carbon atoms), or with atoms of other types.

Carbon can form a vast variety of molecules by bonding with other atoms and by forming chains or rings of carbon atoms with other atoms bonded to the chain.

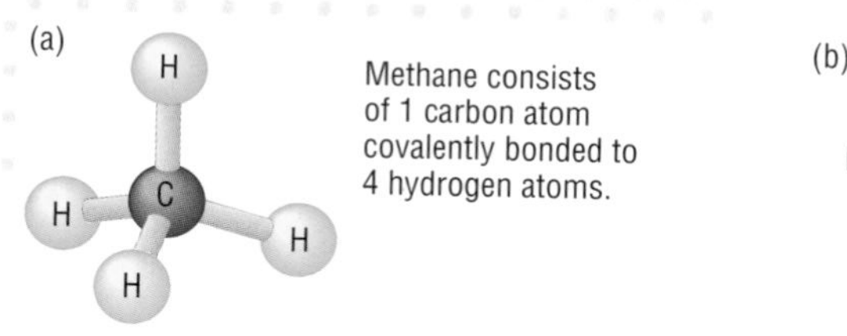

(b)

H H H H
H—C—C—C—C—H
H H H H

Several carbons can form a chain

Figure 1 Methane showing covalent bonding structure; carbon chain of a hydrocarbon

Double bonds also occur

In some cases, carbon forms two bonds with another atom. Key examples in biology include C=C double bonds in **hydrocarbon chains**, and C=O double bonds found in many molecules, including organic acids.

(a) Carbon–carbon double bond C_2H_4

H H
C=C
H H

(b) Carbon–oxygen double bond found in organic acids

O
R—C
O—H

Figure 2 Double bonds in ethene and carboxylic acid (R means a hydrocarbon chain)

Monomers and polymers

Biological molecules are grouped according to their chemical properties. The most important groups of biological molecules are the carbohydrates, proteins, nucleic acids and lipids. In the first three types, large molecules are made by bonding together similar, smaller molecules.

The term **monomer** refers to a single, small molecule, many of which can be joined together to form a **polymer**. Although they are made of smaller molecules bonded together, **lipids** are not polymers because the smaller molecules are very different from each other.

	Monomer	Polymer
Carbohydrates	Monosaccharides (simple sugars)	Polysaccharides
Proteins	Amino acids	Polypeptides and proteins
Nucleic acids	Nucleotides	DNA and RNA

Table 1 Biological polymers

Condensation and hydrolysis

The chemical reaction that links biological monomers together is called a **condensation** reaction. In making a polymer, the same reaction is repeated many times in order to link many monomers together to form a polymer. Condensation reactions also link the different subunits together in lipid molecules.

In condensation reactions:

- a water molecule is released
- a new covalent bond is formed
- a larger molecule is formed by the bonding together of smaller molecules.

The chemical reaction that splits larger molecules to monomers is called a **hydrolysis** reaction. Hydrolysis is the reverse of condensation.

In all hydrolysis reactions:

- a water molecule is used
- a covalent bond is broken
- smaller molecules are formed by the splitting of a larger molecule.

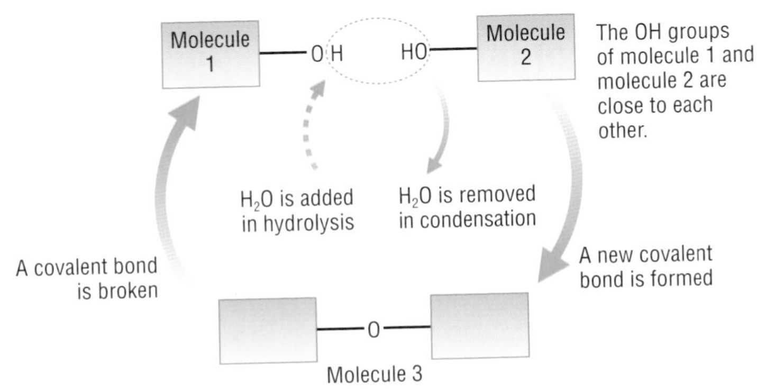

Figure 3 Hydrolysis and condensation

Hydrogen bonds

Polymers can be very large molecules. They often have specific functions that rely on their shape. Hydrogen bonds form when a slightly negatively charged part of a molecule comes close to a slightly positively charged hydrogen atom in the same (or another) molecule. This is most easily seen in water.

Hydrogen bonding makes water a very special substance. You will find out why in spread 2.1.12.

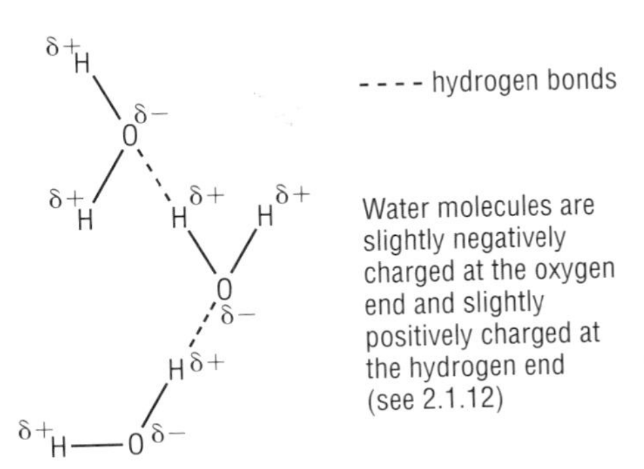

Figure 4 Hydrogen bonds in water

Hydrogen bonds are not strong bonds. They are often described as 'interactions'. However, in some polymers many thousands of hydrogen bonds can form and this helps to stabilise the structure of the molecule.

Examiner tip

Remember that hydrogen bonds are relatively weak so they are easily broken. For example, heating breaks them. Covalent bonds are very difficult to break. Breaking covalent bonds produces different molecules or atoms.

Questions

1. The formula that represents one molecule of carbon dioxide is CO_2. Draw a diagram to show how the atoms in this molecule are bonded together.
2. You can hydrolyse molecules in a laboratory by boiling them with acid. This is not possible inside organisms. What do organisms contain to catalyse hydrolysis reactions in cells?
3. What type of energy can break hydrogen bonds easily?

2.1 (3) Carbohydrates 1: simple sugars

By the end of this spread, you should be able to . . .

* **State the structural difference between α (alpha) and β (beta) glucose.**
* **Describe the formation and breakage of glycosidic bonds in the synthesis and hydrolysis of a disaccharide.**
* **Describe the molecular structure of alph-glucose as an example of a monosaccharide carbohydrate.**
* **Explain how the structure of glucose relates to its functions in living organisms.**

Carbohydrates in living organisms

Carbohydrates make up about 10% of the organic matter of a cell. The functions of carbohydrates in organisms include:

- energy source – released from glucose during **respiration**
- energy store – e.g. starch
- structure – e.g. cellulose.

Some carbohydrates also form part of larger molecules (e.g. **nucleic acids**, **glycolipids**).

Carbohydrates contain the elements carbon, hydrogen and oxygen. The term carbohydrate essentially means 'hydrated carbon', because the elements are found in the proportions $C_n(H_2O)_n$. This means that for every carbon present in a carbohydrate, the equivalent of a water molecule is also present.

Key definition

Carbohydrates make up a group of molecules containing carbon, hydrogen and oxygen in the ratio $C_n(H_2O)_n$.

Simple sugars

The simplest **carbohydrates** are called the **monosaccharides**. These are the **monomers** (basic units) of carbohydrates. All larger carbohydrates are made by joining monosaccharides together.

There are a number of different monosaccharides, containing between three and six carbon atoms. All have very similar properties – they:

- are soluble in water
- are sweet tasting
- form crystals.

The monosaccharides are grouped according to the number of carbon atoms in the molecule.

- 3-carbon monosaccharides (each molecule has three carbon atoms) are known as *triose* sugars.
- 5-carbon monosaccharides are known as *pentose* sugars.
- 6-carbon monosaccharides are known as *hexose* sugars.

The most common monosaccharides are hexoses. These include glucose and fructose. They contain one water molecule for every carbon atom present, so both glucose and fructose contain six carbons, 12 hydrogens and six oxygens. This is written as $C_6H_{12}O_6$.

Pentoses and hexoses tend to occur in nature as ring structures.

Two forms of glucose

Glucose can be drawn as a chain or a ring structure (Figure 1). When the ring structure forms, it can do so in two slightly different ways – the H and OH at carbon 1 may be above or below the plane of the ring (Figure 2). In α-glucose, the OH at C_1 is below the plane of the ring; in β-glucose, the OH at C_1 is above the plane of the ring. Although both are glucose molecules, this difference in structure leads to some very different properties.

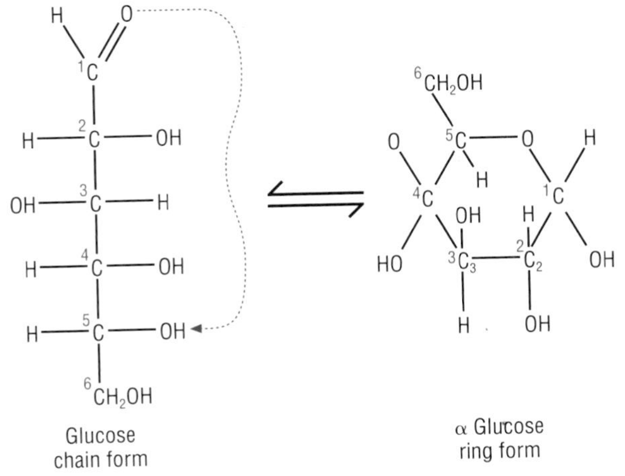

Glucose chain form

α Glucose ring form

Carbons are numbered 1–6 to show how the ring structure is made. The numbering also allows us to note where further bonds are formed

Figure 1 Chain and ring forms of glucose

Ring structure diagrams are simplified to show only the groups considered. The two molecules shown are both glucose. Their constituent atoms are identical.

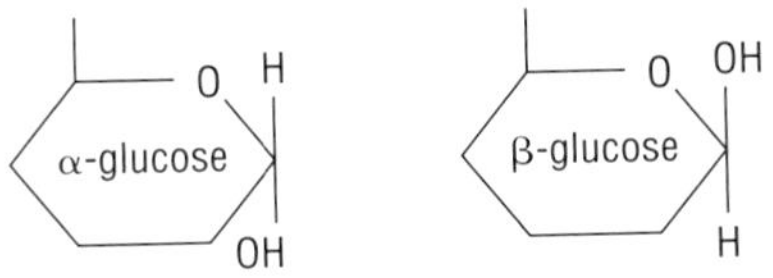

α-glucose

β-glucose

The structural (or shape) difference between them can be seen in a 3-dimensional view.

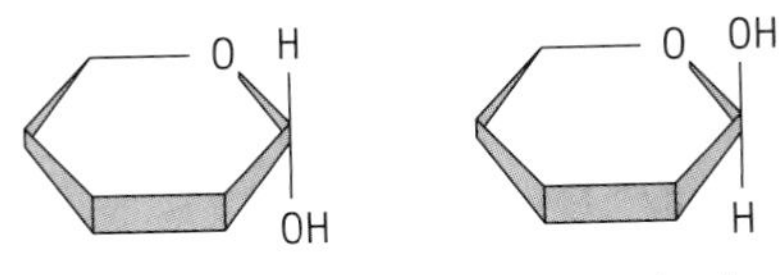

Different shaped forms of the same molecule are called **isomers**.

Figure 2 Alpha and beta glucose; note the different positions of the OH groups on C_1

Joining monosaccharides and splitting disaccharides

Two monosaccharide molecules can be joined together in a **condensation** reaction, forming a **disaccharide** molecule. A new covalent bond called a **glycosidic bond** forms, and water is eliminated. The reverse hydrolysis reaction uses a water molecule to break the glycosidic bond.

Building the **polysaccharides** starch, glycogen and cellulose, and breaking down larger molecules, for example during digestion, involves the making and breaking of glycosidic bonds. Note that disaccharides are still sugars.

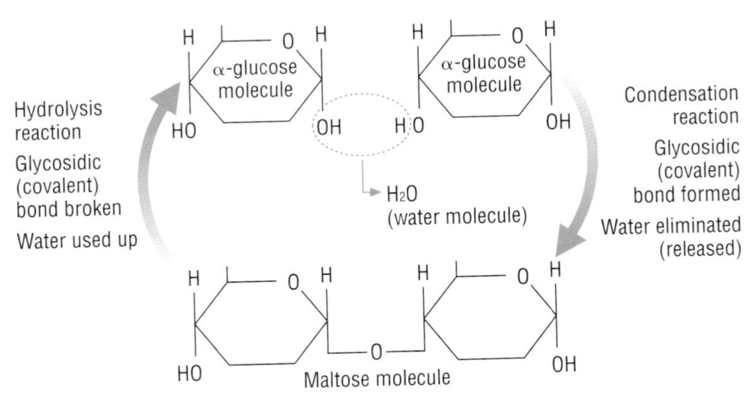

Maltose is a disaccharide sugar. It is sweet and soluble.

Figure 3 Making and breaking bonds

Examiner tip

Remember that there are two products of a condensation reaction – the larger molecule produced as a result of bonding two smaller molecules together, ***and*** a water molecule. Do not forget to write down the water molecule produced.

Questions

1. What is the molecular formula for a triose sugar?
2. List the properties common to all monosaccharides and disaccharides.
3. Write out an equation showing the condensation reaction to produce maltose.
4. Explain why molecules of α-glucose and β-glucose have different shapes.

2.1  Carbohydrates 2: energy storage

By the end of this spread, you should be able to . . .

* **Describe the formation and breakage of glycosidic bonds in the synthesis and hydrolysis of a polysaccharide.**
* **Compare and contrast the structure and functions of starch (amylose) and cellulose.**
* **Describe the structure of glycogen.**
* **Explain how the structures of starch and glycogen relate to their functions in living organisms.**

Energy and structure

Glucose molecules contain a large number of bonds that can be broken to form simpler molecules. The breaking of glucose into simpler molecules of water and carbon dioxide in **respiration** releases energy. This energy can be used to make **ATP** (the molecule that holds small 'packets' of energy for use in cell processes).

Respiration is often written in a word equation as:

glucose + oxygen → carbon dioxide + water + energy that is used to form ATP

The breaking down of glucose in living organisms takes place in a series of many steps. Each step is driven by a specific **enzyme**. In order to be able to use glucose in respiration, an organism must have enzymes that can specifically break the glucose molecule.

Animals and plants have enzymes that break α-glucose only. Plant and animal enzymes cannot break down β-glucose because of its different arrangement of the H and OH at C_1 (carbon 1). This is because enzyme function is based on shape. As shown in Figure 2 on spread 2.1.3, the overall shape of α-glucose is different from that of β-glucose. This means that while α-glucose can be respired, β-glucose cannot.

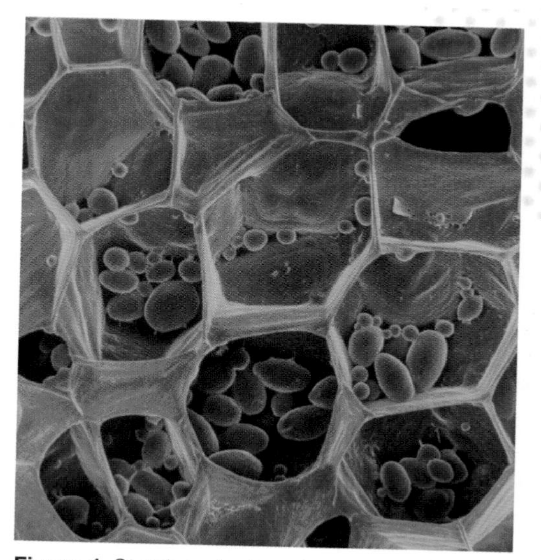

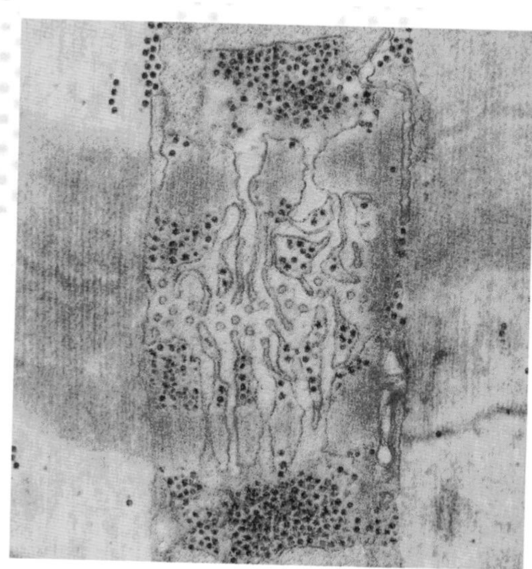

Figure 1 Starch grains in potato (left, ×150) and glycogen granules in muscle cells (right)

Carbohydrate polymers – stores of potential energy

Two α-glucose molecules bonded together form a **disaccharide** called **maltose**. The same condensation reaction can be carried out over and over again to join glucose molecules together, so forming a molecule called **amylose**. Amylose can consist of many thousands of glucose molecules bonded together. The **glycosidic bond** between all the glucose subunits occurs between carbon number 1 of one molecule and carbon number 4 of the next, so it is often called a 1,4–glycosidic bond.

The long chains of amylose coil into a spring because of the shape of the glucose molecules and the formation of the glycosidic bonds. This makes amylose quite compact. Iodine molecules can become trapped in the 'coils' of the spring. This causes iodine (in potassium iodide solution) to change colour from yellow/brown to blue/black. This is the basis of the starch test (spread 2.1.13).

An important feature of these large molecules is that, unlike the glucose molecules from which they are formed, they are not water-soluble.

Starch – the energy-storage polysaccharide in plants

Starch consists of a mixture of long, straight-chain amylose molecules and branched amylopectin. Starch in plants is a mixture of amylose and amylopectin. It is stored in chloroplasts and elsewhere in the plant cell in membrane-bound starch grains. The cells of plant storage organs (e.g. potato tubers) contain a lot of starch grains. Starch can be broken down to glucose molecules, which may then be respired to release energy.

Glycogen – the energy-storage polysaccharide in animals

Glycogen is sometimes referred to as animal starch. It is identical to starch in that it is made up of α-glucose subunits. It is also a large molecule that can be broken down to release the glucose to be respired.

Glycogen differs from starch because the 1–4 linked glucose chains in glycogen tend to be shorter and have many more branches extending from the chain. This means that glycogen is more compact than starch, and forms glycogen granules in animal cells – especially liver and muscle cells.

Features of the energy-storage molecules starch and glycogen

Because both starch and glycogen are made by bonding many thousands of α-glucose molecules together, they are described as energy-storage molecules.

- They do not dissolve, so the stored glucose does not affect the water potential of the cell. This feature is vital in both plants and animals, as glucose stored in a cell as free molecules would dissolve and dramatically reduce the water potential.
- They hold glucose molecules in chains so that they can easily be 'broken off' from the ends to provide glucose for respiration when required.

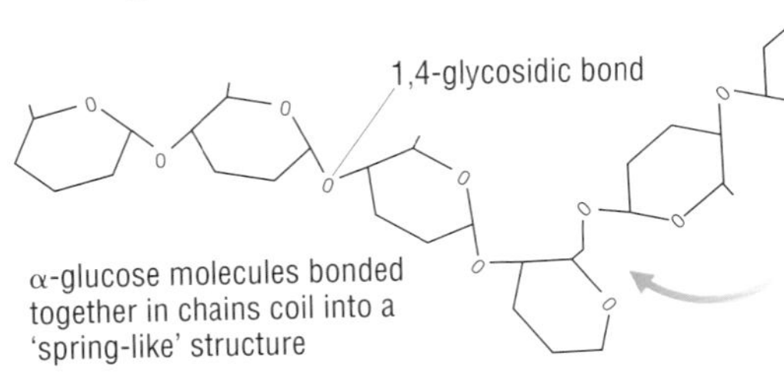

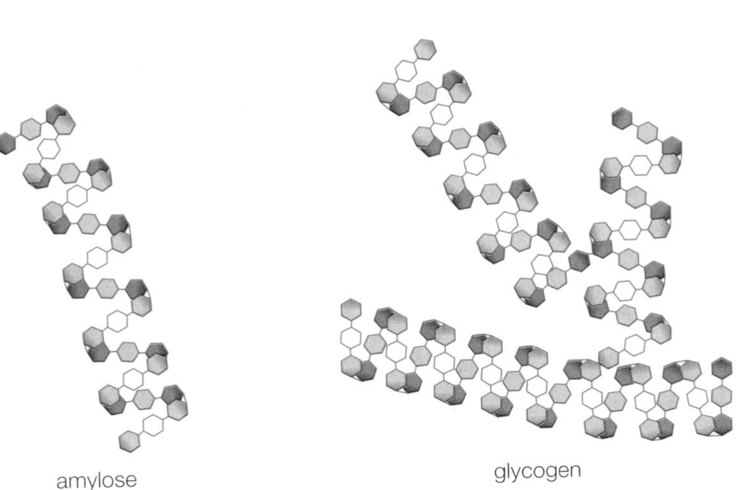

Figure 2 Subunit structure of amylose and glycogen

Key definition

Polysaccharides are **polymers** of **monosaccharides**. They consist of hundreds to thousands of monosaccharide **monomers** bonded together to form a single large molecule.

Examiner tip

Always state that the condensation or hydrolysis reactions in living organisms are catalysed by enzymes. Polysaccharides are stable molecules – they would not simply fall apart in the conditions found within organisms.

Questions

1. Glycogen is significantly more branched than amylopectin. Explain how this difference is important to animals.
2. Why are glycogen and starch called 'storage' molecules, whereas α-glucose is termed an 'energy source'?

Carbohydrates 3: structural units

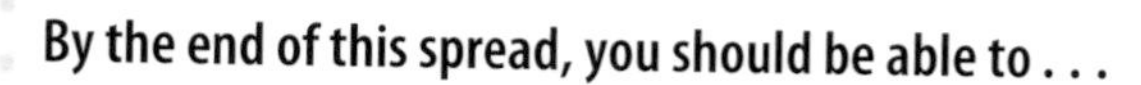

By the end of this spread, you should be able to . . .

* **Compare and contrast the structure and functions of amylose and cellulose.**
* **Explain how the structure of cellulose relates to its function in living organisms.**

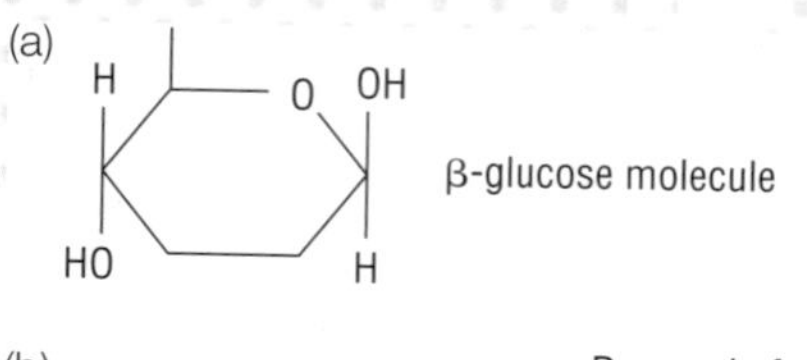

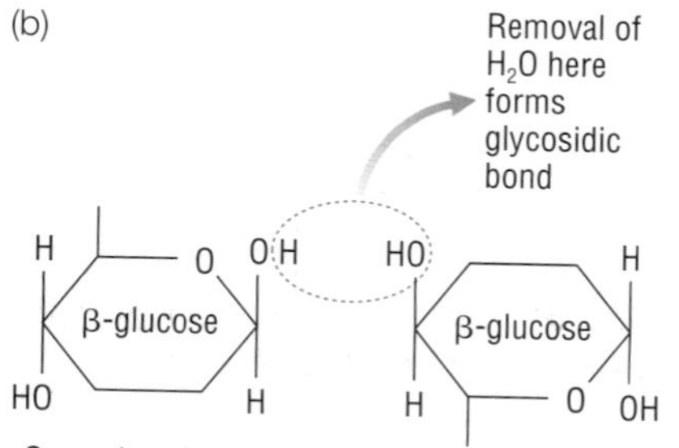

Carbohydrate polymers – structural units

β-glucose molecules can be bonded together in a long chain (**polymer**) through numerous **condensation** reactions. α-glucose molecules can also be condensed together to form coiled, spring-like chains. β-glucose molecules have a slightly different shape from α-glucose molecules – when β-glucose condenses, the resulting chains are long and straight.

(c)

Chains of β-glucose joined by condensation reactions are straight.

Figure 1 Orientation of β-glucose and chain formation

These straight-chain molecules can contain 10 000 β-glucose molecules. They are stronger than the chains found in amylose. These β-glucose polymer chains are called cellulose chains.

Cellulose is found only in plants. It is the most abundant structural **polysaccharide** in nature.

Cellulose

Cellulose fibres are arranged in a very specific way to form plant cell walls. Because the glucose **monomers** contain so many OH groups, many hydrogen bonds can form between them. About 60–70 cellulose molecules become cross-linked by hydrogen bonds to form bundles called microfibrils. These, in turn, are held together by more hydrogen bonds to form larger bundles called macrofibrils.

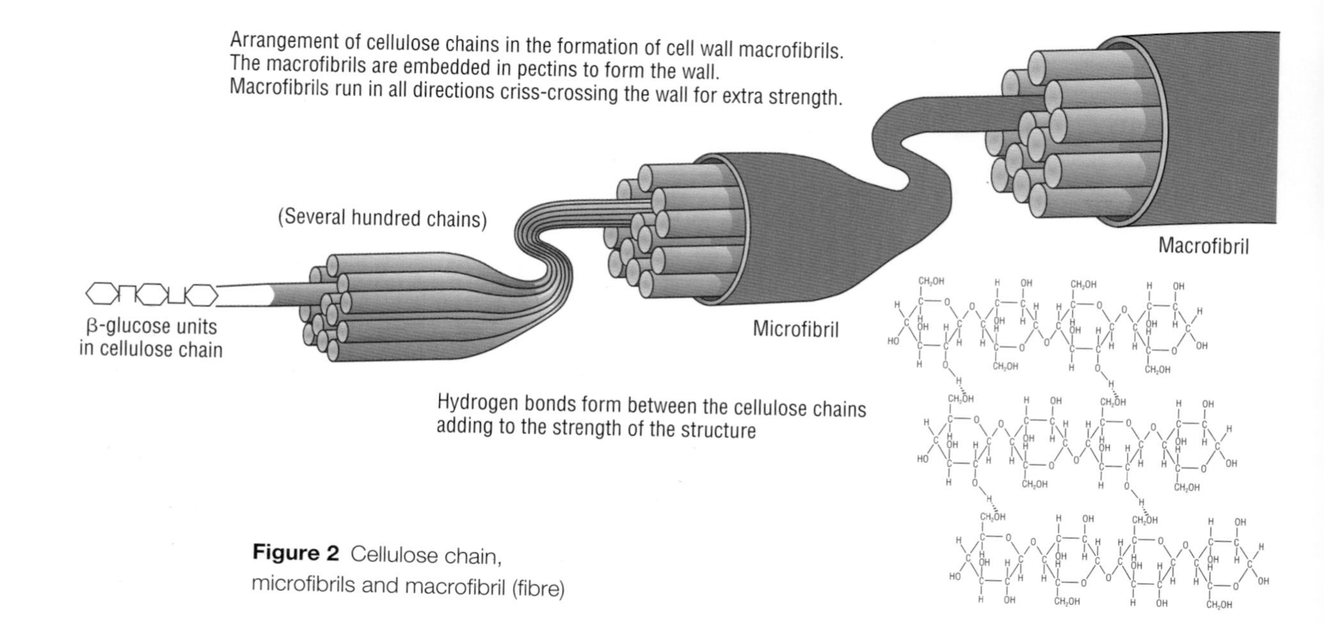

Figure 2 Cellulose chain, microfibrils and macrofibril (fibre)

The macrofibrils have great mechanical strength – close to that of steel. They are embedded in a polysaccharide glue of substances called pectins, to form cell walls.

Structure and function of plant cell wall

- The cell walls around plant cells give great strength to each cell, supporting the whole plant.
- The arrangement of macrofibrils allows water to move through and along cell walls, and water can pass in and out of the cell easily.
- Water moving into plant cells does not cause the cells to burst, as it does in animal cells – the wall prevents bursting, and in **turgid** cells it helps to support the whole plant.
- The arrangement of macrofibrils in cell walls determines how cells can grow or change shape. For example, guard cell walls have arrangements of macrofibrils that result in the opening and closing of stomata (singular: **stoma**) as water moves in or out of the cell (see spread 1.2.13).
- Cell walls can be reinforced with other substances to provide extra support, or to make the walls waterproof.

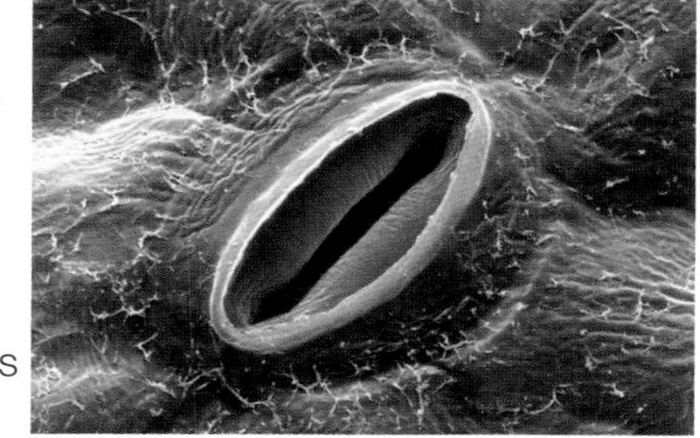

Figure 3 Stomatal pore and guard cells (×650)

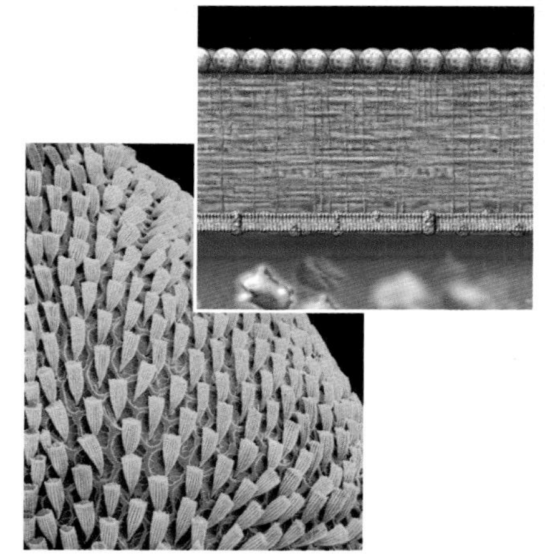

Figure 4 Bacterial cell wall (above) and beetle exoskeleton (below, ×120)

Other structural carbohydrates

A number of organisms use structural **carbohydrate** polymers. The polysaccharide chitin forms the exoskeleton of insects (Figure 4, left). The polysaccharide peptidoglycan is the basis of the cell walls found around most bacterial cells (Figure 4, right).

Key definition

Cellulose is a **carbohydrate polymer** made by bonding many β-glucose molecules together in long chains.

Overview of carbohydrates

Carbohydrate	Examples	Characteristics	Role in organisms
Monosaccharides (monomers)	Glucose (6 carbon)	Small, soluble, sweet and crystalline	Provides energy via respiration
	Deoxyribose (5 carbon)		Part of DNA – information molecule
Disaccharides (dimers)	Maltose (glucose + glucose)	Small, soluble, sweet and crystalline	A sugar obtained when starch is broken down in hydrolysis reactions. It can be split further to glucose for respiration
Polysaccharides (polymers)	Starch and glycogen	Large molecules of many α-glucose molecules joined by condensation reactions. Insoluble in water. Forms grains/ granules.	Energy-storage carbohydrates – starch in plants; glycogen in animals and fungi
	Cellulose	Large molecules of many β-glucose molecules joined by condensation reactions. Insoluble in water. Very strong.	Structural. Found only in plants, where it forms cell walls.

Table 1 Overview of carbohydrates

Examiner tip

The differences between starch and cellulose occur because of the differences between α and β glucose. In alpha glucose, the OH at C_1 is below; in beta glucose, the OH at C_1 is above the plane of the ring.

Questions

1. List the features of cellulose that make it a good structural molecule for plant cell walls.
2. Explain why cellulose is very difficult to break down naturally, and so forms the fibre component of human diets.
3. Explain how turgid cells help to keep green-stemmed plants upright.

2.1  Amino acids – the monomers of proteins

By the end of this spread, you should be able to . . .

* **Describe the structure of an amino acid.**
* **Describe the formation and breakage of peptide bonds in the synthesis and hydrolysis of dipeptides and polypeptides.**

Proteins in living organisms

Proteins make up about 50% of the organic matter of a cell. They are large molecules, made up of the elements carbon, hydrogen, oxygen and nitrogen. Some proteins also contain sulfur.

Proteins have many functions:

- they are structural components, e.g. of muscle and bone
- they are membrane carriers and pores, e.g. for **active transport** and **facilitated diffusion**
- all **enzymes** are proteins
- many **hormones** are proteins
- **antibodies** are proteins.

Overall, proteins provide building materials important for growth and repair in all organisms. They are also crucial to most metabolic activity within all organisms.

All proteins are made from amino acids

Proteins are large molecules because they are **polymers**. They are made by joining together a large number of similar, smaller subunits (**monomers**). The monomers that are joined together to make proteins are called **amino acids**. A protein consists of a long chain of amino acids joined end-to-end.

Amino group
Acid group
H H O
N C C
H H OH
Glycine is the simplest amino acid

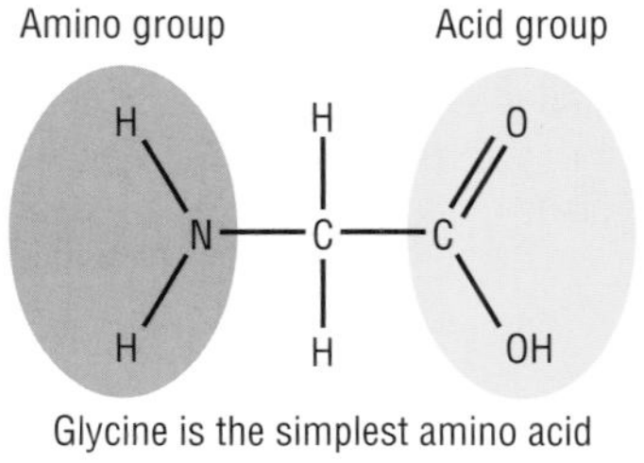

Figure 1 Basic amino acid structure – glycine

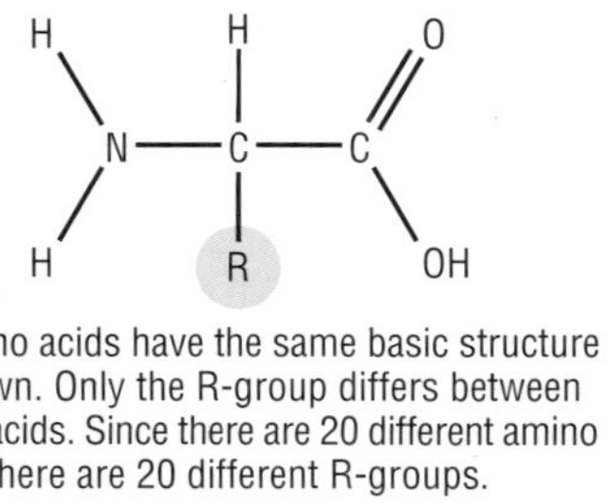

H H O
N C C
H R OH
All amino acids have the same basic structure as shown. Only the R-group differs between amino acids. Since there are 20 different amino acids, there are 20 different R-groups.

Figure 2 Amino acid structure showing R-groups (side chains)

Amino acids – similar but very different

All amino acids have the same basic structure. They all have an amino group at one end of the molecule, an acid group at the other end of the molecule, and a carbon in between.

Amino acids joined end-to-end give a repeating 'backbone'. For example, joining four amino acids together would give a backbone N–C–C–N–C–C–N–C–C–N–C–C.

Not all amino acids are identical. There are 20 types of naturally occurring amino acid. Each one is based on the same structure as shown above, but there are differences between amino acids because they have different R-groups (Figure 2).

The R-group in glycine is a hydrogen atom bonded to the second carbon. Some R-groups are large – larger than the C–C–N part of the molecule. Some are positively charged, some are negatively charged. Some are **hydrophobic**, others are **hydrophilic**.

Amino acids in plants and animals

Plants are able to manufacture the amino acids they need, provided they can obtain nitrate from the soil. This nitrate is converted to amino groups and bonded to organic groups made from the products of **photosynthesis**.

Animals must take in proteins as part of their diet. These proteins are digested to amino acids. Proteins can be built from these amino acids.

There are some types of amino acid (8–10 of the 20 types) that animals cannot build from materials they take into their bodies. These are called 'essential amino acids' – they are an essential part of the diet. Most essential amino acids are found in meat. Plants contain fewer of the essential amino acids. Vegetarians need to be careful to balance their diet to get a supply of all of the essential amino acids. Soya contains all of them.

Animals cannot store excess amino acids

In animals, amino acids that are surplus to the body's requirements cannot be stored. The amino group makes them toxic if too much is present. The amino group is removed in a process called **deamination**. In mammals, this process takes place in the liver. The amino groups removed are converted to a substance called **urea** and removed in the urine.

Joining amino acids together

All amino acids are joined in exactly the same way, no matter which R-group they contain. A **condensation** reaction between the acid group of one amino acid and the amino group of another forms a **covalent bond** between the two amino acids. A water molecule is also produced.

The new bond formed is called a **peptide bond**. The new molecule produced is called a **dipeptide**. The peptide bond can be broken by a **hydrolysis** reaction, which uses a water molecule in order to break the bond.

The making and breaking of peptide bonds is required in the building and rebuilding of all protein molecules in organisms, and in breaking down proteins to amino acids, for example in digestion.

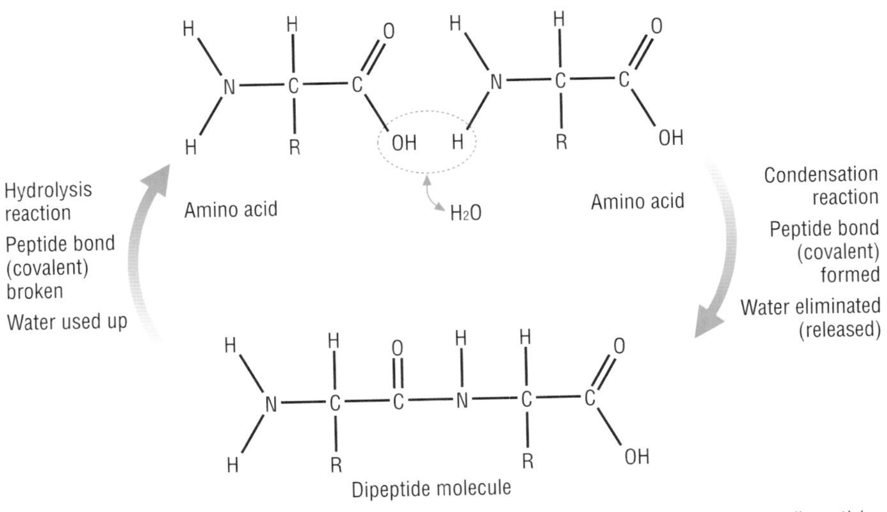

Figure 3 Condensation and hydrolysis. 2 amino acids can be joined to make a dipeptide. The dipeptide can be split back to 2 amino acids

Key definition

Amino acids are the monomers of all proteins. All amino acids have the same basic structure. The 20 different amino acids involved in protein synthesis differ only in the R-group bonded to the central carbon.

Examiner tip

You do not need to learn the structure of different amino acid R-groups. You do need to learn the basic structure of an amino acid, and you must be able to show how a dipeptide is formed in a condensation reaction.

Questions

1 Why are some amino acids described as 'non-essential amino acids'?

2 The backbone for a molecule with four amino acids is N–C–C–N–C–C–N–C–C–N–C–C.
 (a) Draw this out in full, showing the amino acids joined and peptide bonds formed.
 (b) How many molecules of water would be produced in forming this amino acid chain?
 (c) Explain why we might call this chain a 'polypeptide'.

2.1 (7) Proteins from amino acids

By the end of this spread, you should be able to . . .

* **Explain the term primary structure.**

Dipeptides, polypeptides and proteins

Two **amino acids** joined together form a **dipeptide**. As more and more amino acids are joined together by peptide bonds, a **polypeptide** is formed.

A protein may consist of a single polypeptide chain of hundreds of amino acids. Some proteins consist of more than one polypeptide chain bonded to form an even larger molecule.

The amino acids in a polypeptide chain are sometimes referred to as amino acid residues, because part of the molecule is lost in the **condensation** reaction that produces the **peptide bond**.

Making polypeptides and proteins

Polypeptides and proteins are made (synthesised) in cells on **ribosomes**. This process is known as protein synthesis. It uses information in the form of a molecule called **messenger RNA** (**mRNA**; see spreads 2.1.16–17) to put amino acids in the right order to make a specific polypeptide chain.

As the mRNA passes through the ribosome, amino acids are joined together one at a time. As each new amino acid joins, a condensation reaction occurs, forming a new peptide bond. This produces a longer and longer chain of amino acids. The sequence of amino acids produced is determined by the mRNA. To make different proteins, different mRNA molecules must pass through ribosomes.

Primary structure

An organism may contain 10 000 or so different proteins. Each protein has its own function within the organism. Each one is formed from amino acids joined by peptide bonds in a chain. A protein may be hundreds of amino acids long, but all proteins will have an amino group at one end and an acid group at the other.

Key definition

The **primary structure** of a protein is given by the specific sequence of amino acids that make up the protein.

The function of each protein is determined by its structure. The structure of each protein is determined firstly by its amino acid sequence. The unique amino acid sequence of a polypeptide or protein is called its **primary structure**.

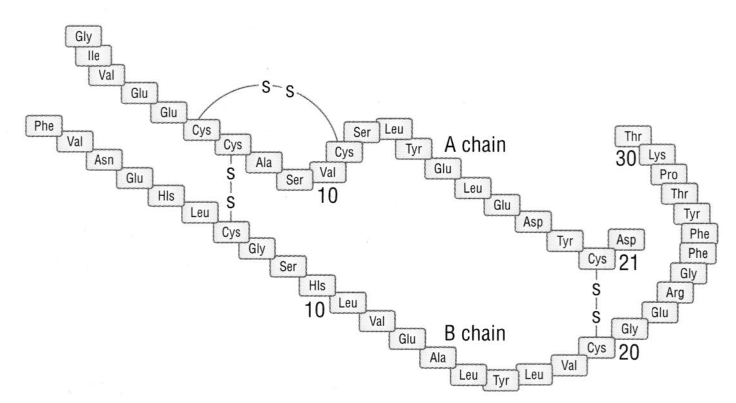

Figure 1 Primary structure of insulin

Different amino acids have different properties. This means that the sequence of amino acids found in a protein will have an effect on its properties. For example, if the protein contains a number of amino acids with **hydrophobic** R-groups, then the final protein will be a particular shape and may be found embedded in a membrane.

Forming different proteins

Consider a peptide chain that is four amino acids long.

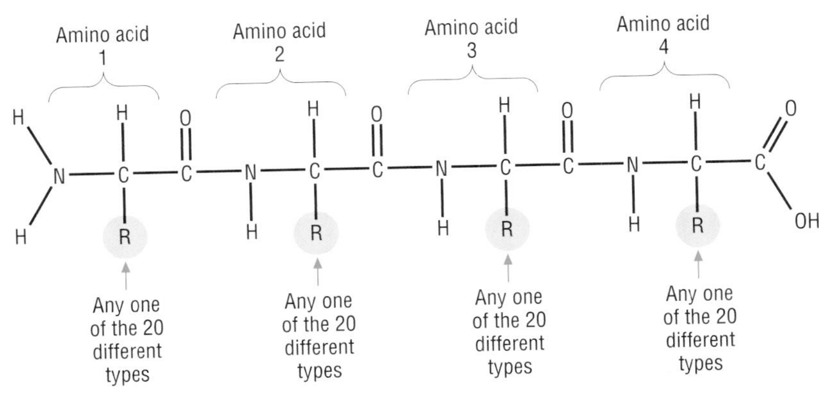

Figure 2 Four-amino-acid structure

At position 1, the R-group of the amino acid may be any one of the 20 different ones available. At position 2, again the R-group may be any one of the 20 different ones available – and so on.

To calculate the total number of different possibilities (rather than trying to write them all down), we need to multiply the number of possibilities at each point. In this case $20 \times 20 \times 20 \times 20 = 160000$, which means that 160 000 different sequences of four amino acids are possible.

Given that even a small protein may be 100 amino acids long, this means that, in theory, the number of different possible proteins is extremely large.

Figure 3 Differences between old and young skin due to loss of collagen

Breaking down proteins and polypeptides

It is important to remember that the formation and breakage of peptide bonds (and other types of covalent bond) in organisms is catalysed by **enzymes**. Covalent bonds are very strong and do not simply 'appear' or 'fall apart' in the conditions found within cells.

Enzymes that catalyse the breaking of peptide bonds are known as **protease** enzymes. These are not only found in the parts of an organism where food is digested – organisms continually break down and rebuild proteins. Two good examples are as follows.

- Hormone regulation – it is vital that hormones are broken down so that their effects are not permanent and can be controlled. Any cell that is targeted by a hormone contains enzymes that can break down that hormone.
- Ageing – one of the features of ageing is that the skin loses elasticity and becomes wrinkled. This occurs because older skin is less able to rebuild the protein collagen and other proteins that give young skin its smooth and elastic properties.

Examiner tip

Always state that condensation or hydrolysis reactions are *enzyme-catalysed* reactions if you are referring to the process within organisms.

Questions

1. Why do all proteins have an amino group at one end and an acid group at the other?
2. If an amino acid residue is found in the middle of a polypeptide chain, how much of the original amino acid molecule has been 'lost'?
3. Suggest where the enzymes responsible for breaking down a polypeptide hormone are located in the cell.

Levels of protein structure

By the end of this spread, you should be able to . . .

* **Explain the term secondary structure with reference to hydrogen bonding.**
* **Explain the term tertiary structure with reference to hydrophobic and hydrophilic interactions, disulfide bonds and ionic interactions.**

Key definition

Secondary structure refers to coiling and pleating of parts of the polypeptide molecule; **tertiary structure** refers to the overall 3D structure of the final polypeptide or protein molecule.

Examiner tip

Never refer to proteins as being 'killed' by heat. State that heating *denatures* proteins because their structure is changed by the breaking of hydrogen bonds and other weak interactions. This means the protein can no longer function, and its function cannot be restored by cooling.

Levels of protein structure

To make a specific protein, **amino acids** must be bonded together in a specific sequence. This sequence is determined by the **DNA**. As amino acids are bonded, the amino acid chain becomes longer. To avoid tangling and breaking, parts of the chain are stabilised by being coiled up or pleated as they are made. The coils and pleats are held in place by numerous **hydrogen bonds**. The amount of coiling or pleating depends on the types of amino acids being added to the chain, and so depends on the **primary structure**.

Once the whole chain is complete and a **polypeptide** is formed, the coils and pleats come together in a specific way to form a specific overall three-dimensional shape. Again, the overall shape is determined by the original sequence of amino acids (R-groups in the primary structure), and is held in place by a number of different bonds. The levels of protein structure are as follows.

Primary structure

This is the sequence of amino acids that forms the protein.

Secondary structure

The protein's secondary structure is formed when the chain of amino acid coils or folds to form an **alpha helix** or a beta pleated sheet. Hydrogen bonds hold the coils in place. Although hydrogen bonds are quite weak, many are formed, so overall they give great stability to parts of the protein molecule.

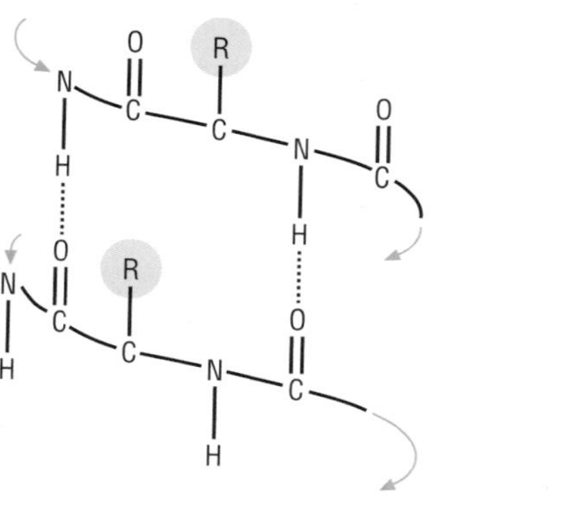

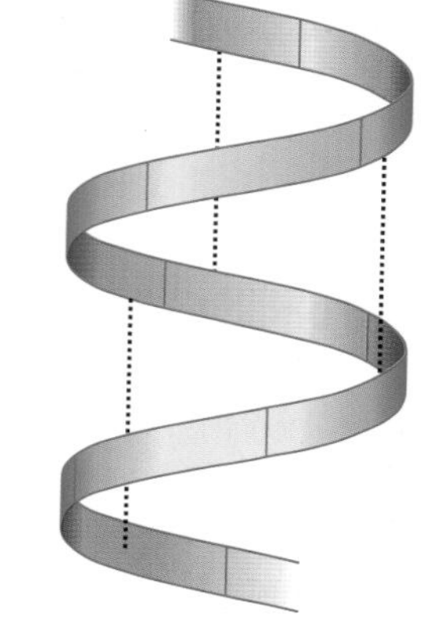

Figure 1 Alpha helix and polypeptide chain

An α-helix has 36 amino acids per 10 turns of the coil. H-bonds form between one amino acid and the one 'four places' along the chain.

Tertiary structure

The final three-dimensional shape of a protein is formed when these coils and pleats themselves coil or fold, often with straight runs of amino acids in between. This three-dimensional shape is held in place by a number of different types of bonds and interactions. A protein's tertiary structure is vital to its function. For example, a **hormone** must be a specific shape in order to fit into the hormone **receptor** of a target cell. A structural protein, such as collagen, is shaped to be strong, with protein chains wound around each other in a specific way. An enzyme must have an **active site**, the shape of which is **complementary** to that of its **substrate**.

Tertiary structure in proteins is stabilised by a number of bonds.

1. Disulfide bonds
   The amino acid cysteine contains sulfur. Where two cysteines are found close to each other a covalent bond can form.

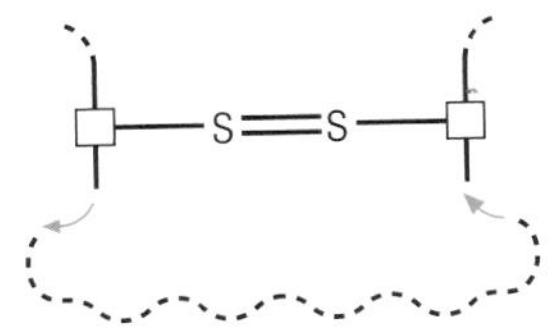

2. Ionic bonds
   R-groups sometimes carry a charge, either +ve or −ve. Where oppositely charged amino acids are found close to seach other an ionic bond forms.

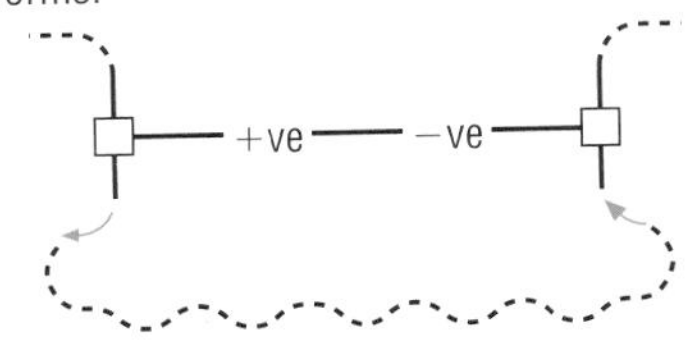

3. Hydrogen bonds
   As in secondary structure. Wherever slightly positively charged groups are found close to slightly negatively charged groups hydrogen bonds form.

4. Hydrophobic and hydrophilic interactions
   In a water-based environment, hydrophobic amino acids will be most stable if they are held together with water excluded. Hydrophilic amino acids tend to be found on the outside in globular proteins, with hydrophobic amino acids in the centre.

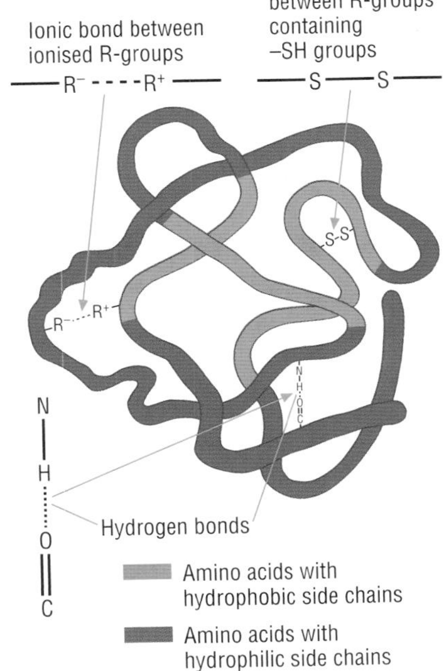

Figure 2 The bonds responsible for maintaining tertiary structure

Tertiary structure, protein function and heat

Heating a protein increases the **kinetic energy** in the molecule. This causes the molecule to vibrate and breaks some of the bonds holding the tertiary structure in place. As most of the bonds holding the tertiary structure in place are quite weak (they are not **covalent bonds**), they are easily broken. If enough heat is applied, the whole tertiary structure can unravel and the protein will no longer function. This process is called **denaturation**. The protein is said to be denatured. Even if the protein is then cooled, it will not reform the original complex structure.

Globular and fibrous proteins

The three-dimensional shape of proteins fall into two main categories.

Globular proteins tend to roll up into a compact globe- or ball-shaped structure. Any **hydrophobic** R-groups are turned inwards towards the centre of the structure, while the **hydrophilic** R-groups tend to be on the outside. This makes the proteins water-soluble, because water molecules can easily cluster around them.

Fibrous proteins form fibres. Most have regular, repetitive sequences of amino acids and are usually insoluble in water.

Protein type	3D feature	Solubility in water	Role	Examples
Globular	Roll up to form balls	Usually soluble	Usually have metabolic roles	Enzymes found in all organisms. Plasma proteins and antibodies found in the blood of mammals.
Fibrous	Form fibres	Usually insoluble	Usually have structural roles	Collagen found in bone and cartilage. Keratin found in fingernails and hair.

Table 1 Structure and function in proteins

Questions

1. Some organisms can survive in extremely hot locations (some above 80 °C). Suggest what types of bond are found in greater numbers holding the tertiary structure of the proteins in such organisms, and explain why.
2. Go back to the section on membranes (spread 1.1.8) – give examples of how structure is related to function in proteins found in membranes.
3. Draw a transmembrane protein pore in outline and add labels showing the types of amino acid R-group that would be found in the different parts of the protein molecule.

2.1 (9) Proteins in action

By the end of this spread, you should be able to . . .

* **Explain the term quaternary structure with reference to the structure of haemoglobin.**
* **Describe the structure of a collagen molecule.**
* **Compare the structure and function of haemoglobin and collagen.**

Quaternary structure

Quaternary structure refers to the fact that some proteins are made up of more than one **polypeptide** subunit joined together, or a polypeptide and an inorganic component. Such proteins only function if all the subunits are present. Quaternary structure may involve two identical polypeptides coming together to form the final working protein, or it may involve a number of different polypeptide subunits coming together. Examples of proteins with quaternary structure include **haemoglobin** and insulin.

Haemoglobin – a transport protein

Figure 1 Red blood cells

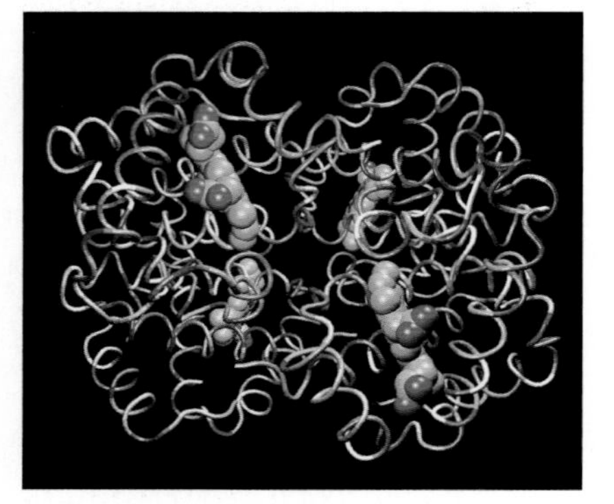

Figure 2 Haemoglobin molecule

Haemoglobin's quaternary structure consists of four polypeptide subunits. Two are called α-chains, the other two are β-chains. The four subunits together form one haemoglobin molecule, which is a water-soluble **globular protein**.

As with all proteins, the **tertiary structure** of each subunit is held in place by a number of bonds and interactions. These interactions give the subunits (and hence the complete molecule) a very specific shape. This shape is vital for the molecule to carry out its function.

Haemoglobin's function is to carry oxygen from the lungs to the tissues. It binds oxygen in the lungs and releases it in the tissues. A specialised part of each polypeptide, called a haem group, contains an iron (Fe^{2+}) ion. The haem group is responsible for the colour of haemoglobin.

haemoglobin + oxygen → oxyhaemoglobin
(purple-red) (bright red)

An oxygen molecule can bind to the iron in the haem group. This means that one complete haemoglobin molecule can bind up to four oxygen molecules.

The haem group is not made of **amino acids**, but it is an essential part of the molecule. Such groups are found in a number of proteins. They are called **prosthetic groups**.

Key definitions

Haemoglobin is a **globular** transport protein. **Collagen** is a **fibrous** structural protein.

Collagen – a structural protein

Collagen is a fibrous protein. A collagen molecule is made up of three polypeptide chains wound around each other. It looks like a twisted rope. Each of the three chains is itself a coil, made up of around 1000 amino acids. Hydrogen bonds form between the chains. This gives the structure strength.

The strength of the molecule is increased further because each collagen molecule forms **covalent bonds**, called cross-links, with other collagen molecules next to it. The cross-links that form are staggered along the collagen molecules, adding to the strength of the molecule. This results in a structure called a collagen fibril. Many fibrils together form a collagen fibre.

The function of collagen is to provide mechanical strength in many areas.

- In the walls of arteries, a layer of collagen prevents blood that is being pumped from the heart at high pressure from bursting the walls.

- Tendons connect skeletal muscles to the bones. Tendons are mostly collagen, and form a strong connection that allows muscles to pull bones for movement.
- Bones are formed from collagen, reinforced with materials such as calcium phosphate to make them hard.
- Cartilage and connective tissue are also made of collagen.
- Cosmetic treatments using collagen are becoming increasingly popular. For example, collagen can be injected into the lips to give them a fuller appearance.

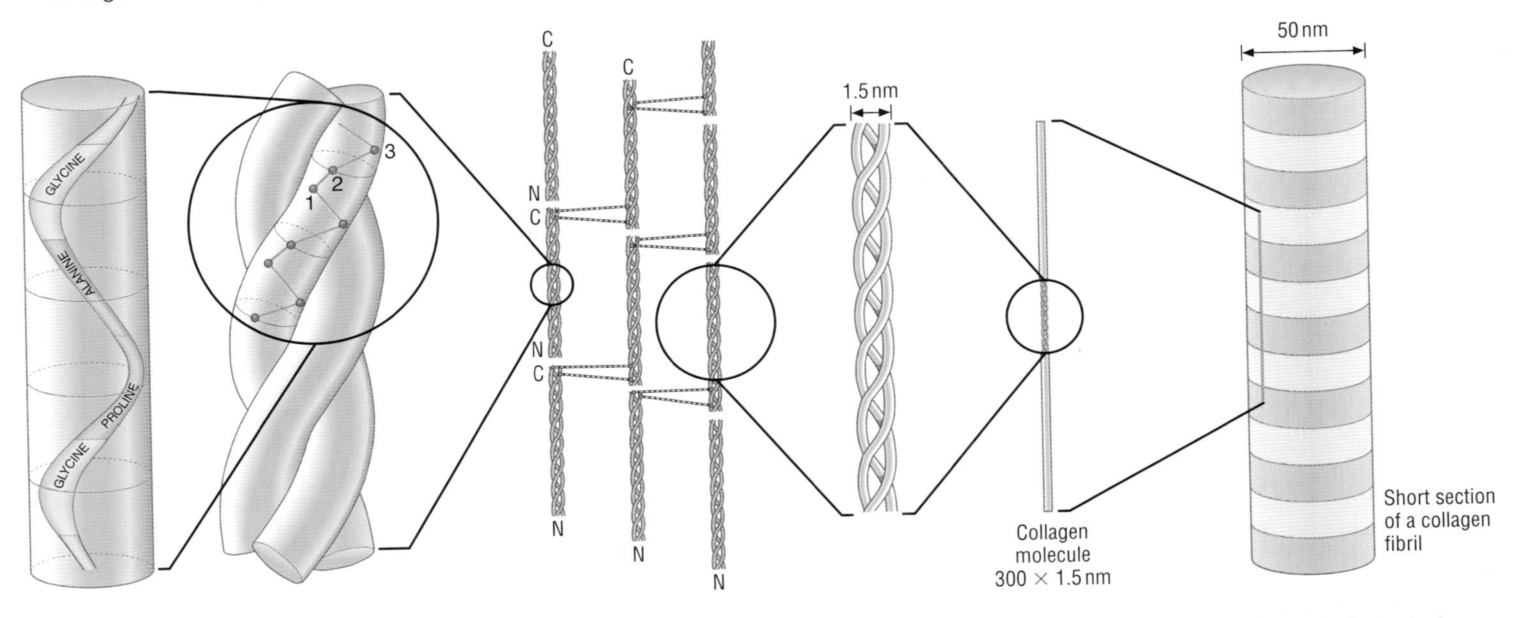

Figure 3 Collagen structure

Haemoglobin	Collagen
Globular protein	Fibrous protein
Soluble in water	Insoluble in water
Wide range of amino acid constituents in primary structure	Approximately 35% of the molecule's primary structure is one type of amino acid (glycine)
Contains a prosthetic group – haem	Does not have a prosthetic group
Much of the molecule is wound into alpha helix structures	Much of the molecule consists of left-handed helix structures

Table 1 Key similarities and differences between haemoglobin and collagen

Examiner tip

You must be able to compare the structure *and* function of haemoglobin with that of collagen. State the features in pairs: 'haemoglobin is soluble while collagen is insoluble', etc.

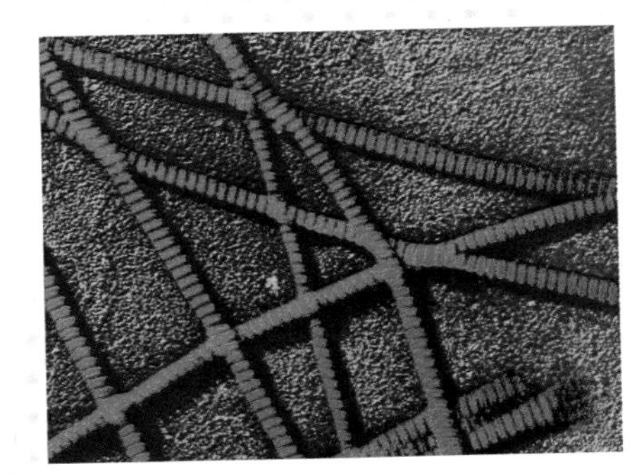

Figure 4 Collagen fibrils

Figure 5 Collagen injection into lips

Questions

1. Describe the similarities and differences between haemoglobin and collagen.
2. Collagen shares many features with cellulose, both functionally and structurally. List and explain these similarities.
3. What is the single major difference between collagen and cellulose?

2.1 (10) Lipids are not polymers

By the end of this spread, you should be able to . . .

* **State that lipids (fats and oils) are a range of biological molecules including triglycerides.**

> **Key definition**
>
> **Lipids** are a diverse group of chemicals that dissolve in organic solvents, such as alcohol, but not in water. They include fatty acids, triglycerides and cholesterol.

Lipids in living organisms

Lipids make up about 5% of the organic matter of a cell. At room temperature, a solid lipid is called a *fat* and a liquid lipid is called an *oil*. Lipids perform a number of functions within living organisms:

- a source of energy – lipids can be **respired** to release energy to generate **ATP**
- energy storage – lipids are stored in adipose cells (cells that store lipid) in 'fat stores' in organisms
- all biological membranes are made from lipids
- insulation – e.g. the blubber in whales is lipid that reduces heat loss; lipids also provide 'electrical insulation' around long nerve cells
- protection – e.g. the surface (cuticle) of plant leaves is protected against drying out by a layer of lipid
- some **hormones** (steroid hormones) are lipids.

Lipids contain the elements carbon, hydrogen and oxygen. The proportion of oxygen in lipids is very low, much lower than that found in carbohydrates. Lipids are insoluble in water.

Glycerol and fatty acids

Glycerol and fatty acids (Figure 1) are found in all the fats and oils that perform roles in energy storage and supply, and those found in membranes. While the glycerol molecule is always the same, the fatty acid molecules found in lipids can differ significantly. As with amino acids, animals cannot make some of the fatty acids they need from the raw materials taken into their bodies. These are called essential fatty acids, and they must be taken in 'complete' as part of the diet.

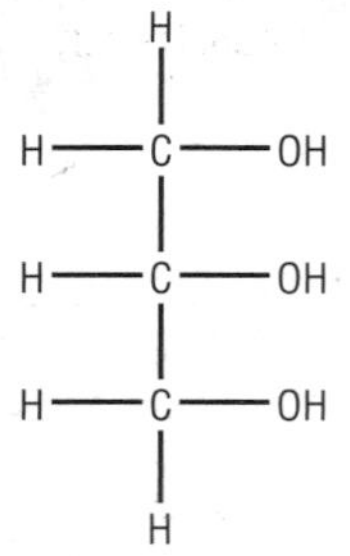

Glycerol, a 3-carbon molecule with three OH groups

Figure 1 Molecular structure of glycerol

Fatty acids – similar, but very different

All fatty acids have an acid group at one end. This acid group is the same as that found on an amino acid. The rest of the molecule consists of a **hydrocarbon chain** (a chain containing only carbon and hydrogen). The hydrocarbon chain can be anything from two to 20 carbons long. The fatty acids found most commonly have around 18 carbons in the hydrocarbon chain.

Saturated or unsaturated

The term unsaturated or polyunsaturated is often used when promoting the healthier aspects of foods. It is becoming common knowledge that too much saturated fat is a feature of a poor diet. These terms refer to the hydrocarbon chain, and whether it is 'saturated' with hydrogen or not. Put simply, if all the possible bonds are made with hydrogen, the fatty acid is saturated.

(a) A fatty acid molecule

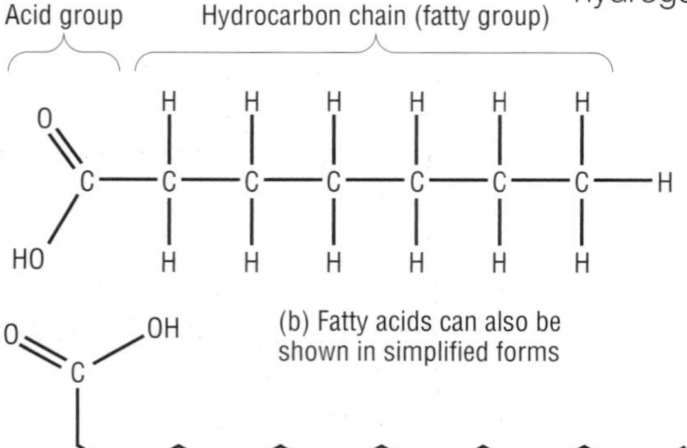

Figure 2 Fatty acid molecules showing acid group

Unsaturated fatty acids have C=C bonds, so fewer hydrogen atoms can be bonded to the molecule. A single C=C double bond gives a mono-unsaturated fatty acid. Two or more C=C double bonds give a polyunsaturated fatty acid.

Introducing C=C double bonds changes the shape of the hydrocarbon chain. This shape change makes the molecules in a lipid push apart and so makes them more fluid. This means that lipids containing many unsaturated fatty acids are often oils, but those with mainly saturated fatty acids are often fats.

Many animal lipids contain a great deal of saturated fatty acids. They are solid at room temperature and are called fats. Many plant lipids contain a great deal of unsaturated fatty acid. They are liquid at room temperature and are called oils. The animal lipid lard is solid, but the plant lipid olive oil is liquid.

Palmitic acid is a 16-carbon fatty acid. It is saturated, meaning all available bonds in its fatty chain have a hydrogen attached.

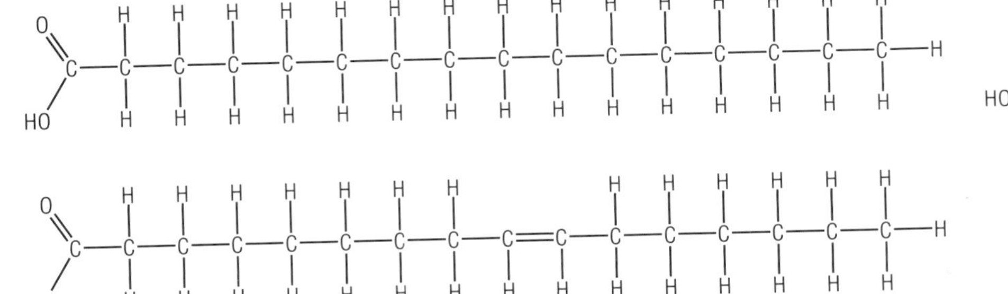

Palmitoleic acid is a 16-carbon fatty acid. It is unsaturated, having a double bond between carbons 9 and 10. This double bond makes the chain 'kink' at that position.

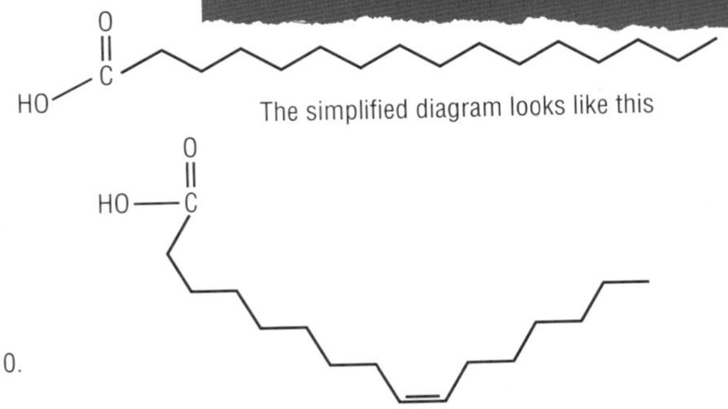

The simplified diagram looks like this

Figure 3 Saturated and unsaturated fatty acids

Triglycerides

A triglyceride consists of one glycerol molecule bonded to three fatty acid molecules. All fatty acid molecules are joined to glycerol molecules in exactly the same way. A **condensation** reaction between the acid group of a fatty acid molecule and one of the OH (hydroxyl) groups of the glycerol molecule forms a covalent bond. A water molecule is also produced.

The new bond formed is called an **ester bond**. The new molecule produced is a monoglyceride. Condensation reactions between acid groups of two more fatty acid molecules with the two remaining OH groups on the glycerol form a **triglyceride** molecule.

Triglycerides are insoluble in water, so they are described as **hydrophobic**. This is because the charges on the molecule are distributed evenly around the molecule. This means that hydrogen bonds cannot form with water molecules, so the two types of molecule do not mix together easily.

Examiner tip

If you are asked what triglycerides are made from, or what are the products of hydrolysis of a triglyceride, don't say 'glycerol and fatty acids', say 'one glycerol molecule and three fatty acids'.

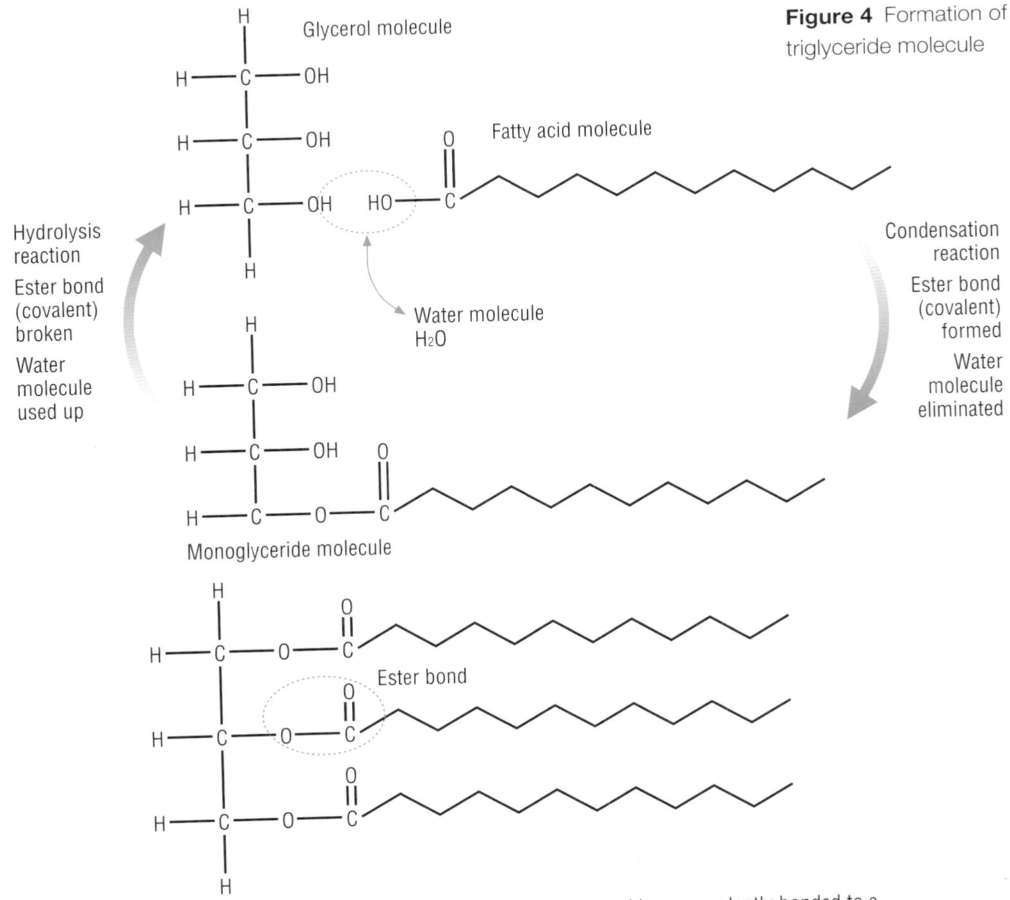

Glycerol molecule
Fatty acid molecule
Hydrolysis reaction
Ester bond (covalent) broken
Water molecule used up
Water molecule H_2O
Condensation reaction
Ester bond (covalent) formed
Water molecule eliminated
Monoglyceride molecule
Ester bond

A triglyceride molecule is formed when three fatty acids are covalently bonded to a glycerol molecule

Figure 4 Formation of triglyceride molecule

Questions

1. Explain why lipids are not polymers.
2. What is the difference between a saturated and an unsaturated fat?
3. Explain why triglyceride is suitable for its function in living organisms.
4. How many molecules are produced when a triglyceride molecule is hydrolysed?

2.1 (11) Essential oils?

By the end of this spread, you should be able to . . .

* **Compare the structure of triglycerides and phospholipids.**
* **Explain how the structure of triglyceride, phospholipid and cholesterol molecules relates to their functions in living organisms.**

Phospholipids – the basis of biological membranes

Essentially, a phospholipid molecule is almost identical to a **triglyceride** molecule. It consists of a **glycerol** molecule with **fatty acid** molecules bonded by **condensation** reactions to produce **ester bonds**.

In phospholipids, the third fatty acid is not added to the glycerol molecule. Instead, a phosphate group is covalently bonded to the third OH group on the glycerol. The bonding of the phosphate group occurs by a condensation reaction and so a water molecule is released.

The phosphate 'head' of the molecule is **hydrophilic**, but the **hydrocarbon chain** fatty acid 'tails' are **hydrophobic**. The majority of a phospholipid molecule is insoluble in water, like all lipid molecules. The water-solubility of the head group gives phospholipids their characteristics in terms of the capacity to form membranes.

A phospholipid molecule consists of a glycerol molecule with two fatty acids and a phosphate group.

Hydrophilic head (where X can be a variety of chemical groups)

Hydrophobic tails

Phospholipids are usually shown in a simplified form

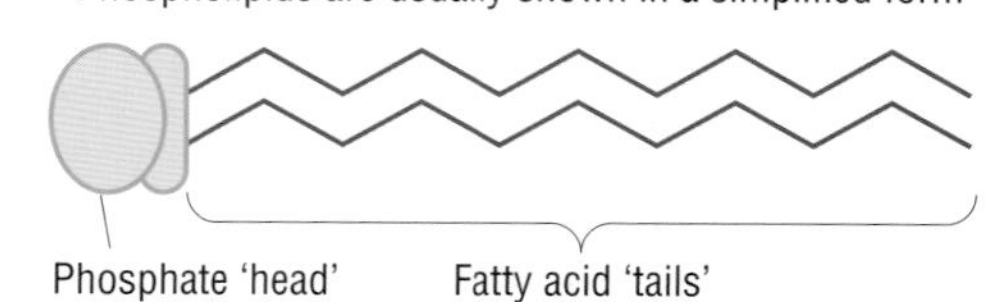

Figure 1 Formation of a phospholipid molecule

Figure 2 A kangaroo rat never needs to drink because of lipid respiration and water-saving features

Phospholipids and membrane fluidity

The fatty acids that make up a phospholipid may be saturated or unsaturated. Organisms can control the fluidity of membranes using this feature. For example, organisms living in colder climates have an increased number of unsaturated fatty acids in their phospholipid molecules. This ensures that membranes remain fluid, despite the low temperatures.

Lipids and respiration

Respiration of lipids first requires the **hydrolysis** of the ester bonds holding the fatty acids and glycerol together. This is the reverse of the condensation reaction that joins them together. Both the glycerol and fatty acids can then be broken down completely to carbon dioxide and water. This releases energy, which is used to generate **ATP** molecules (spread 2.1.4).

The respiration of one gram of lipid gives out about twice as much energy as the respiration of one gram of **carbohydrate**. Because lipids are insoluble in water, they can be stored in a compact way and they do not affect the **water potential** of the cell contents. These features make triglyceride an excellent energy-storage molecule.

The respiration of lipid gives out a great deal more water than the respiration of carbohydrate. This metabolic water is vital to some organisms (Figure 2).

Cholesterol and steroid hormones

Cholesterol is a class of lipid. It is not formed from fatty acids and glycerol like triglycerides and phospholipids. It is a small molecule made from four carbon-based rings. Cholesterol is found in all biological membranes. Its small, narrow structure and hydrophobic nature allow it to sit between the phospholipid hydrocarbon tails and help to regulate the fluidity and strength of the membrane.

The steroid hormones **testosterone**, **oestrogen** and vitamin D are made from cholesterol. The lipid nature of the steroid hormones means they can pass directly through the phospholipid bilayer in order to reach their target receptor (or site). This is usually inside the nucleus, so they also pass through the lipid bilayer of the **nuclear envelope**.

Cholesterol is vital to living organisms, so many cells (especially in the liver) can make it. Excess cholesterol may be a problem in humans because:

- in bile, cholesterol can stick together to form lumps called gallstones
- in blood, cholesterol can be deposited in the inner linings of blood vessels causing **atherosclerosis**, which can result in a number of circulatory problems.

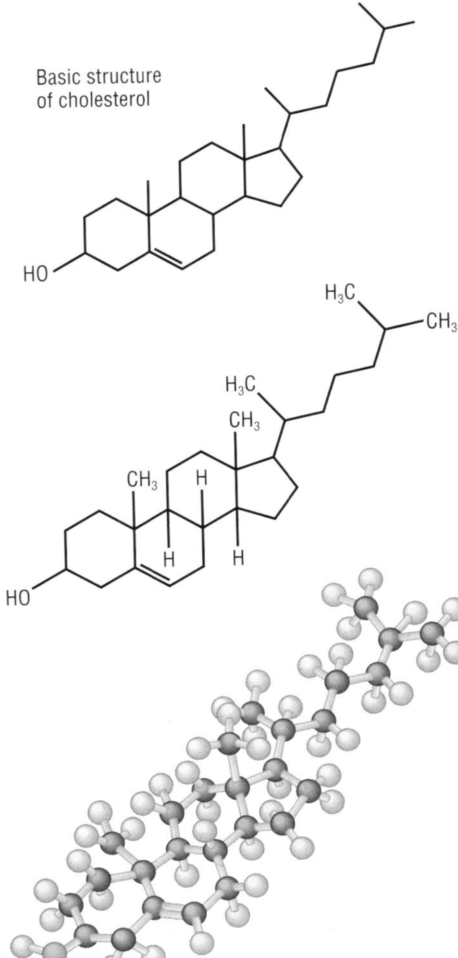

Figure 3 Cholesterol structure

Essential – but potentially deadly

A condition known as familial hypercholesterolaemia (FHC: high blood cholesterol levels that run in families) is a genetic disorder, where cells manufacture and secrete cholesterol even though there is already sufficient in the blood to provide for the organism's requirements. This happens because the cells do not obey the signals to stop cholesterol production, as they lack a particular cell surface **receptor**.

Individuals with this condition in its worst form can suffer heart attacks (myocardial infarction) and strokes by the age of 2 years.

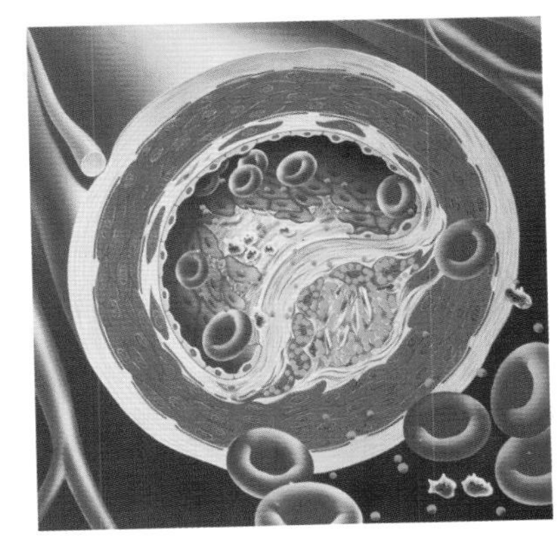
Figure 4 Atherosclerosis

Lipid	Structure	Main role	Other features
Triglyceride	Glycerol plus three fatty acids	Compact energy store, insoluble in water so doesn't affect cell water potential	Stored as fat, which also has thermal insulation and protective properties
Phospholipid	Glycerol plus two fatty acids and a phosphate group	Forms a molecule that is part hydrophobic, part hydrophilic, ideal for basis of cell surface membranes	Phosphate group may have carbohydrate parts attached – these glycolipids are involved in cell signalling
Cholesterol	Four carbon-based ring structures joined together	Forms a small, thin molecule that fits into the lipid bilayer giving strength and stability	Used to form the steroid hormones

Table 1 The main types of lipid

Questions

1. Lipid is stored in animals as fat under the skin, and around delicate organs such as the kidneys. Explain why both an excess of lipid and a severe lack of lipid may be problems in animals.
2. Camels' humps are mainly lipid – list and explain the advantages of this.
3. The vitamins A, D, K and E are fat-soluble. Where are these vitamins likely to be stored?
4. Explain how the formation of ester bonds is similar to, and different from, the formation of (a) glycosidic bonds and (b) peptide bonds.

2.1 (12) Water – a vital biological molecule

How Science Works

By the end of this spread, you should be able to . . .

* **Describe how hydrogen bonding occurs between water molecules.**
* **Relate this, and other properties of water, to the roles of water in living organisms.**

Key definition

A **hydrogen bond** is a weak interaction that can occur wherever molecules contain a slightly negatively charged atom bonded to a slightly positively charged hydrogen. Water molecules hydrogen-bond with each other extensively.

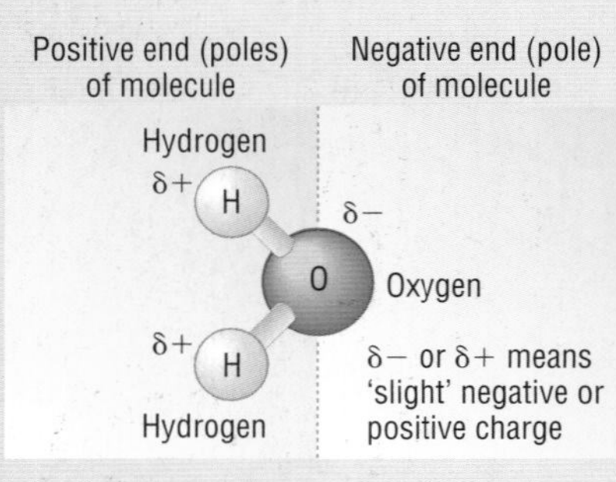

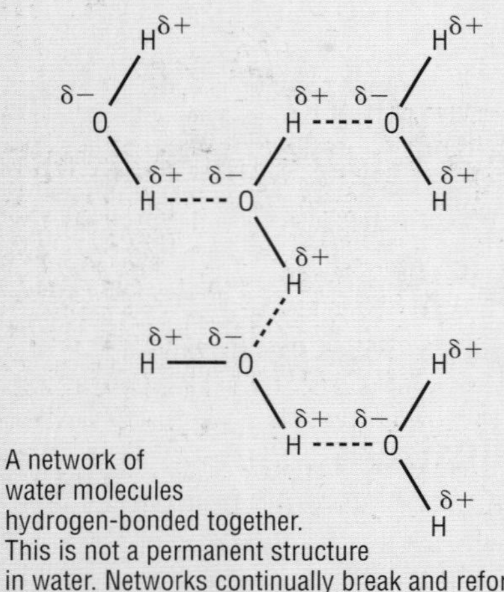

Figure 1 Water molecule showing δ+ and δ– ends and hydrogen bond formation

Hydrogen bonds explain water's unusual properties

Water is a very small molecule consisting of two hydrogen atoms **covalently bonded** to one oxygen atom. The shared electrons in water are not shared evenly. The oxygen atom is capable of pulling the shared electrons towards it and so away from the hydrogen atoms in the molecule.

This means that water molecules become slightly negatively charged at the oxygen end and slightly positively charged at the hydrogen ends. Water is described as a polar molecule.

Hydrogen bonds in liquid water and ice

In pure liquid water, the molecules form **hydrogen bonds** with each other. They form a network that allows the molecules to move around, continually making and breaking hydrogen bonds as they do so. This makes it much more difficult for water molecules to 'escape' the liquid to become gas. It explains why water must be heated to 100 °C before it boils. Molecules such as hydrogen sulfide (H_2S), which are similar in size to water, are gases at normal environmental temperatures.

As temperature is reduced, water molecules move less because they have reduced **kinetic energy**. More hydrogen bonds form, but they do not break so easily. As water becomes solid, the hydrogen bonds formed hold the structure in a semi-crystalline form. This form is less dense than liquid water, so ice forms on the surface of water as it cools.

Hydrogen bonds and temperature changes

The hydrogen bonds in liquid water restrict the movement of the water molecules, so a relatively large amount of energy is needed to increase the temperature of water. This keeps the temperature of large bodies of water, such as lakes and oceans, stable even when the temperature changes dramatically.

The evaporation of water uses up a relatively large amount of energy. This means that water evaporating from a surface 'removes' heat from the surface. So heat energy is used in evaporation.

Density and freezing

Water is unusual because its solid form – ice – is less dense than its liquid form. As water cools, its density increases until the temperature drops to 4 °C, then its density begins to decrease again. This means that ice floats on liquid water, so it insulates the water below. This allows living organisms to survive the winter, and even to live under the ice.

Hydrogen bonds, cohesion and surface tension

A drop of water on a waxy surface, such as a plant leaf, looks almost spherical. It hardly wets the leaf at all. This is because the hydrogen bonds pull water molecules in at the surface. This property of water molecules sticking to each other is termed **cohesion**. Cohesion also results in the **surface tension** seen at the surface of a body of water.

Water as a solvent

The solubility of a substance in water depends on whether water molecules can interact with it. Any molecule that (like water) is polar will dissolve in water. This happens as follows:

- The substance to be dissolved (the **solute**) has slightly negative and slightly positive parts. These will interact with water molecules.
- The water molecules cluster around the slightly charged parts of the solute molecule.

- This keeps solute molecules apart, so they are dissolved.
- Once in solution, molecules can move around and react with other molecules. This is the basis of most metabolic reactions, which take place in solution in the cytoplasm.
- Ions are charged particles that also dissolve very easily because water molecules can cluster around them and separate them.

Water as a transport medium

Water remains liquid over a large temperature range and can act as a solvent for many chemicals. This makes it an ideal transport medium in living organisms.

Properties of water

Property	Importance	Examples
Solvent	Metabolic processes in all organisms rely on chemicals being able to react together in solution.	70–95% of cytoplasm is water. Dissolved chemicals take part in processes such as **respiration** and **photosynthesis** in living organisms.
Liquid	The movement of materials around organisms, both in cells and on a large scale in multicellular organisms, requires a liquid transport medium.	Blood in animals and the vascular tissues in plants use water as a liquid transport medium.
Cohesion	Water molecules stick to each other creating surface tension at the water surface. Cohesion also makes long, thin water columns very strong and difficult to break.	Transport of water in the **xylem** relies on water molecules sticking to each other as they are pulled up the xylem in the transpiration stream. Some small organisms make use of surface tension to 'walk on water'.
Freezing	Water freezes, forming ice on the surface. Water beneath the surface becomes insulated and less likely to freeze.	Organisms such as polar bears live in an environment of floating ice packs. Lakes tend not to freeze completely, so aquatic organisms are not killed as temperatures fall.
Thermal stability	Large bodies of water have fairly constant temperatures. Evaporation of water can cool surfaces by removing heat.	Oceans provide a relatively stable environment in terms of temperature. Many land-based organisms use evaporation as a cooling mechanism, for example in panting or sweating.
Metabolic	Water takes part as a reactant in some chemical processes.	Water molecules are used in **hydrolysis** reactions and in the process of photosynthesis.

Table 1 Properties of water related to its role in living organisms

Figure 2 Water – something that all organisms have in common is that they cannot survive without it

Examiner tip

Hydrogen bonds are not just found in water. You will be expected to know how hydrogen bonds contribute to water's properties, as well as to the stability of other molecules.

Questions

1. Explain why water is described as a 'polar molecule'.
2. What type of molecules will not dissolve in water? Explain why.
3. Why is it important that water is liquid over a large range of temperatures?
4. Name the cellular process involved in moving liquids into the cell.
5. Name the process by which water moves into and out of cells.

(13) Practical biochemistry – 1

By the end of this spread, you should be able to . . .

* **Describe how to carry out chemical tests to identify the presence of proteins, carbohydrates and lipids.**

Finding out what is present

There are a number of simple chemical tests that can confirm the presence of various biological molecules within a sample. These tests indicate only the presence of the type of molecule, not how much is present, so they are known as **qualitative** tests. These tests are often referred to as food tests, because they can be used to detect the presence of the various biological molecules in food samples.

The tests all rely on the biological molecules going into solution. It may be necessary to grind or break up the sample in order to carry out the tests properly.

Tests for the presence of carbohydrates

Starch

To show the presence of starch, you can add a solution of iodine (in potassium iodide) to the sample. If starch is present, the iodine solution changes colour from yellow-brown to blue-black.

Add a few drops of iodine solution

Iodine solution

Place a sample of food to be tested in a well on a spotting tile or in a test tube

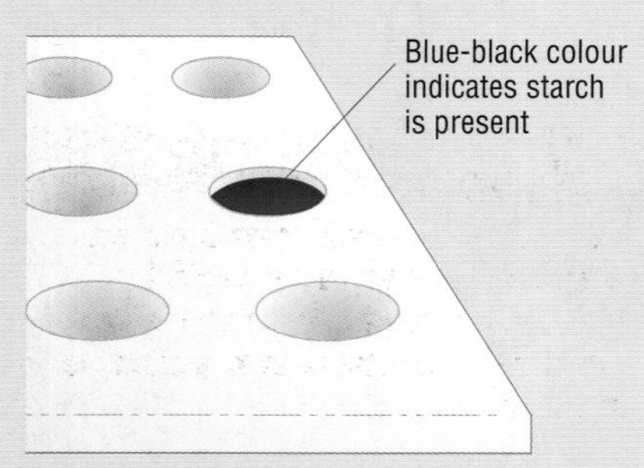

Figure 1 Starch test

Reducing sugars

All **monosaccharide** and many **disaccharide** sugars are known as **reducing sugars**. Put simply, this means that a molecule can react with other molecules by giving electrons to them. (In chemistry, adding electrons to another molecule or particle is called **reduction**.)

When a reducing sugar is heated with Benedict's solution (alkaline copper sulfate), the solution changes colour from blue to orange-red. The orange-red is described as a **precipitate** because the orange-red substance comes out of solution and forms solid particles dispersed in the solution. This is **Benedict's test**.

Non-reducing sugars

Some sugars do not react with Benedict's solution at all, so a reducing sugar test would show up as negative (no colour change).

Sucrose is a non-reducing sugar. It is formed by a **condensation** reaction forming a **glycosidic bond** between a glucose molecule and a fructose molecule. Fructose and glucose are monosaccharides, and sucrose is a disaccharide.

The formation of the glycosidic bond in sucrose is slightly different from that in maltose (which is a reducing sugar). This difference prevents the sucrose from reacting with Benedict's solution.

If your substance does not react with Benedict's solution (the reducing sugar test), you have to use a different test.

- First you must make sure there are no reducing sugars in the sample.
- Boil the sample with hydrochloric acid.
- This hydrolyses any sucrose present, splitting sucrose molecules to give glucose and fructose monosaccharides.
- Then cool the solution and neutralise it by adding sodium hydrogencarbonate solution or sodium carbonate solution (both are alkalis).
- Carry out the reducing sugar test again.
- If sucrose is present in the original sample, the test will now give a positive result because the monosaccharides glucose and fructose are present.

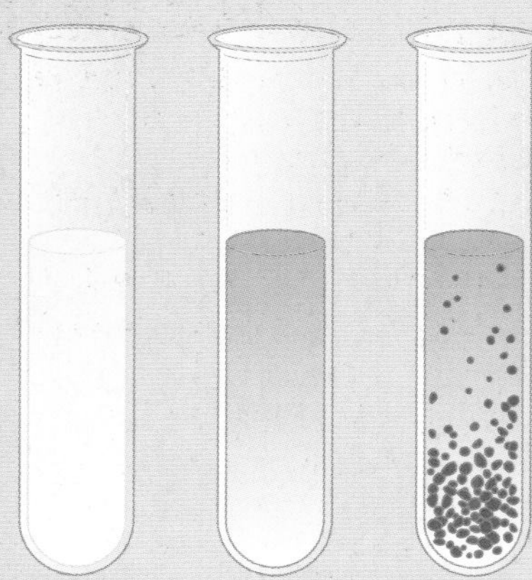

Figure 2 Benedict's test

Test for the presence of proteins

You can test for the presence of **proteins** by adding biuret reagent to a sample. This is called the **biuret test**. Biuret reagent is a pale blue colour, and contains sodium hydroxide and copper sulfate. These chemicals react with the **peptide bonds** found in protein, which results in a colour change to lilac. This test does not require any heating.

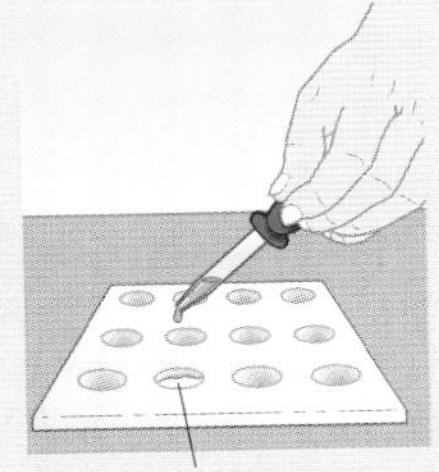

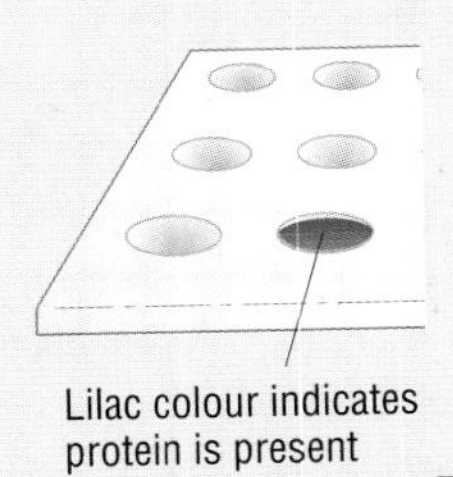

Figure 3 Biuret test

Figure 4 Ethanol emulsion test

Test for the presence of lipids

You can test for the presence of **lipid** by using the **ethanol emulsion test**.

- Mix the sample with ethanol (an alcohol). This dissolves any lipid present (lipids are soluble in alcohols).
- Then pour the liquid (alcohol with dissolved fat) into water contained in another clean test tube.
- If lipid is present, a cloudy white **emulsion** will form near the top of the water – the lipid comes out of solution and becomes dispersed as tiny droplets in the water.

Food tests

Testing for	Description	Result (colour change)
Starch	Add a few drops of iodine solution	Brown to blue/black
Reducing sugar	Add Benedict's solution and heat to 80 °C in a water bath	Blue to orange-red
Non-reducing sugar	If reducing sugar test is negative, boil with hydrochloric acid, cool and neutralise with sodium hydrogencarbonate solution or sodium carbonate solution; repeat Benedict's test	Blue to orange-red (on second test)
Protein	Add biuret reagent	Blue to lilac
Lipid	Add ethanol to extract (dissolve) lipid and pour alcohol into water in another test tube	White emulsion forms near top of water

Table 1 Summary of simple biochemical tests

Key definition

Food tests are simple tests that show the presence of various biological molecules in samples or structures. Iodine can stain plant tissue sections to show where starch is.

Examiner tips

In answering recall questions about the biochemical tests for biological molecules, you must:
- state that heat to 80 °C needs to be applied in Benedict's test
- state the colour change you would see in a positive result for any of the tests.

It is worth learning these by heart:
reducing sugars (Benedict's, heat) – red-orange
proteins (biuret, cold) – purple
lipids (ethanol and water, cold) – look cloudy.

Questions

1. Explain what is meant by chemical reduction.
2. Why would a solution of amino acids give a negative result for the biuret test?
3. Explain what is meant by the terms 'precipitate' and 'emulsion'.
4. Hydrolysis of cellulose gives beta-glucose molecules. What would happen if cellulose were hydrolysed then heated with Benedict's solution?

2.1 How Science Works

(14) Practical biochemistry – 2

By the end of this spread, you should be able to . . .

* **Describe how the concentration of glucose in a solution may be estimated by using colour comparisons.**

Finding out how much is present

The simple tests described in spread 2.1.13 detect only the *presence* of the various molecules. If you want to determine *how much* is present, you need to carry out a **quantitative** test (a test that gives a value for the amount present).

Benedict's test and precipitation

- **Benedict's test** reveals the presence of **reducing sugars**.
- It results in an orange-red precipitate.
- The more reducing sugar there is present, the more precipitate will be formed, and the more Benedict's solution (copper sulfate) will be 'used up'. If the precipitate is filtered out, then the concentration of the remaining solution can be measured.
- This will tell you how much Benedict's solution has been used up, so you can then estimate the concentration of reducing sugar in the original sample.

Comparing solutions with a colorimeter

- A colorimeter is a device that shines a beam of light through a sample.
- A photoelectric cell picks up the light that has passed through the sample. It will give you a reading showing how much light has passed through.
- Place the solution in a sample chamber between the light and the photoelectric cell in a small clear plastic container called a cuvette or in a special test tube.
- The more copper sulfate that is used up in Benedict's test in a sample, the less light will be blocked out and the more light will be transmitted. This means the readings taken give a measure of the Benedict's reaction.

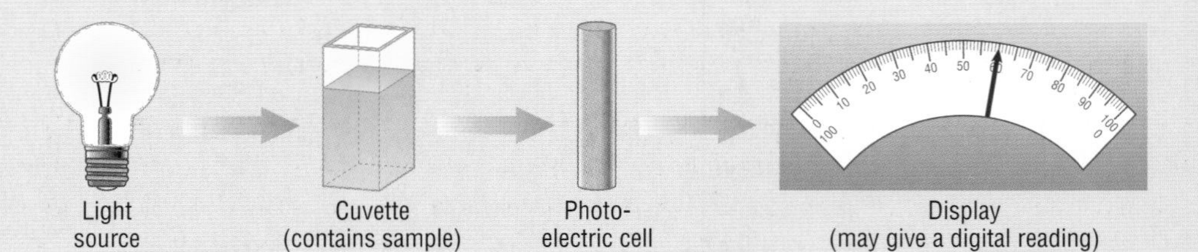

- When using a colorimeter, the device is usually zeroed between each reading by placing an appropriate 'blank' sample to reset the 100% transmission/0% absorption. In this case, the blank used would be water.
- Colour filters are often used for greater accuracy. In this case, a red filter would be used.

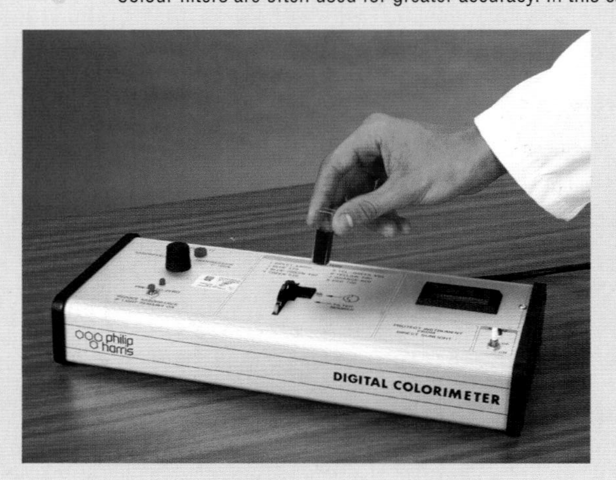

Figure 1 You could compare the reducing sugar content of orange juice and apple juice, for example, in this way. When using a colorimeter, you must zero the device between each reading with an appropriate blank sample

Calibration – making precise measurements

Using a colorimeter to take readings of different samples that contain reducing sugar does not tell us *how much* reducing sugar is present – it simply tells us which sample contains more.

In order to quantify (measure) the amount, a **calibration** curve must be made.

- Prepare a calibration curve by taking a range of known concentrations of reducing sugar.
- Carry out Benedict's test on each one. Then filter the precipitate out of the solution.
- Use a colorimeter to give readings of the amount of light passing through the solutions.
- Plot the readings in a graph to show light getting through (transmission) against reducing sugar concentration.
- You are now ready to take a reading from an unknown sample. From the graph, you can read off the equivalent reducing sugar concentration for the new reading.

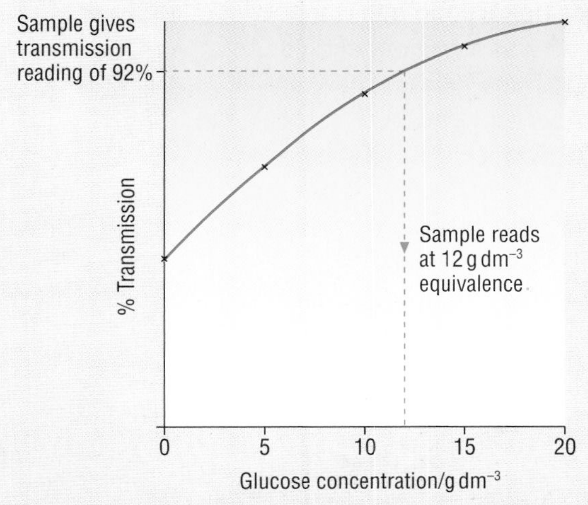

Figure 2 A range of glucose solutions was made up. After applying Benedict's test and filtering off the precipitate, the samples were placed in a colorimeter to measure the amount of light passing through. The graph shows the results of a test carried out on a sample that had a transmission reading of 92%. Looking at the calibration curve, we can see that the sample contained the equivalent of 12 g dm^{-3} reducing sugar

The use of calibration curves is common in biology. Finding the concentration of a substance in a sample by comparing it with known standards in this way is often referred to as an **assay**.

Examiner tip

Be very careful when reading graphs in examination questions. The information shown in Figure 2 could be recorded as 'percentage transmission' (how much light gets through) or 'absorption' (how much light is absorbed). Unless you work out what the axes are showing, you may make basic mistakes.

Key definitions

A **quantitative** test gives a measure of a substance in units, not simply an indication of its presence.

Assay techniques are often used to compare measurements with known samples so that quantitative measurements can be made.

Questions

1. Colorimeters can be used to measure both transmission (how much light passes through the sample) and absorption (how much light is absorbed by the sample). Sketch the graph for absorption that would be given by the calibration curve readings shown in Figure 2.
2. Explain why cuvettes must be transparent and must be handled carefully.
3. Outline how you could compare the reducing sugar concentration in apple juice and orange juice.

2.1 (15) Nucleotides – coding molecules

By the end of this spread, you should be able to . . .

* **State that DNA is a polynucleotide, usually double stranded, made up of nucleotides containing the bases adenine, thymine, cytosine and guanine.**
* **State that RNA is a polynucleotide, usually single stranded, made up of nucleotides containing the bases adenine, uracil, cytosine and guanine.**

Nucleic acids in living organisms

Nucleic acids come in two forms: **DNA** (deoxyribonucleic acid) and **RNA** (ribonucleic acid). In living organisms these molecules hold the coded information to build that organism.

Almost all the DNA in a **eukaryotic cell** is found in the nucleus, where it acts as the information store. RNA is found in three different forms. These three forms are needed to read and translate the information in order to produce the various proteins required to make the living, functioning organism.

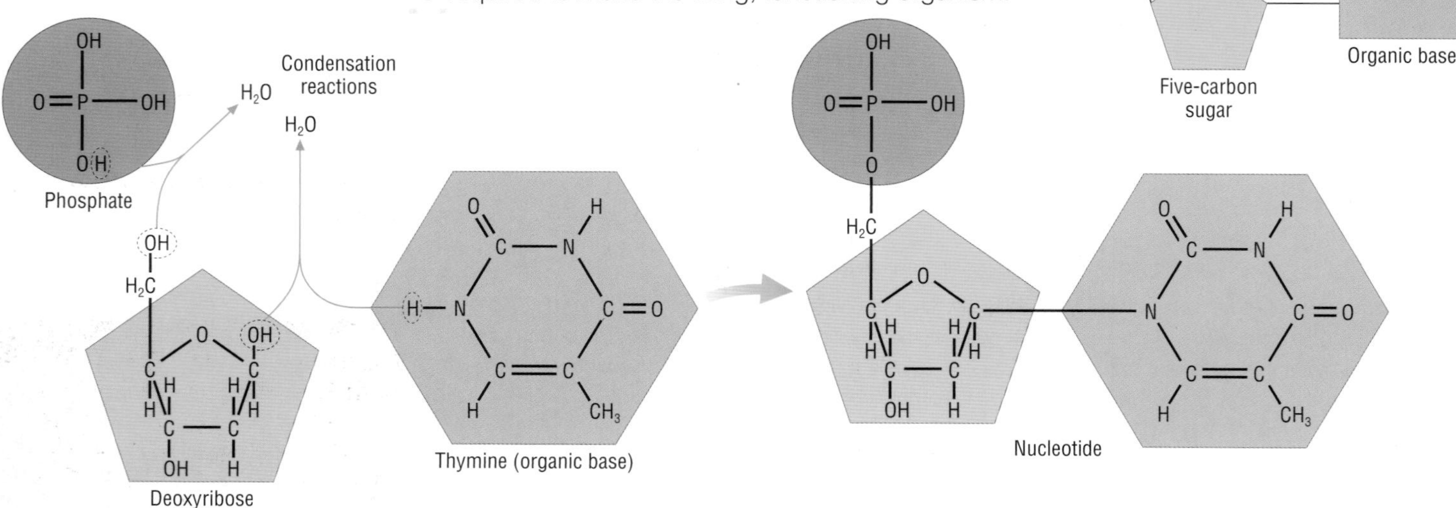

Figure 1 A single nucleotide

A phosphate, sugar and base are joined by condensation reactions to form a single nucleotide. In this case the sugar is deoxyribose and the base is thymine

Nucleotides are monomers of nucleic acids

The **monomer** of all nucleic acids is called a **nucleotide**.

Each single nucleotide is itself made by the joining of three subunits:

- one phosphate group
- one sugar molecule
- one **organic nitrogenous base**.

The three subunits are joined by **covalent bonds** to form a single nucleotide molecule.

Key definition

Nucleotides are the monomers of all nucleic acids. Each nucleotide is formed by bonding together a phosphate group, a sugar molecule and a nitrogenous base.

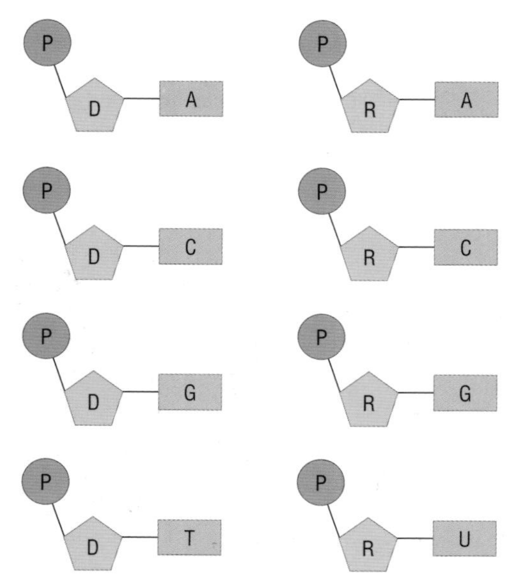

Figure 2 Structure diagrams of DNA and RNA nucleotides

Nucleotides – common features

As nucleic acids code for the building of all living organisms, you might expect there to be a large number of different monomers available to produce such a code. In fact, just four different nucleotides make up the code carrying the instructions to make proteins in all living things.

- The phosphate group in all nucleic acids is always the same.
- The sugar molecule is a 5-carbon sugar, either deoxyribose (in DNA) or ribose (in RNA).
- There are five possible bases: adenine, thymine, guanine, cytosine and uracil (Figure 2).

Joining nucleotides together

A **condensation** reaction between the phosphate group of one nucleotide and the sugar of another nucleotide joins the two together. Repeating this bonding gives a long chain of nucleotides.

As nucleotides are bonded together to form chains, the 'backbone' of the molecule consists of a repeating sugar–phosphate chain. Organic bases project from the backbone. It is the sequence of these bases that forms coded information in nucleic acids.

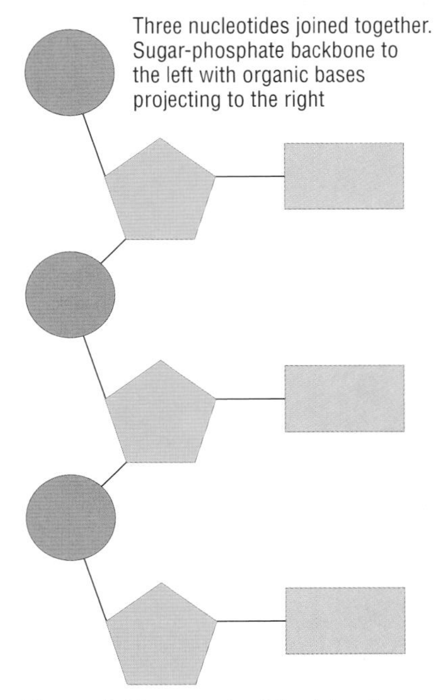

Figure 3 Three nucleotides joined together

From nucleotides to nucleic acids

Chains of nucleotides bonded together are called nucleic acids. When nucleotides are joined together to form long-chained polymers, only nucleotides carrying the same sugar molecule bind together. This means that nucleic acids can be described as either ribonucleic acids (RNA) when the sugar is **ribose**, or deoxyribonucleic acids (DNA) when the sugar is **deoxyribose**.

Organic bases are either purines or pyrimidines

The five organic bases are grouped. Three are called **pyrimidines** and two are called **purines**. Pyrimidines are smaller than purines. You will see how important these differences are when we look at the structure of DNA later (spread 2.1.16).

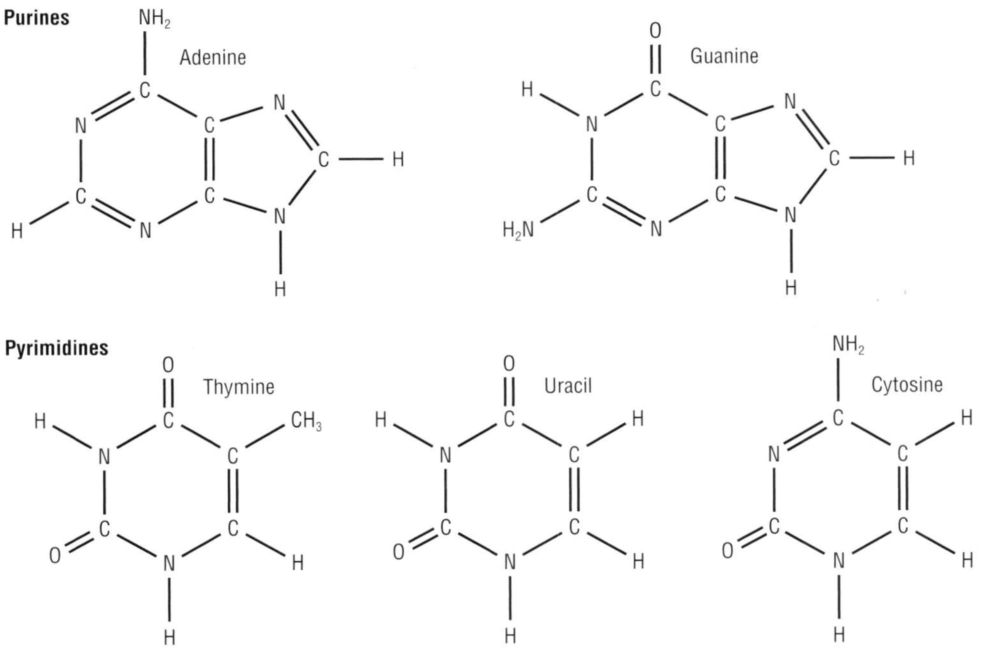

Purines: Adenine, Guanine

Pyrimidines: Thymine, Uracil, Cytosine

Figure 4 Purine and pyrimidine molecules

Examiner tip

Organic bases contain nitrogen. Always call them 'nitrogenous bases', not just 'bases'. Remember that both nucleic acids *and* amino acids always contain nitrogen.

Too much nucleic acid causes gout

Uric acid is produced when excess purines are broken down in the liver. It is excreted in the urine. Some people have too much uric acid in their blood. The uric acid is insoluble at lower temperatures, and forms crystals that are deposited in joints at the extremities, such as the toes. The toe joint becomes very painful and swollen, a condition called gout.

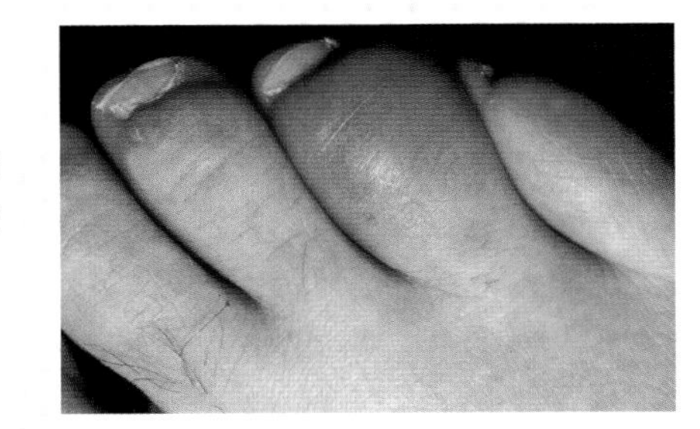

Figure 5 A gouty toe

Questions

1. How many molecules of water will be produced when a nucleotide is made from its separate component parts?
2. Explain why nucleic acids are named after the sugar they contain.
3. In a nucleic acid polymer, which component groups are at the ends of the molecule?

2.1 (16) DNA – information storage

By the end of this spread, you should be able to . . .

* **State that DNA is a double-stranded polynucleotide.**
* **Describe how a DNA molecule is formed by hydrogen bonding between complementary base pairs on two antiparallel DNA strands.**
* **Explain how twisting of the DNA molecule produces the double helix shape.**
* **Outline how DNA replicates semi-conservatively, with reference to the role of DNA polymerase.**

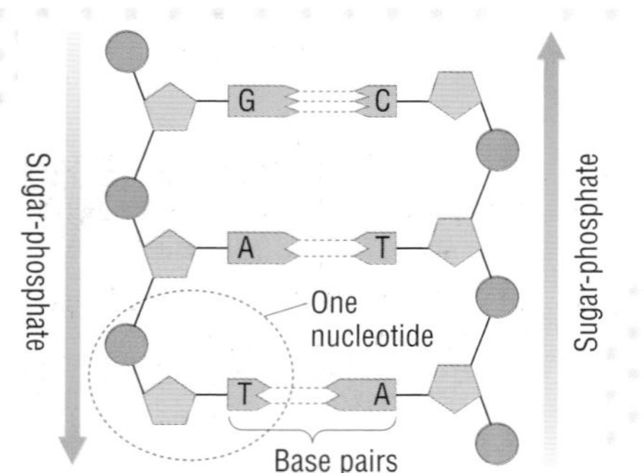

Part of a DNA molecule. The two antiparallel strands are held together by hydrogen bonds

Figure 1 Bonding in a DNA molecule

DNA is a stable polynucleotide

DNA is a long-chain polymer of nucleotide monomers. This polymer is called a **polynucleotide**. A DNA molecule forms when two polynucleotide strands come together. They form what looks like a ladder – the sugar phosphate backbones of the two chains form the uprights, and the bases project towards each other to form the rungs.

Hydrogen bonds between the bases in opposite uprights strengthen the rungs of the ladder. This makes DNA a very stable structure, which is vital as it carries the instructions to make an organism. If it were unstable, the instructions could go wrong too easily.

Hydrogen bonding and base pairing – getting it right

The two DNA strands run parallel to each other because the space between them is taken up by the nitrogenous bases projecting inwards. The term 'antiparallel' is used because the strands run in opposite directions to each other. The sugars are pointing in opposite directions.

Nucleotides with adenine as the base can make two hydrogen bonds with nucleotides with thymine as the base

P D A T D P

Nucleotides with guanine as the base can make three hydrogen bonds with nucleotides with cytosine as the base

P D G C D P

Figure 2 A–T, C–G and number of hydrogen bonds

The chains are always the same distance apart because the bases pair up in a specific way. Where a **pyrimidine** appears on one side, a **purine** appears on the other. Even more important is that where the purine is **adenine** (A), the pyrimidine is always **thymine** (T). Where the purine is **guanine** (G), the pyrimidine is always **cytosine** (C). As the strands come together, hydrogen bonds form between the bases. The different structure of the bases means that the **base-pairing rules** always apply. The base pairing is described as **complementary**. So A is complementary to T, and C is complementary to G.

In a complete DNA molecule, the antiparallel chains twist, like twisting a rope ladder, to form the final structure. This is known as a **double helix**.

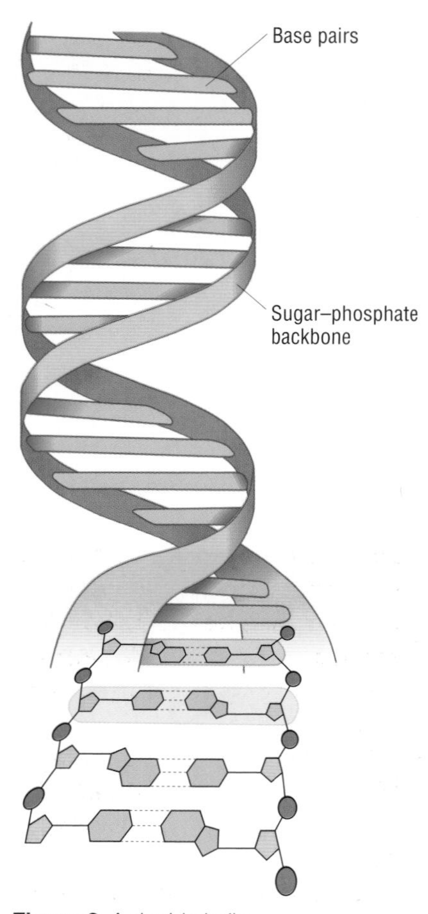

Figure 3 A double helix

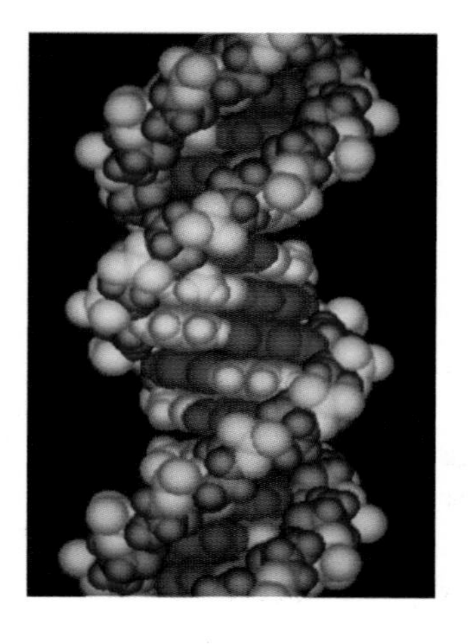

Making copies

When a cell divides, each new cell must receive a full set of instructions. Each cell must have a full copy of all the DNA for that organism (see spread 1.2.2). This means there has to be a way of copying DNA strands precisely. This DNA replication takes place in **interphase** of the cell cycle (see spread 1.1.13), and is the process that creates identical sister **chromatids**.

In order to make a new copy of a DNA molecule:

* the double helix is untwisted
* hydrogen bonds between the bases are broken apart to 'unzip' the DNA – this exposes the bases

- free DNA nucleotides are hydrogen-bonded onto the exposed bases according to the base-pairing rules (A–T, C–G)
- covalent bonds are formed between the phosphate of one nucleotide and the sugar of the next to seal the backbone.

This continues all the way along the molecule until two new DNA molecules (double helices) are formed, each an exact replica of the original DNA molecule because of the base-pairing rules.

This process of DNA replication is described as **semi-conservative replication**. Each new DNA molecule consists of one conserved strand plus one newly built strand. Pioneering work by the scientists Meselsohn and Stahl showed how semi-conservative DNA replication was responsible for the production of new DNA molecules.

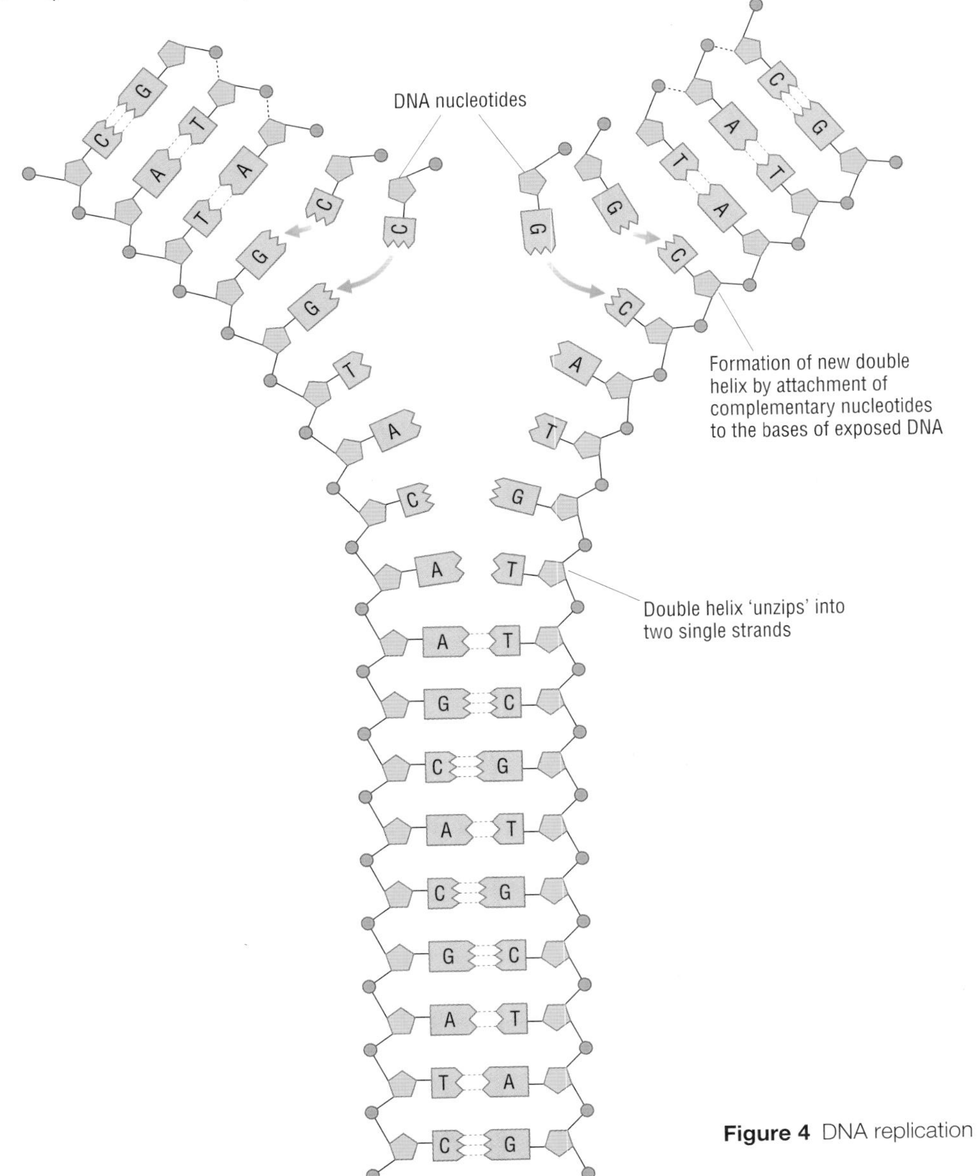

Figure 4 DNA replication

Key definition

DNA is a stable **polynucleotide** molecule. It acts as an information store because the bases projecting from the backbone act as a coded sequence. Organisms differ in their DNA only because they contain different sequences of bases in the DNA.

Examiner tip

Learn the base-pairing rules. A–T, C–G. It is common to ask students to write out the complementary sequence to a given DNA sequence.

Structure to function in DNA

- The sequence of bases is an example of information storage. The information is in the form of codes to build proteins.
- The molecules are long, so a large amount of information can be stored.
- The base-pairing rules mean that complementary strands of information can be replicated.
- The double helix structure gives the molecule stability.
- Hydrogen bonds allow easy unzipping for copying and reading information.

Questions

1. Define the term 'antiparallel'.
2. A DNA strand has the base sequence ATTAGGCTAT. Write down the complementary strand sequence.
3. A DNA molecule is 20% thymine (T). What percentage of each of the other types of base would it contain?
4. What type of diseases can result from DNA copying going wrong?

2.1 (17) Reading the instructions

By the end of this spread, you should be able to . . .

* **State that a gene is a sequence of DNA nucleotides that codes for a polypeptide.**
* **Outline the roles of DNA and RNA in the cells of living organisms.**

RNA is different

RNA is structurally different from DNA in a number of important ways:

- the sugar molecule that makes up the **nucleotides** is **ribose**
- the nitrogenous base **uracil** (U) is found instead of the organic base **thymine** (T)
- the **polynucleotide** chain is usually single-stranded
- three forms of RNA molecule exist.

The base-pairing rules apply

RNA nucleotides never contain the nitrogenous base thymine. This means that RNA is made up of nucleotides containing the purines **adenine** (A) and **guanine** (G), and the pyrimidines **cytosine** (C) and uracil (U).

As in DNA, cytosine can form hydrogen bonds with guanine. Uracil is very similar to thymine and can form hydrogen bonds with adenine. So base pairing of A–U, C–G can take place.

The base-pairing rules mean that molecules of RNA can be made that are **complementary** to molecules of DNA. This is because exposed DNA nucleotides (see spread 2.1.15) can have free RNA nucleotides **hydrogen-bonded** to them and then the sugar–phosphate backbone is sealed up to form a chain of RNA nucleotides. This is the basis of copying the genetic code of the DNA base sequence (**transcription**).

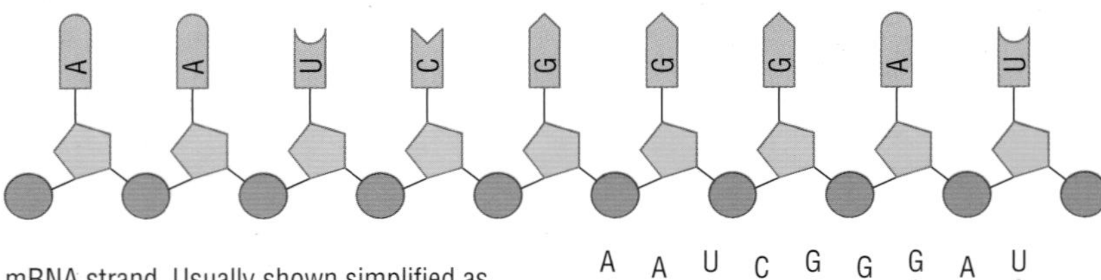

mRNA strand. Usually shown simplified as

A A U C G G G A U

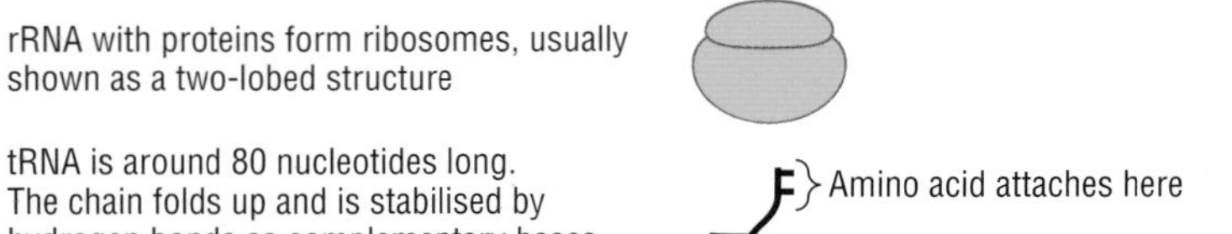

rRNA with proteins form ribosomes, usually shown as a two-lobed structure

tRNA is around 80 nucleotides long. The chain folds up and is stabilised by hydrogen bonds as complementary bases come near each other. The structure is usually shown simplified as a hairpin.

Amino acid attaches here

Figure 1 RNA types and functions

Three forms of RNA

RNA molecules exist in three forms:

- **messenger RNA** (**mRNA**) is made as a strand complementary to one strand of a DNA molecule (the template strand) – it is therefore a copy of the other DNA strand (the coding strand) of the double helix
- **ribosomal RNA** (**rRNA**) is found in **ribosomes**
- **transfer RNA** (**tRNA**) carries amino acids to the ribosomes, where they are bonded together to form **polypeptides**.

What are the instructions for?

- The sequences of bases on DNA make up codes for particular protein molecules. They code for the sequence of amino acids in the protein.
- The sequence coding for a particular protein (a **gene**) can be exposed by splitting the hydrogen bonds that hold the double helix together in that region.
- RNA nucleotides form a complementary strand (mRNA). This mRNA molecule is a copy of the DNA coding strand or gene.
- The mRNA peels away from the DNA and leaves the nucleus through a nuclear pore.
- The mRNA attaches to a ribosome.
- Then tRNA molecules bring amino acids to the ribosome in the correct order, according to the base sequence on the mRNA.
- The amino acids are joined together by **peptide bonds** to give a protein with a specific **primary structure** (this primary structure then gives rise to the **secondary** and **tertiary structures**).

Key definition

A **gene** is a length of DNA (part of a DNA molecule) that codes for one (or more) polypeptides. Each gene occupies a specific place (locus) on a chromosome. Different versions of the same gene are called alleles.

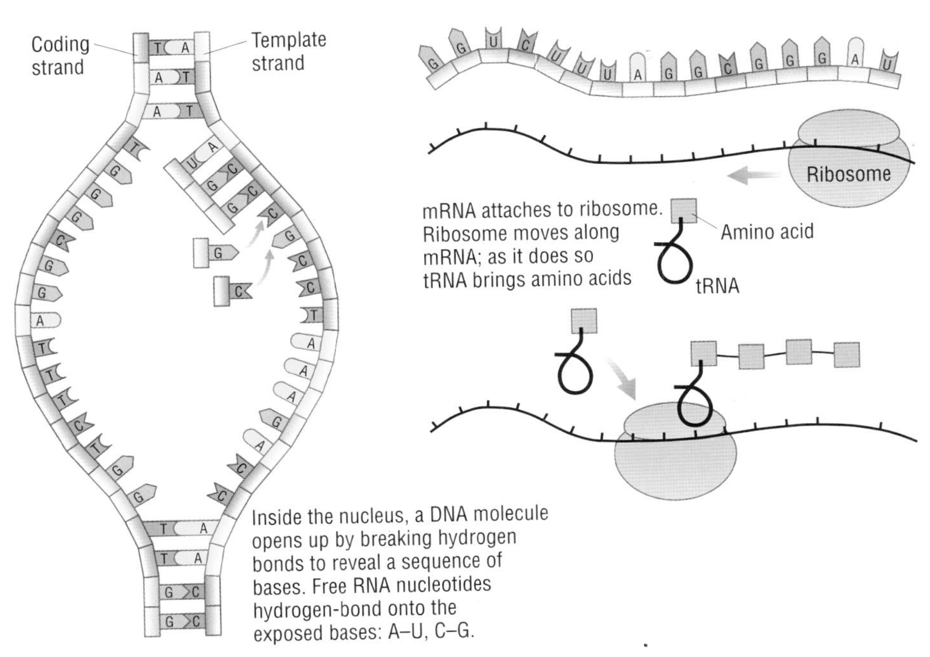

Figure 2 Overview of protein synthesis

An idea whose time has come

In 1953, in Cambridge, James Watson and Francis Crick discovered the structure of DNA and worked out how it might form the instructions for life. But their work on DNA, and the key moment of discovery, owed much to work being done at the same time by Rosalind Franklin, Raymond Gosling and Maurice Wilkins at King's College, London. The X-ray patterns (diffraction patterns) they produced using DNA crystals enabled Watson and Crick finally to piece together a full model of the structure of DNA that also explained how copies might be made from it (DNA replication).

Figure 3 Crick, Watson, Wilkins and Franklin

Questions

1 Explain why the mRNA strand produced in the nucleus is complementary to the template strand, and a copy of the coding strand.
2 If a DNA template strand code reads ATTCGCGTTAAT, what would the complementary mRNA strand read? What is the code on the DNA coding strand?
3 Suggest why mRNA is less stable than DNA, and explain why this is a necessary feature of mRNA.
4 Make a table to compare and contrast the structure of DNA with that of RNA.

2.1 (18) Enzymes are globular proteins

By the end of this spread, you should be able to . . .

* **State that enzymes are globular proteins with a specific tertiary structure.**
* **State that enzymes catalyse metabolic reactions in living organisms.**

All enzymes are proteins

A **globular protein** has a specific three-dimensional shape or **tertiary structure**. The shape comes from the protein's **primary structure** (the specific sequence of **amino acids** that form the protein) and its **secondary structure**. In globular proteins, the final 3D shape usually has **hydrophobic** amino acid R-groups in the centre of the 'ball', and **hydrophilic** amino acid R-groups around the outside of the ball. The amino acid chain spirals, pleats and turns to form the overall structure.

All enzymes are very similar in many ways:

- they are globular proteins – generally soluble in water
- they act as **catalysts** – speeding up chemical reactions, but not being 'used up' as part of the reaction
- they are specific – catalysing a reaction involving only one type of **substrate**
- the globular structure contains a 'pocket' or cleft area called an **active site**
- their activity is affected by temperature and pH.

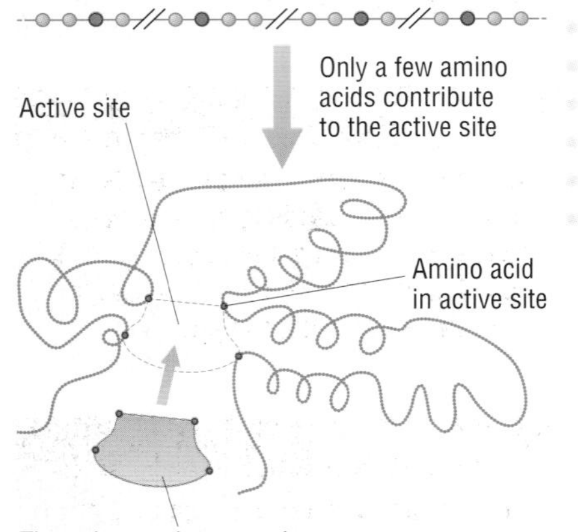

Figure 1 Primary structure to tertiary structure in enzymes and the identification of amino acids as part of the active site

The active site is a tiny part of an enzyme

Enzymes are relatively large molecules, consisting of hundreds of amino acids. The vast majority of the amino acids are involved in maintaining the specific tertiary structure of the enzyme. Enzyme function, like the function of all proteins, is related to shape. For any enzyme to work properly, the tertiary structure must be maintained in a very specific way.

Essentially, the whole primary, secondary and tertiary structure of enzymes is involved in achieving a very specific shape for the active site. The active site is the area of the enzyme where the catalytic activity of the enzyme occurs.

Each enzyme has a very specific, individual active site shape, maintained by a very specific overall tertiary structure. Very few (often fewer than 10) amino acids form the actual active site. As the active site shape of an enzyme is very specific and individual, this means the reaction an enzyme can catalyse is also very specific and individual.

Catalysts in nature

Enzymes are often referred to as biological catalysts. Like all catalysts, they speed up chemical reactions. Industrial and commercially used catalysts include some metals and other chemicals, but in comparison with enzymes they are often rather slow. This has led to a whole branch of biotechnology to investigate and use enzymes in industrial processes to replace such inorganic catalysts. The added advantage of enzymes is that, unlike inorganic catalysts, they are specific to one catalytic reaction and do not produce a range of unwanted by-products.

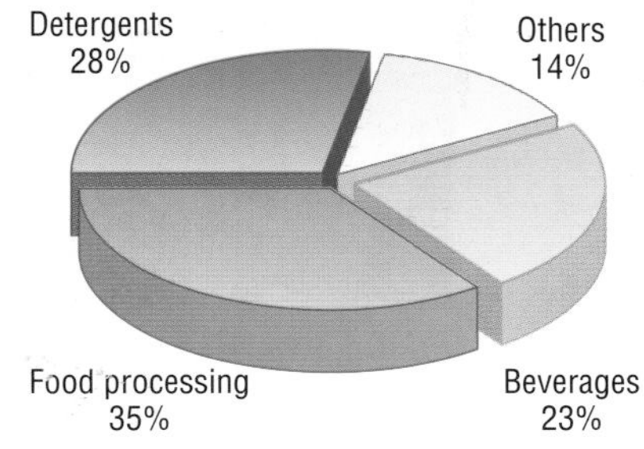

Figure 2 Commercial applications of enzymes

How many enzymes?

Very many chemical reactions occurring within (and sometimes outside) a cell are catalysed by enzymes. Metabolism can be described as enzyme-driven.

In looking at biological molecules, for example, we have discussed the formation and breakage of **glycosidic bonds**, **ester bonds** and **peptide bonds** (spreads 2.1.3, 2.1.10 and 2.1.6). Each of these processes requires at least one specific enzyme in order to catalyse the reaction. The processes of protein synthesis, digestion, **respiration** and **photosynthesis** each require a number of different enzymes. Each catalyses a specific reaction in the sequence of events that make up the process.

It is estimated that an individual cell may contain over 1000 different enzymes.

Substrate and product

In a chemical reaction that is catalysed by an enzyme, substrate is turned into product. For example, if we use an enzyme, in this case maltase, to catalyse the conversion of maltose to glucose, then the substrate maltose is converted to the product glucose.

$$\text{maltose} \xrightarrow{\text{maltase}} \text{glucose} + \text{glucose}$$

The name given to an enzyme is usually derived from the substrate of the reaction that is catalysed, with the suffix '-ase'.

Enzyme	Reaction catalysed
Lactase	The breakdown of milk sugar, lactose, into glucose and galactose monomers. Lactose-intolerant individuals do not produce lactase and suffer stomach cramps, bloating and diarrhoea if they take in milk products.
Catalase	The breakdown of hydrogen peroxide to water and oxygen gas. Almost all organisms produce catalase because hydrogen peroxide is a toxic by-product of some metabolic reactions.
Ribulose bisphosphate carboxylase (rubisco)	Plants need carbon dioxide for photosynthesis. The enzyme rubisco catalyses the binding of carbon dioxide to a molecule called ribulose bisphosphate.
ATP-ase	The breakdown of ATP to produce ADP and a phosphate group. This reaction releases a small amount of energy that is used to drive energy requiring processes such as active transport.
Glycogen synthetase	The building up of glycogen by catalysing the joining together of glucose molecules. Glycogen is the storage carbohydrate of animals.

Table 1 Some common enzymes and the reactions they catalyse

Questions

1 Explain why enzymes are so specific in the reactions that they catalyse.
2 Suggest why all enzymes are protein molecules.
3 The name given to enzymes that catalyse the breakage of peptide bonds is 'proteases'. What name would be given to enzymes that catalyse the breakage of:
   (a) glycosidic bonds
   (b) ester bonds?

Examiner tip

Don't confuse the enzyme *maltase* with its substrate *maltose*.

Key definition

A **catalyst** is defined as a molecule (or element) that speeds up a chemical reaction but does not get used up in the reaction. At the end of the reaction, the catalyst remains unchanged.

Examiner tip

Any feature of an enzyme-controlled reaction can affect the rate of the reaction only if it affects either the number of collisions that can take place between enzyme and substrate *or* the shape of the enzyme's active site. Answer exam questions on enzyme-controlled reactions in relation to these two features.

2.1  Inside and out – where enzymes work best

By the end of this spread, you should be able to . . .

* **State that enzyme action may be extracellular or intracellular.**
* **State that enzymes catalyse metabolic reactions in living organisms.**

Figure 1 Organisms have developed enzymes to function at extremes: top, red algae growing on ice; middle, bacteria in thermal pools; bottom, flamingoes feeding on algae growing at pH 10 in Lake Nakuru, Kenya

Organisms vary considerably

There is a huge range of organisms on the planet, with a variety of forms, from single-celled organisms to huge multicellular organisms. All of the organisms on Earth make use of **enzymes** to **catalyse** metabolic reactions.

As all enzymes are **protein** molecules, they have the same weaknesses as all protein molecules. For an enzyme to work, its shape must remain intact. Organisms are able to live in a wide variety of environments only because they are adapted to deal with different environments because their enzymes continue to function.

Life in extremes

A few organisms can survive and grow in some of the most extreme environments on Earth (Figure 1).

Enzymes and endotherms

Endothermic animals are able to maintain their internal body temperature independently of the environment. This ability has allowed endothermic animals (birds and mammals) to live in most of the different environments in the world. Regulating body temperature means that enzymes can function at near-optimum temperature inside the organism. This ability comes at a very high energy cost to the organism. Birds and mammals require a much greater food supply than similarly sized reptiles. But the advantages of having enzyme activity at a continuous and optimum level have allowed endothermic animals to survive very well, both on land and in water.

Nutrition and digestion – enzyme locations

Organisms that obtain their nutrients by consuming other organisms are known as **heterotrophs**. These organisms need to break down the body of the organisms they are consuming in order to extract the nutrient molecules they need for their own growth and energy requirements.

Digestion involves breaking down larger molecules into their subunits. This requires the breaking of bonds including **glycosidic**, **peptide** and **ester bonds**, and is catalysed by different types of enzymes.

Some organisms secrete (or release) enzymes outside themselves onto the food source. The enzymes digest the molecules into their **monomers**, which the organism can then take in and use.

Other organisms have an internal digestive system. As food taken into the organism passes through the digestive system, various enzymes are mixed with it in order to digest the nutrients it contains. Many enzymes in digestive systems are also **extracellular** in that they are released from the cells that make them, onto food within the digestive system spaces.

Many enzymes, including some involved in parts of the digestive process, are found in the cytoplasm of cells or attached to cell membranes. These enzymes are described as **intracellular** because their action takes place inside cells.

Key definition

Extracellular enzymes catalyse reactions *outside* the cell.

Intracellular enzymes catalyse reactions *inside* the cell.

Examiner tip

Read the information given in each question carefully. You need to be able to explain the action of enzymes in relation to the question.

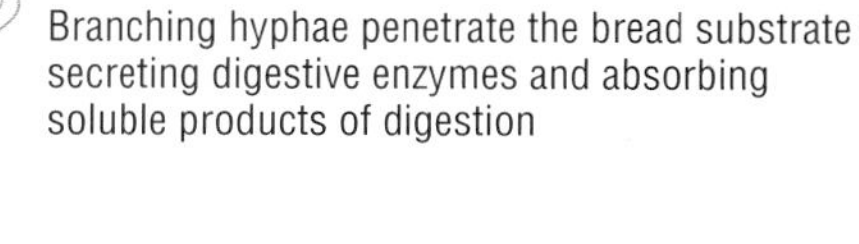

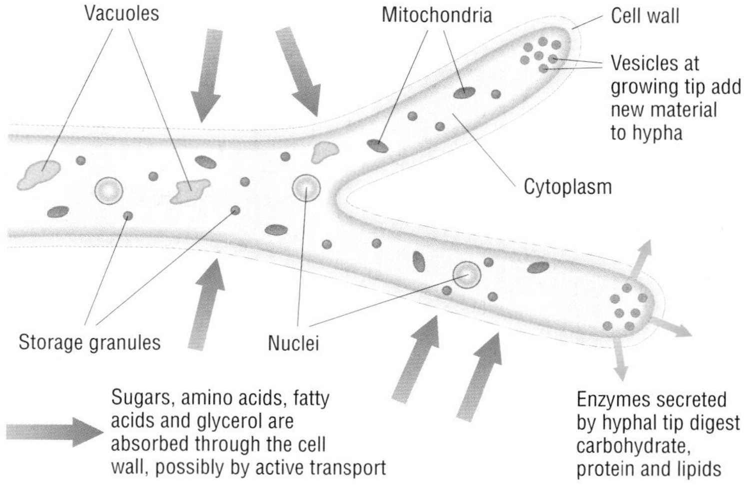

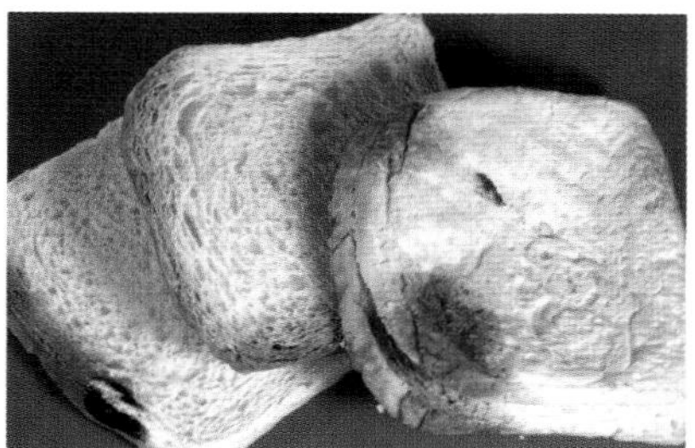

Figure 2 Mould produces extracellular enzymes to digest bread

Enzymes and protection

Enzymes that can catalyse reactions to break down molecules are useful tools in protecting an organism. Many organisms make use of enzymes as a defence mechanism.

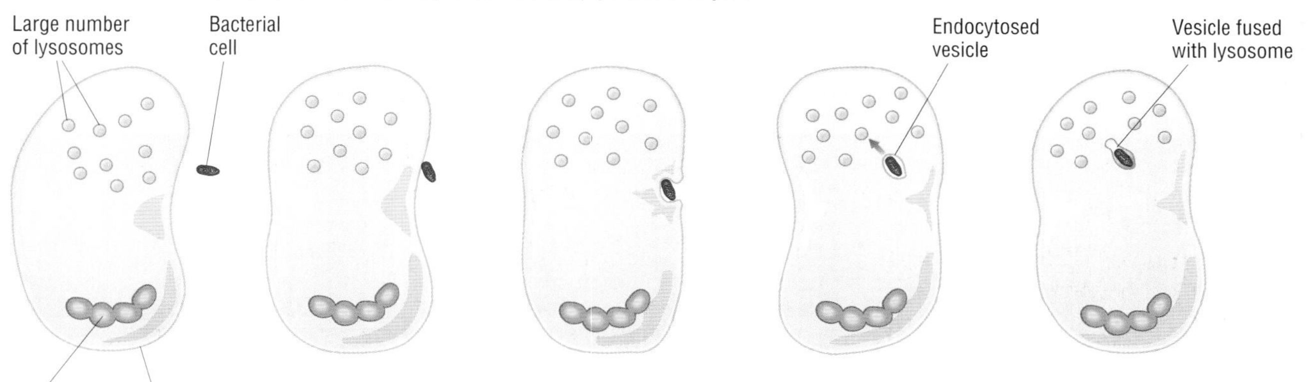

Figure 3 Phagocytosis

Questions

1 Suggest the advantages of having an internal digestive system compared with secreting enzymes outside the organism.
2 Explain why many white blood cells involved in phagocytosis contain a high concentration of enzymes.
3 Suggest why all organisms, no matter what sort of environment they live in, have enzymes in their cells.

2.1 (20) Enzyme action

By the end of this spread, you should be able to . . .

* **Describe the mechanism of enzyme action with reference to specificity, active site, lock-and-key and induced-fit hypotheses.**
* **Explain what is meant by enzyme–substrate complex and enzyme–product complex.**
* **Describe how enzymes lower the activation energy of a reaction.**

Key definition

Activation energy is the amount of energy that must be applied for a reaction to proceed. Different reactions require different levels of activation energy. Enzymes reduce the amount of activation energy needed to allow a reaction to proceed.

Falling apart and staying together

Covalently bonded molecules (like the biological molecules discussed in spread 2.1.7) do not just assemble or break up – they are far too stable.

Let us consider a **maltose** molecule. Maltose is made up of two glucose molecules joined by a **glycosidic bond**. In order to split maltose (in a **hydrolysis** reaction) into two glucose molecules, the glycosidic bond needs to be broken. At the same time, a water molecule must split. The parts of the split water molecule must then bond back onto the split parts of the maltose molecule to re-form glucose molecules.

In order to make this reaction happen in a test tube, maltose can be boiled in acid. This provides the right conditions for maltose molecules to collide with water molecules 'energetically' enough to achieve hydrolysis. The extra energy that is needed to enable the reaction to take place is called the **activation energy**.

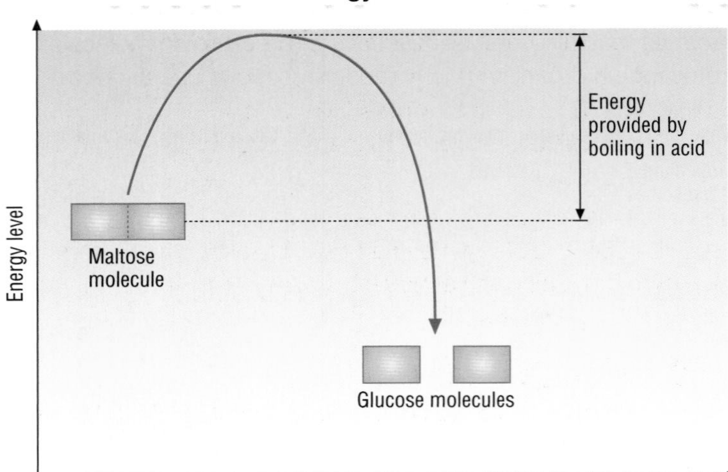

The activation energy required to carry out the conversion of maltose to glucose

Figure 1
Activation energy (maltose to glucose)

From one stable molecule to another

We can describe biologically important molecules as being too stable to break up or self-assemble under conditions found in the cell. Boiling in acid gives molecules a great deal of extra energy. The energy is now enough to destabilise the structure of the molecule, so that when it collides with another molecule the reaction proceeds.

Clearly, the cell is unlikely to survive boiling in acid in order to drive reactions, so it needs **catalysts** (enzymes) to drive its metabolic reactions. Without catalysts, metabolic reactions could not occur at anything like the speed needed to maintain life.

Getting over the hump – reducing activation energy

Enzymes work by reducing the amount of activation energy required. This means that reactions can proceed quickly at temperatures much lower than boiling point. They can do this because of the way the **active site** is shaped to fit the **substrate** molecule (or molecules).

Optimum pH varies between enzymes

All enzymes have their own **optimum pH**. This is the pH at which the rate of reaction is highest. For many enzymes, but by no means for all of them, the optimum pH is around neutral (pH 7).

At the optimum pH, the concentration of hydrogen ions in solution gives the tertiary structure of the enzyme the best overall shape. This shape holds the active site in the shape that best fits the substrate (the shape that is **complementary** to the shape of the substrate).

Enzymes usually work in a fairly narrow pH range. Changes to pH, altering the concentration of hydrogen ions even very slightly, result in a fall in the reaction rate because the shape of the enzyme molecule is disrupted and so the shape of the active site is changed.

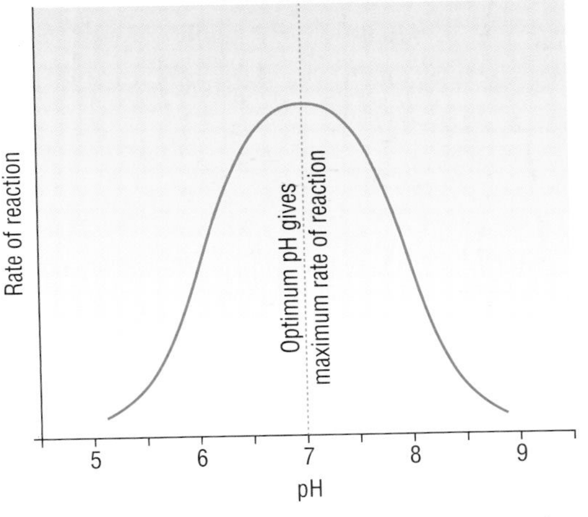

Figure 4 Optimum pH graph versus rate of reaction

Reducing or increasing the pH away from the optimum pH reduces the rate of reaction because the concentration of H^+ in solution affects the tertiary structure of the enzyme molecule.

Extremes of pH can lead to denaturation

Measuring the effect of pH on enzyme action usually involves carrying out enzyme-controlled reactions at different pH values using **buffer** solutions. The production of a product, or the disappearance of a substrate, can be measured in a variety of ways (see spreads 2.1.27 and 2.1.28).

Key definition

Optimum pH is the pH value at which the rate of an enzyme-controlled reaction is at its maximum. Each enzyme has a specific optimum pH.

Examiner tip

Minor changes in pH do not denature enzymes. The bonds disrupted by the pH change can re-form if the pH returns to optimum. Only refer to denaturation at *extreme* changes of pH away from the optimum.

pH and location

The wide range of environments in which enzymes have to work includes a range of pH levels. You can see a clear example of enzymes operating in different pH conditions in the human digestive system. In the stomach, the protein-digesting enzyme pepsin works well. Pepsin has an optimum pH of around pH 2. This is ideal for working in the stomach, which contains hydrochloric acid, giving it a pH of around 2.

Partly digested food leaving the stomach moves into the small intestine. Intestinal secretions neutralise the food, so it can be digested further. The protein-digesting enzyme trypsin works in the small intestine. It has an optimum pH of around 7, ideal for the environment of the small intestine.

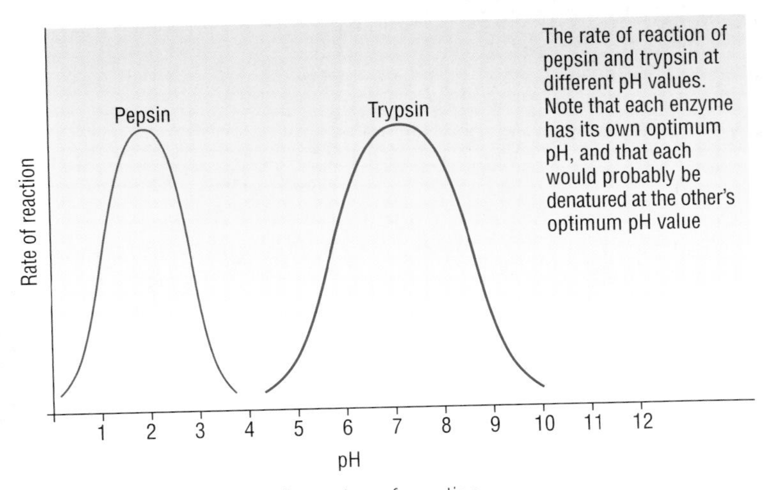

Figure 5 Pepsin and trypsin – rates of reaction

Questions

1. Explain why both increasing *and* decreasing pH away from the optimum results in a reduction in the reaction rate.
2. Individual hydrogen bonds are fairly weak. Explain how such weak bonds can be responsible for holding the tertiary structure of an enzyme molecule in place.
3. Explain why the optimum pH for pepsin is different from the optimum pH for trypsin, even though both enzymes are proteases.
4. Why do enzymes usually work only within very narrow pH ranges?

2.1 Enzymes at work – concentration effects

By the end of this spread, you should be able to . . .

* **Describe and explain the effects of enzyme concentration and substrate concentration on enzyme activity.**

Increasing the substrate concentration

In an experimental situation, the concentration of **substrate** can be varied for a fixed concentration of **enzyme** molecules. If there is no substrate, the reaction cannot proceed because **enzyme–substrate complexes** cannot form.

As the concentration of substrate increases, collisions between enzyme and substrate molecules occur more often. More enzyme–substrate complexes form, so more product is formed. The reaction rate increases.

If the concentration of substrate increases further, a point will be reached where the reaction rate reaches a maximum value.

At this point, all the enzyme molecules present are forming enzyme–substrate complexes as fast as possible. In effect, all the **active sites** are occupied at all times, so any further increase in substrate concentration will have no effect on the reaction rate.

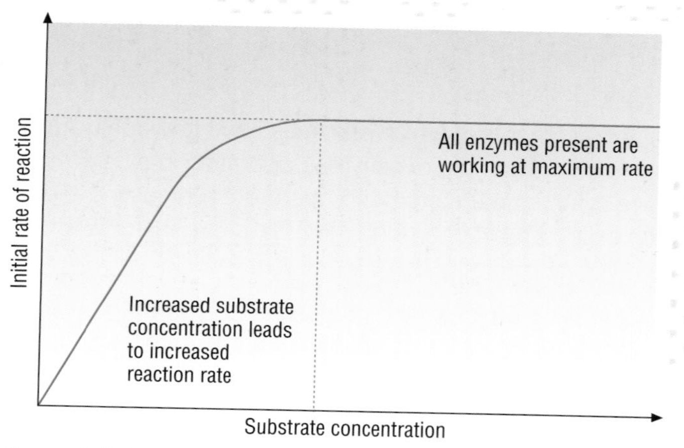

Figure 1 Effect of substrate concentration on the rate of an enzyme-controlled reaction

Increasing the enzyme concentration

Reversing the situation above, so that the concentration of enzyme is varied for a fixed concentration of substrate molecules, follows a similar pattern.

- As the enzyme concentration increases, more active sites become available.
- More enzyme–substrate complexes (and so more products) form, so the reaction rate increases.
- If the enzyme concentration increases further, a point will be reached where all substrate molecules are occupying enzyme active sites.
- The reaction rate is the maximum possible for the fixed substrate concentration.
- Repeating the experiment with even more enzyme cannot give a higher reaction rate.

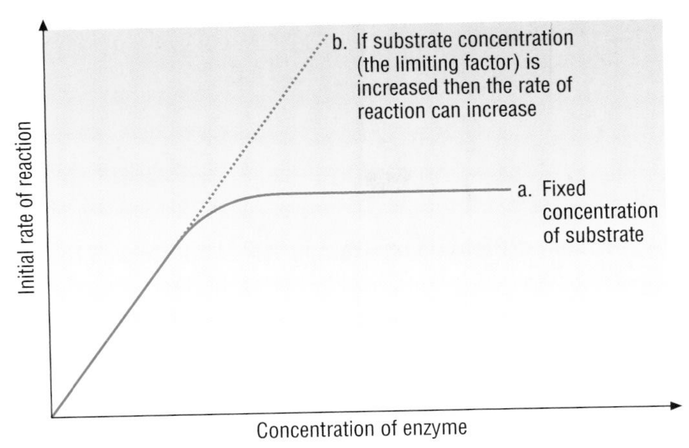

Figure 2 Effect of enzyme concentration on the rate of an enzyme-controlled reaction

Initial reaction rate

- In an experimental situation where enzyme and substrate are mixed together, the rate of the enzyme-controlled reaction will be highest at the point when enzyme and substrate are mixed.
- As the reaction proceeds, product molecules are formed and increase in number.
- At the same time, substrate molecules are used up and decrease in number.
- This means the frequency of collisions between enzyme and substrate (and so the reaction rate) goes down as an enzyme-controlled reaction proceeds.
- The highest reaction rate, known as the initial reaction rate, gives the maximum possible reaction rate for an enzyme under a particular experimental situation.

Limiting factors

In Figures 1 and 2, the levelling off (known as a plateau) occurs because either enzyme concentration (Figure 1) or substrate concentration (Figure 2) prevents any further increase in the reaction rate. They are limiting the reaction, and are described as **limiting factors**. If the concentration of the limiting factor is increased, the reaction rate increases.

Key definition

A factor is described as a **limiting factor** in a situation where, if all other conditions are kept constant, increasing the concentration of that factor alone will increase the reaction rate.

Enzyme and substrate concentrations in cells

Enzyme concentrations in cells are usually maintained at a relatively low level. This is partly because enzymes, being **catalysts**, work over and over again, driving the same reaction. The control of **metabolism**, as we shall see later (spread 2.1.29), is based on the control of enzyme activity. One way of regulating enzyme activity is by adjusting the concentrations of enzymes and/or substrates.

Examiner tip

When describing a graph showing enzyme activity, you should use the information on the axes and in the question to add detail to your answer. *Describe* what is happening in each region of the graph – if you are asked to *explain*, say *why* these things are happening.

Questions

1. Sketch a graph to show the effect of increasing the substrate concentration for a fixed enzyme concentration (as in Figure 1). Add to the sketch to show how the reaction rate would change if the enzyme concentration was increased.
2. Sketch a graph to show reaction rate against enzyme concentration where substrate concentration is continually increased so that the substrate never becomes limiting.
3. Suggest why enzymes are usually maintained at low concentrations in cells.

2.1  Enzymes at work – inhibitors of action

By the end of this spread, you should be able to . . .

* **Explain the effects of competitive and non-competitive inhibitors on the rate of enzyme-controlled reactions, with reference to both reversible and non-reversible inhibitors.**

Key definition

An enzyme **inhibitor** is any substance or molecule that slows down the rate of an enzyme-controlled reaction by affecting the enzyme molecule in some way.

Inhibitors of enzyme-controlled reactions

Inhibitors are defined as substances that reduce the reaction rate in an **enzyme**-controlled reaction because they have some effect on the enzyme molecule. A wide range of enzyme inhibitors exists. Some have their effect on the **active site** of just one enzyme. Others affect another part of the molecule, indirectly causing a change in the shape of the active site. Some inhibitors can affect a number of different enzymes.

Competitive inhibitors

Competitive inhibitor molecules have a similar shape to that of the **substrate** molecules. This means they can occupy the active site, forming enzyme–inhibitor complexes. These complexes do not lead to the formation of products because the inhibitor is not identical to the substrate. The enzyme does not catalyse a reaction.

Enzyme inhibition occurs because whenever an inhibitor molecule is occupying an enzyme's active site, a substrate molecule cannot enter. So the number of enzyme–substrate complexes is reduced and the reaction rate slows down.

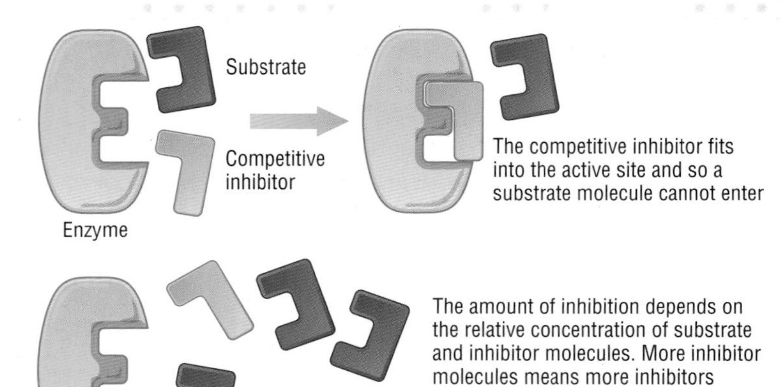

Figure 1 How a competitive inhibitor works

The level of inhibition depends on the concentrations of inhibitor and substrate. Where the numbers of substrate molecules are increased, the level of inhibition decreases because a substrate molecule is more likely than an inhibitor molecule to collide with an active site.

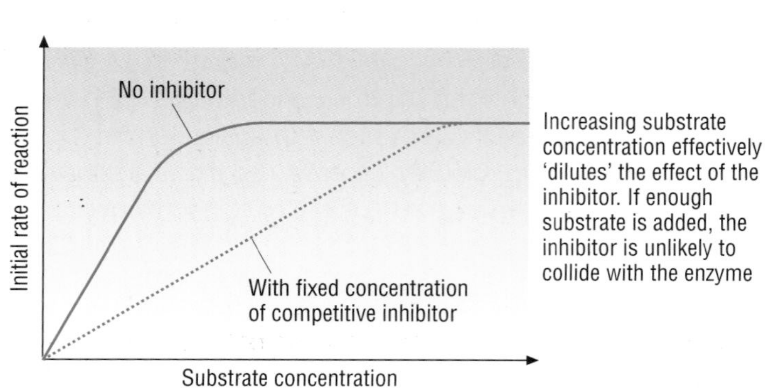

Figure 2 Effect of concentrations of inhibitor and substrate on the rate of an enzyme-controlled reaction

Non-competitive inhibitors

Non-competitive inhibitors do not compete with substrate molecules for a place in the active site. Instead, they attach to the enzyme molecule in a region away from the active site. The attachment of a non-competitive inhibitor molecule to an enzyme molecule distorts the **tertiary structure** (3D shape) of the enzyme molecule. This leads to a change in shape of the active site. This means the substrate no longer fits into the active site, so enzyme–substrate complexes cannot form and the reaction rate decreases.

The level of inhibition depends on the number of inhibitor molecules present. If there are enough inhibitor molecules to bind to all the enzyme molecules present, then the enzyme-controlled reaction will stop. Changing the substrate concentration will have no effect on this form of inhibition.

Examiner tip

Don't assume that non-competitive inhibition is irreversible. Many reversible non-competitive inhibitors are vital to controlling metabolic functions by regulating enzyme activity in cells.

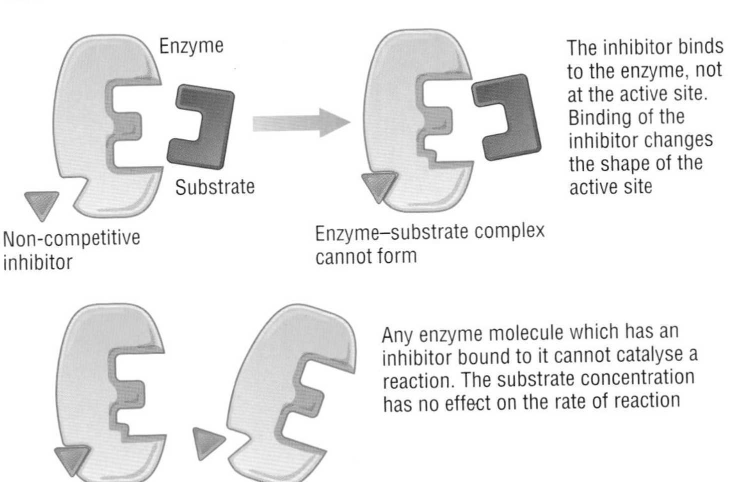

Figure 3 How non-competitive enzyme inhibitors work

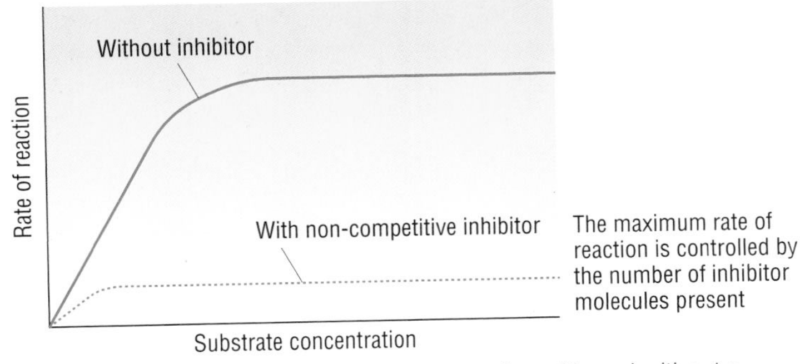

Figure 4 Rate of an enzyme-controlled reaction with and without a non-competitive inhibitor

Permanent inhibitors

Most competitive inhibitors do not bind permanently to the active site. They bind to the active site for a short period and then leave. Their action is described as reversible, as removal of the inhibitor from the reacting mixture leaves the enzyme molecules unaffected.

Many non-competitive inhibitors bind permanently to enzyme molecules. The inhibition is not reversible (it is irreversible), and any enzyme molecules bound by inhibitor molecules are effectively **denatured**.

Inhibition is not always a bad thing

The regulation of a number of metabolic pathways involves the inhibition of enzymes to control the reaction rates (see spread 2.1.29).

Questions

1. Suggest a method of determining whether the inhibition of an enzyme-controlled reaction is competitive or non-competitive.
2. Competitive inhibitor molecules can be much larger than the substrate molecule they 'compete' with. Suggest how this is possible.
3. Explain why inhibition of enzyme activity is important in controlling metabolic processes.

2.1 

Enzymes at work – coenzymes and prosthetic groups

By the end of this spread, you should be able to . . .

* **Explain the importance of cofactors and coenzymes in enzyme-controlled reactions.**

Key definition

A **cofactor** is any substance that must be present to ensure enzyme-controlled reactions take place at the appropriate rate. Some cofactors are a part of the enzyme (prosthetic groups); others affect the enzyme on a temporary basis (coenzymes and inorganic ion cofactors).

Enzymes shaping up

Some **enzymes** can only catalyse a reaction if another **non-protein** substance is present. There are a number of different substances that help to ensure that enzyme-controlled reactions can take place at an appropriate rate. These substances are known as **cofactors**.

Coenzymes

These are small, organic, non-protein molecules that bind for a short period to the active site. They may bind either just before, or at the same time as, the **substrate** binds. In many reactions, **coenzymes** take part in the reaction and, like substrate, are changed in some way. Unlike substrates, coenzymes are recycled back to take part in the reaction again. The role of coenzymes is often to carry chemical groups between enzymes so they link together enzyme-controlled reactions that need to take place in sequence.

Vitamin B_3 (nicotinamide) plays an important role in helping the body break down carbohydrates and fat to release energy. This vitamin is used in the body to make a coenzyme that is required for the enzyme pyruvate dehydrogenase to function properly. Pyruvate dehydrogenase catalyses one of the reactions in the sequence involved in **respiration**. Normal growth and development cannot proceed without vitamin B_3, and a disease known as pellagra results if vitamin B_3 is absent from the diet.

Prosthetic groups

A coenzyme that is a permanent part of an enzyme molecule is called a **prosthetic group**. Prosthetic groups are also found in other protein molecules. Prosthetic groups are vital to the function of the enzyme (or protein) molecule. They contribute to the final 3D shape, and to other properties of the molecule, including its charges.

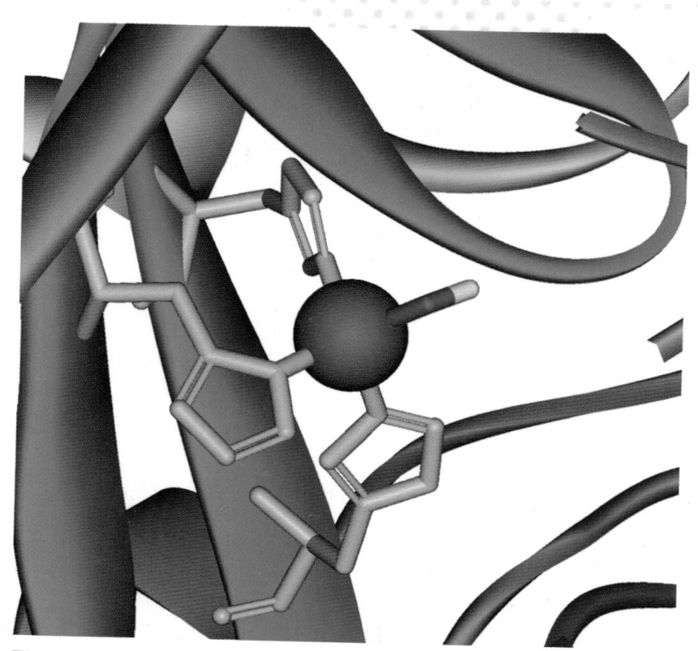

Figure 1 Zinc prosthetic group in the enzyme carbonic anhydrase

The enzyme carbonic anhydrase contains a zinc-based prosthetic group. This enzyme is a vital component in red blood cells, where it is involved in catalysing the combining of carbon dioxide and water to produce carbonic acid. This is an important reaction that enables carbon dioxide to be transported in the blood.

Inorganic ion cofactors

In some enzyme-controlled reactions, the presence of certain ions can increase the reaction rate. Ions may combine with either the enzyme or the substrate. The binding of the ion makes the **enzyme–substrate complex** form more easily, because it affects the charge distribution and, in some cases, the shape of the enzyme–substrate complex.

The enzyme **amylase** catalyses the breakdown of starch to **maltose** molecules. This enzyme will function properly only if chloride ions are present. Amylase (originally called diastase) was the first enzyme to be identified and isolated, by the scientist Anselme Payen in 1833.

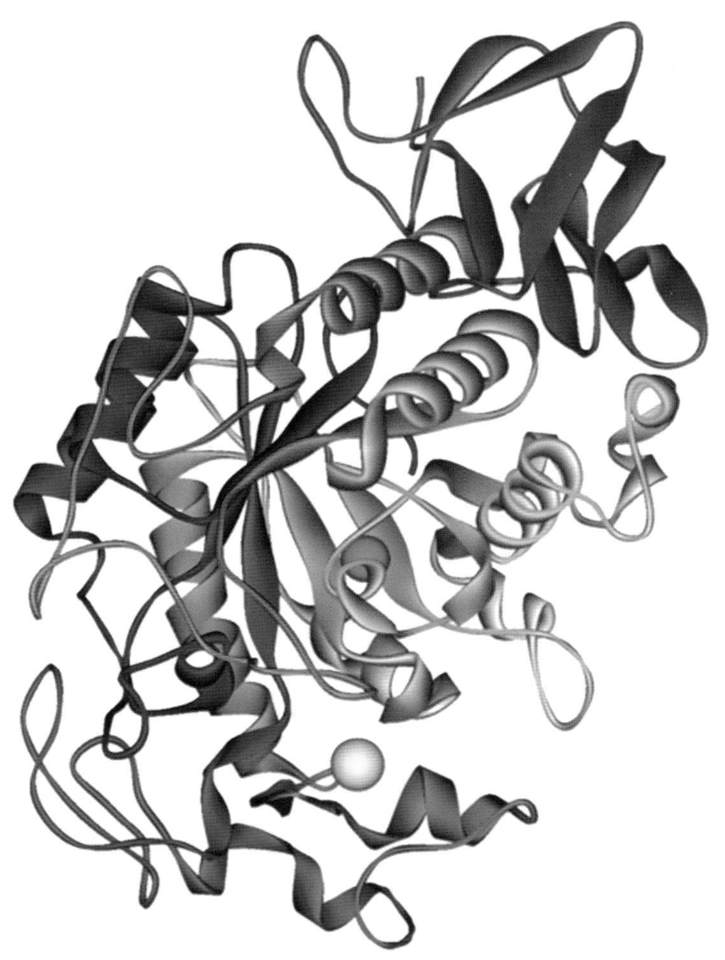

Figure 2 Amylase does not function unless chloride ions (dark green) are present

Examiner tip

Examiners will often use tables of data to show the reaction rate with and without other substances. Be very careful *not* to assume that only the substrate and enzyme are needed to ensure an enzyme-controlled reaction takes place.

Questions

1 The recommended daily dietary allowance for nicotinamide is 18 mg. Suggest why the recommended daily amount is very low.
2 Name the prosthetic group found in the haemoglobin molecule.
3 Prosthetic groups are bound to enzymes as the enzyme is being made, as part of the processing of the molecule to make it active. Suggest where in the cell the addition of prosthetic groups to the enzyme molecules takes place.

Interfering with enzymes – poisons and drugs

By the end of this spread, you should be able to . . .

* **State that metabolic poisons may be enzyme inhibitors, and describe the action of one named poison.**
* **State that some medicinal drugs work by inhibiting the activity of enzymes.**

Deadly poisons

Many poisonous substances have their effects because they inhibit or even overactivate **enzymes**. Potassium cyanide, for example, inhibits cell **respiration**. It does this because it is a **non-competitive inhibitor** for a vital respiratory enzyme called cytochrome oxidase, found in mitochondria. Inhibition of this enzyme decreases the use of oxygen, so **ATP** cannot be made. The organism can only respire anaerobically, which leads to a build up of **lactic acid** in the blood.

Only 100–200 mg of cyanide must be absorbed in order to make an adult lose consciousness. This can occur in as little as 10 seconds. If untreated, the body goes into a coma in around 45 minutes, and death results after around 2 hours.

Key definition

A **biosensor** uses enzyme-controlled reactions to detect the presence of substances in a highly sensitive and specific way. If the substance is present, the enzyme-controlled reaction takes place. The biosensor has a mechanism for revealing whether product is made.

Examiner tip

The same enzyme inhibitor may act as *both* a poison *and* a drug, depending on the amount of inhibitor and its location. The question stem will guide you in what to look for. For example, excess alcohol taken over a period of time is very damaging to the liver – but alcohol is also a treatment for ethylene glycol poisoning.

Enzymes and medicines

Infection by viruses, including HIV, is treated using chemicals that act as **protease** inhibitors. These chemicals prevent viruses from replicating by inhibiting the activity of protease, which the viruses need in order to build new virus coats. The protease inhibitors specifically inhibit the viral protease enzymes, often as **competitive inhibitors**.

Replacement enzymes and cystic fibrosis

One of the symptoms of the inherited disease cystic fibrosis is that the passage of digestive enzymes, normally secreted from the pancreas into the gut, is blocked. This means that individuals with the condition have problems digesting their food. Doctors prescribe enzymes in tablet form to individuals with cystic fibrosis in order to overcome this problem. The enzymes are packaged in an acid-resistant coat so that they are not destroyed by the acid and protein-digesting enzymes in the stomach.

Ethylene glycol poisoning

Ethylene glycol is found in the antifreeze used in car engines. Ethylene glycol is not itself poisonous but, if taken into the body, it is broken down in the liver by the enzyme alcohol dehydrogenase. The breakdown product, oxalic acid, is extremely toxic, and ethylene glycol poisoning can lead to death.

If ethylene glycol intake is suspected, the individual can be given a massive dose of ethanol (alcohol). This leads to severe – though less likely to be fatal – alcohol intoxication. The ethanol acts as a competitive inhibitor of alcohol dehydrogenase. This reduces the rate of production of oxalic acid, allowing the ethylene glycol to be excreted harmlessly.

Ethylene glycol is similar in shape to ethanol, which acts as a competitive inhibitor of the above reaction. Mixing large amounts of ethanol with ethylene glycol slows the above reaction

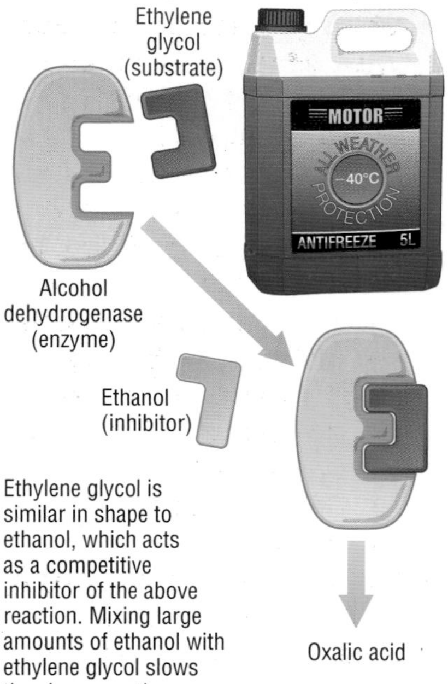

Figure 1 How ethylene glycol is metabolised, by ethanol (alcohol) dehydrogenase, to oxalic acid, which is poisonous

Antibiotics and bacterial enzymes

Antibiotics can kill or inhibit the growth of microorganisms. They are used widely in treating diseases caused by bacterial infections. Penicillin is an inhibitor of a bacterial enzyme that forms cross-links in the bacterial cell wall of some bacteria. This means the walls of growing bacteria are not made, so bacterial reproduction is halted.

Resistance to antibiotics in bacteria is increasingly a problem in treating bacterial infections. Many forms of bacterial resistance occur because, among populations of bacteria, there may be an individual with a mutation and altered enzymes. These enzymes may be capable of inactivating antibiotics. This bacterium is naturally selected when the bacterial population is exposed to antibiotics. It will survive and then reproduce. Many strains of bacteria are resistant to penicillin because they produce an enzyme (beta-lactamase) that breaks down the penicillin molecule.

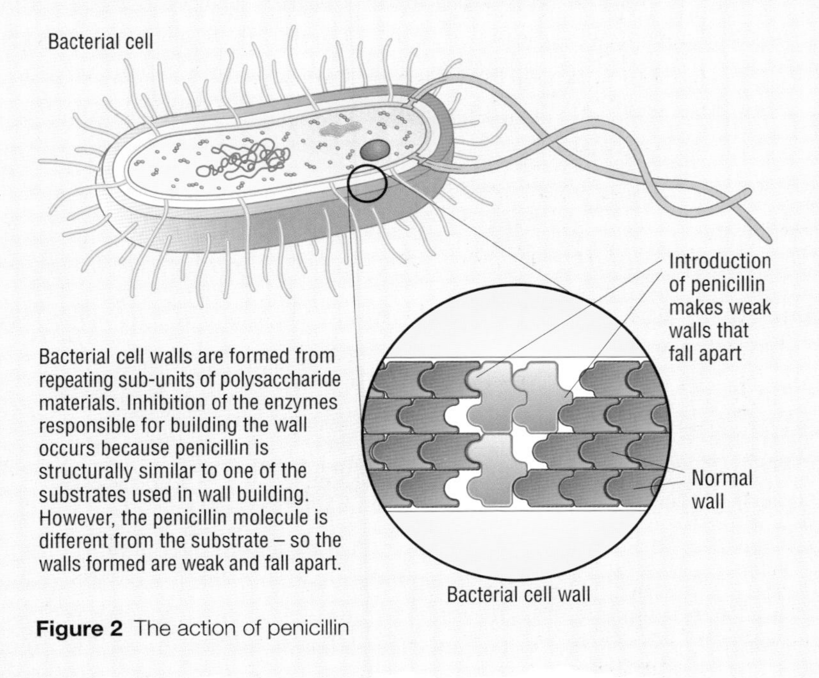

Figure 2 The action of penicillin

Snake venom – an enzyme and toxin cocktail

Snake venom is a mixture of toxins and different enzymes. Phosphodiesterases are present in most snake venoms. These interfere with the working of the prey's heart, causing a fall in blood pressure.

Snake venom also contains an inhibitor of the enzyme acetyl cholinesterase. This enzyme is involved in nerve transmission, and its inhibition results in paralysis.

The enzyme hyaluronidase is a digestive enzyme that breaks down **connective tissue** and so helps the toxins to penetrate tissues quickly. Snake venom also often contains ATP-ases, which are used for breaking down **ATP** to disrupt the prey's use of energy.

Figure 3 Snake biting

Questions

1 Why would the enzymes in tablets given to individuals with cystic fibrosis be destroyed in the stomach if they were not packaged properly?
2 Suggest why glucose does not normally appear in urine, but it does appear in individuals with diabetes mellitus.
3 Many antibiotics are chemicals naturally produced by fungal organisms and released into their environment. What is the advantage to a fungus of producing and releasing antibiotics?
4 Suggest why protease inhibitors can inhibit viral proteases, but do not affect the human protease enzymes in the cell.

2.1 How Science Works

(27) Investigating enzyme action – 1

By the end of this spread, you should be able to . . .

* **Describe how the effects of pH, temperature, enzyme concentration and substrate concentration on enzyme activity can be investigated experimentally.**

Enzymes and practical assessment

Investigations into **enzyme** activity are often used in practical assessments. There are many possible investigations, but they all have many features in common.

What can we investigate?

Figure 1 Catalase reaction

In an enzyme-controlled reaction, enzymes **catalyse** a reaction where **substrate** is turned into product. An investigation into enzyme activity can follow the rate at which substrate is used up, or the rate at which product is formed, in order to give a measure of the reaction rate.

For example, hydrogen peroxide (H_2O_2) is a toxic by-product of a number of metabolic reactions and needs to be removed. The enzyme catalase is present in a wide range of tissues. This enzyme catalyses the conversion of H_2O_2 to water and oxygen gas:

$$2H_2O_2 \xrightarrow{\text{catalase}} 2H_2O + O_2$$

Investigations into the activity of catalase could measure the rate at which H_2O_2 is used up, or the rate at which either water or oxygen gas is produced.

Dependent and independent variables

Investigations into enzyme activity measure one variable, for example the rate at which product is formed, against another, for example variation in temperature.

The variable that is set by the investigator is called the **independent variable**. The variable that is measured by the investigator is called the **dependent variable**. In the example above, if we want to investigate the effect of temperature on enzyme action, the temperatures set in the investigation are the independent variable, and the readings taken for the rate of formation of product are the dependent variable.

It is vital that the independent variable is set precisely, otherwise readings of the dependent variable are meaningless. For example, if we are investigating the effect of temperature, there must be a reliable and accurate way of setting and maintaining the temperature range selected.

Investigations should show a good range for the independent variable. You should use a minimum range of five values of the independent variable, for example five different temperatures.

Equilibration

It is essential that enzymes and substrates are at the same temperature before the investigation begins. Adding cold substrate and enzyme together, then placing them in a water bath, would not give a precise temperature in the reacting mixture. Placing enzyme and substrate separately in a water bath so that they reach the required temperature before the investigation begins is known as equilibration.

What needs to be controlled?

Whatever the variables being investigated, all other variables must be kept constant. The key variables in enzyme-controlled reactions are temperature, enzyme concentration, substrate concentration and pH. So if an investigation into the effect of temperature is being carried out, then the pH, enzyme concentration and substrate concentration must be kept constant.

Key definition

A **variable** is any factor that may change and therefore affect the reaction rate. Enzyme investigations require all variables to be controlled.

Examiner tip

Questions about enzyme activity often need detail about why the reaction rate is changing. Always explain the reaction rate in enzymes in terms of how factors affect collisions and thus formation of the enzyme–substrate complex, or how they affect the shape of the enzyme, thus the shape of the active site and thus formation of the enzyme–substrate complex.

Variable	Method of keeping constant	Reasons
Temperature	Usually kept constant by carrying out enzyme-controlled reactions in a water bath with a thermostat. If not possible to use a water bath, a polystyrene sleeve used as insulation can help keep temperature constant.	Room temperature is too variable. Fluctuations in temperature will affect the enzyme-controlled reaction so readings taken will not reflect the action of the independent variable being tested.
Enzyme concentration	Use accurately measured volumes of enzyme in solution.	Reaction rate depends on the concentration of enzyme molecules present. Using accurately measured volumes of enzyme solution gives a constant concentration of enzyme molecules.
	If you are using enzyme in a living tissue, accurate measurements of the mass of tissue are important.	In living tissue you must assume that all pieces of tissue contain the same number of enzyme molecules.
	If you are using whole pieces of tissue, you must keep the surface area of the tissue constant as well as the mass.	The number of enzyme molecules in contact with substrate molecules will affect the reaction rate. If the surface areas of the pieces of tissue are different, the number of enzyme molecules exposed directly to substrate will be different.
Substrate concentration	Use accurately measured volume or mass of substrate.	Reaction rate depends on the concentration of substrate molecules.
pH value	Use pH buffers – solutions that maintain pH at a set level by keeping the H^+ concentration in solution constant.	Reaction rate depends on the pH because of its effect on the shape of the active site of the enzyme.

Table 1 Commonly used methods for controlling variables

Time scales, rate and control tests

There are two main ways of measuring the reaction rate:

- start the reaction, then measure the concentration of product (or of substrate used up) after a fixed period
- monitor the reaction by taking readings of product formation (or of substrate used up) at a number of time intervals.

A suitable time scale must be used. Enzyme-catalysed reactions are often very rapid. Time scales in such an investigation will be over a period of seconds or perhaps minutes in many cases.

Calculations of rate can be made from measurements over time, where:

reaction rate = 1/time

The rate calculations can then be plotted on a graph, with reaction rate on the *y*-axis and the inependent variable on the *x*-axis.

It is often necessary to include a **control** test in enzyme investigations. Usually you would set up the investigation using water in place of the enzyme, as a control. The control shows that it is the enzyme action, rather than any other factor, that affects the reaction.

Questions

1 Explain why using living tissue as a source of an enzyme can lead to difficulties in controlling enzyme concentration as a variable.

2 Why would we use a thermometer to measure the temperature of a reaction mixture, even if the experiment is being carried out in a thermostatically controlled water bath?

3 Explain the difference between a dependent variable and an independent variable.

(28) Investigating enzyme action – 2

By the end of this spread, you should be able to . . .

* **Describe how the effects of pH, temperature, enzyme concentration and substrate concentration on enzyme activity can be investigated experimentally.**

Following an enzyme-controlled reaction

As an **enzyme**-controlled reaction proceeds, product molecules are produced. These product molecules will also collide with enzyme molecules, so the actual reaction rate will slow as the reaction progresses. If you want to show the effect of any **dependent variable** on the action of an enzyme, it is best to compare the **initial reaction rates** at each value for the dependent variable.

You can do this by taking several readings over the course of a reaction and plotting them on a graph of product formed (or **substrate** used) against time. Take a tangent to the steepest portion of the graph for a measure of the initial reaction rate.

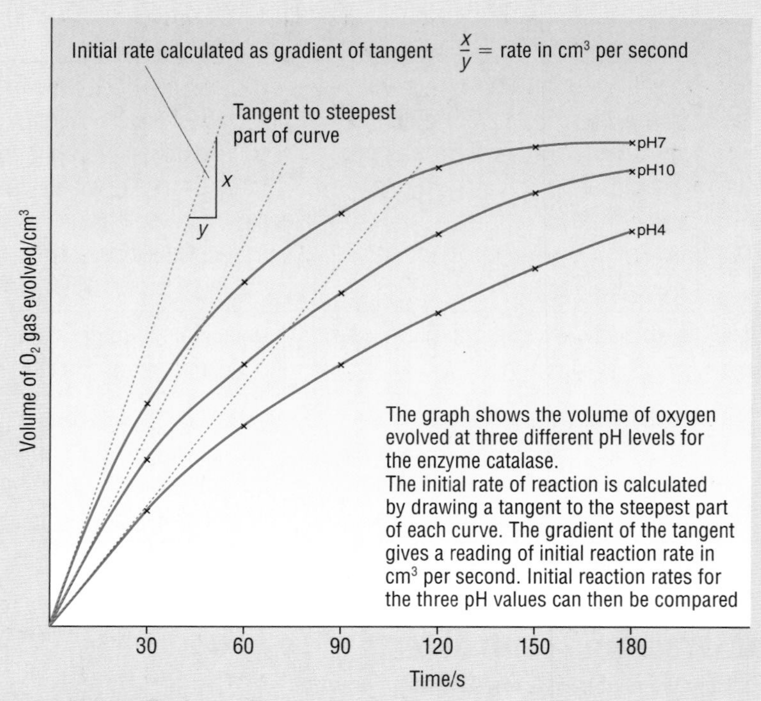

Figure 1 Initial reaction rate comparisons

Examples of enzyme investigations

Effect of pH on amylase action

Amylase hydrolyses starch to **maltose**. You can investigate the effect of pH on the activity of amylase with a simple **assay** technique using a starch–agar plate, as follows.

- Cut wells into the plate using a cork borer.
- Into each well place the same volume of one of a range of pH buffer solutions.
- Place an identical volume of a stock amylase solution into each well except one.
- To that well, add an equal volume of distilled water as a **control**.
- Incubate for 24 h in a dry oven at 35 °C. During this time, the amylase will diffuse through the agar and **catalyse** the conversion of starch to maltose.
- Flood the plate with iodine solution and rinse with water (iodine indicates the presence of starch by turning blue-black).
- Measure the diameter of the cleared zone – this gives a measure of how much substrate has been turned into product.

A starch–agar plate is made up by mixing starch with molten agar. The mixture is poured into a petri-dish and left to set. It forms a semi-rigid gel in the plate

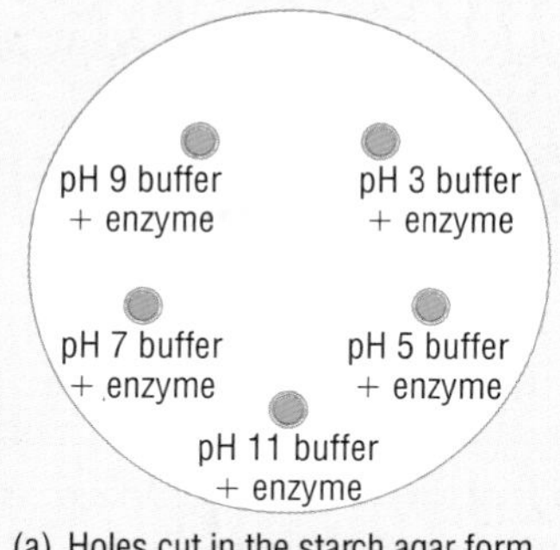

(a) Holes cut in the starch agar form 'wells'. pH buffer and enzyme are placed in each well

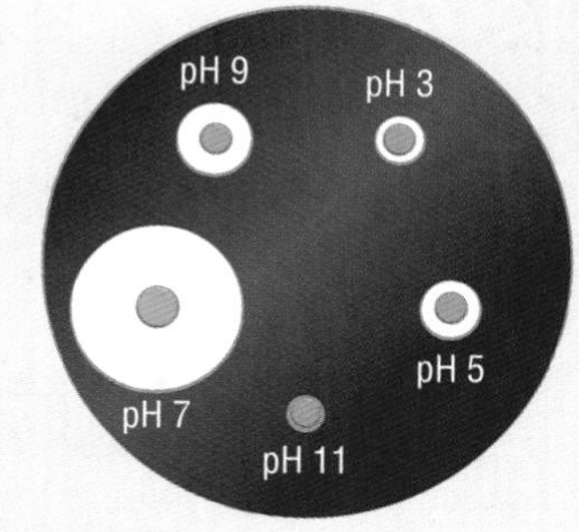

(b) After 24 hours incubation, the plate is flooded with iodine solution and rinsed. The diameter of the clear (non-blue/black) zone is measured

Figure 2 Using starch plates to measure enzyme activity

Effect of temperature on catalase action

Catalase is present in the cells of potato tubers, where it catalyses the breakdown of hydrogen peroxide to water and oxygen gas.

- Take samples of potato tissue using a cork borer, then slice into discs of equal thickness.
- Place an equal number of discs into each of seven test tubes, and place one in each of a range of water baths at 20, 30, 40, 50, 60, 70 and 80 °C.
- Place an equal volume of pH 7 buffer and hydrogen peroxide solution into each of seven separate test tubes, and place one in each water bath. Allow to equilibrate (see spread 2.1.27).
- Taking each in turn, add peroxide/buffer mixture to potato discs, then fix a stopper and side arm into the tube. Close the clip.
- Time how long in seconds it takes for the water bubble in the side arm tube to move 5 cm along the tube (as oxygen gas is produced in the reaction, it pushes the water bubble along the arm).

You can calculate the reaction rate in the range of temperatures using the equation:

rate = 1/time

You can make the numbers friendlier by using 100/time or even 1000/time.

Plotting the rate of reaction (*y*-axis) against temperature (*x*-axis) on a graph allows you to estimate the optimum temperature for the enzyme.

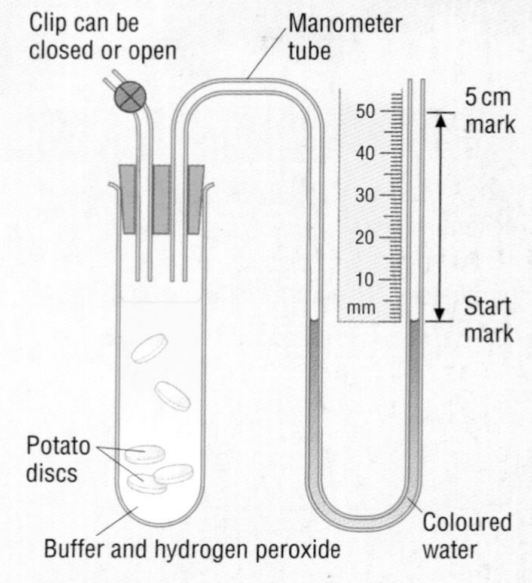

Figure 3 Apparatus for measuring catalase activity

Temperature /°C	Time taken to move coloured water 5 cm/s	Rate calculation 1/t
20	40	0.025
30	14	0.071
40	6	0.167
50	12	0.083
60	55	0.018
70	Did not move	

Table 1 The effect of temperature on catalase activity

The importance of repeats

When carrying out any investigation into the effect of enzymes, you should repeat the investigation at least three times. Repeating investigations helps to identify **anomalous** results. Averaging results for an investigation assumes that wherever there is inaccuracy, some repeats will be inaccurate below the real value, while others are inaccurate above the real value. An investigation with many repeats should give the most accurate result possible.

Key definition

An **anomalous** result or reading is one that looks out of place with other results or readings.

Anomalous results

You should also compare your results with those obtained by the rest of the class. This is because if you are doing something wrong, you may well do it wrong for all your repeats. Results that don't fit the general pattern may be anomalous – if you know that your apparatus is not very accurate, or if you have done something wrong. Even so, you should re-run the experiment to see if you still get the 'anomalous' results, and you should see if anyone else gets them too. The results may not be anomalous – they may be a new discovery!

Examiner tip

Exam questions often give experimental procedures and ask for comments on how they might be improved. Read the information about how variables were controlled, whether repeat readings were taken, and whether a control was used.

Questions

1 The enzyme trypsin breaks down the protein component of some milk powders. Describe how you could use this information to investigate the effect of pH on trypsin activity.
2 Describe a control test for the starch–agar investigation shown in Figure 2 above.

2.1 (29) Enzymes and metabolism – an overview

Key definition

The **turnover number** of an enzyme is the number of reactions an enzyme molecule can catalyse in one second. In catalase, the turnover number is up to 200 000.

Catalytic power

Without **enzymes**, metabolic reactions could not continue at a rate high enough to support any living processes. Enzymes increase reaction rates by a staggering 10^7 times. That's 10 million times faster. In fact, in many enzyme-controlled reactions the limit to the reaction rate is set by how fast the **substrate** molecule collides with the **active site**.

The power of enzymes must be controlled, as the product of any enzyme-controlled reaction is needed in very specific amounts by the organism. Uncontrolled enzyme activity is just as dangerous to an organism as a lack of enzymes. An example of this is seen in the disease multiple sclerosis. The immune system of sufferers of this disease wrongly sets destructive enzymes against parts of nerve cells. This results in nervous system breakdown leading to a number of symptoms, including paralysis.

Control of metabolic sequences

Many metabolic processes involve a series of enzyme-controlled reactions. The processes of **respiration** and **photosynthesis** are examples of complex metabolic sequences. In these metabolic processes, the product of one enzyme-controlled reaction is the substrate for the next enzyme-controlled reaction in the sequence. Such chains of enzyme reactions are often called metabolic pathways.

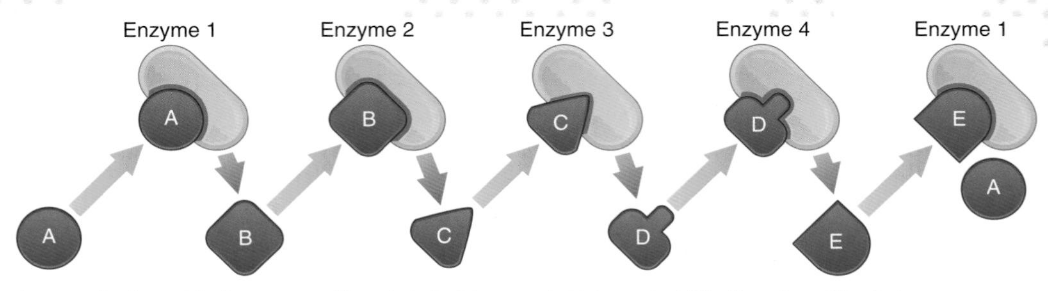

An example of a metabolic sequence. The substrate A is converted into product B by the action of the first enzyme (enzyme 1). Product B is the substrate for enzyme 2, and is converted to product C, and so on.

The end product E can bind to enzyme 1, and acts as a non-competitive inhibitor. This means that the end product E will not build up in the cell, because as E is made, it slows the formation of itself by inhibiting the first enzyme in the sequence.

Figure 1 A metabolic pathway

Examiner tip

If enzyme and substrate are present, a reaction will go ahead. In questions about control of metabolism, remember that control is often about regulating enzyme activity by inhibiting enzymes. This is a natural and necessary process.

In controlling metabolic processes, it is vital to the organism that the end product of the sequence does not build up if it is not required. In many cases, the end product can attach to one of the enzymes early in the sequence. This is exactly the same as reversible **non-competitive inhibition** (see spread 2.1.24).

The end product binds to the enzyme in a part of the enzyme molecule away from the active site. Binding of the end product molecule changes the shape of the active site, so it reduces the reaction rate of the enzyme.

Some enzymes are crucial to life

Some enzymes are found in all organisms, so they appear to be vital for life itself. The enzyme **ATP** synthase, for example, **catalyses** the addition of an inorganic phosphate group to an ADP molecule:

$$\text{ADP} + \text{inorganic phosphate} \xrightarrow{\text{ATP synthase}} \text{ATP}$$

ATP synthase is found in the cells of all organisms, from bacteria to flowering plants and humans. The ATP generated in this reaction acts as a short-term energy supply in cells.

Inborn errors of metabolism

All enzymes are **proteins**. As with all proteins, any organism must have the instructions to make the enzymes it needs, in the form of **DNA**.

If the DNA forming the instructions for an enzyme is faulty (mutated), then the enzyme may not be made correctly. Many diseases are caused by the lack of a functioning specific enzyme in a metabolic sequence. These are called inborn errors of metabolism.

The condition phenylketonuria is an inborn error of metabolism. Individuals with this condition lack a functioning version of the enzyme phenylalanine hydroxylase, which breaks down excess phenylalanine (an amino acid in the diet) to tyrosine. The phenylalanine builds up and, in young individuals, causes irreversible damage to the nervous system and to brain development. Affected individuals must have a diet with very little phenylalanine.

Under normal conditions excess phenylalanine (an amino acid) is converted to the amino acid tyrosine by the enzyme phenylalanine hydroxylase (PAH)

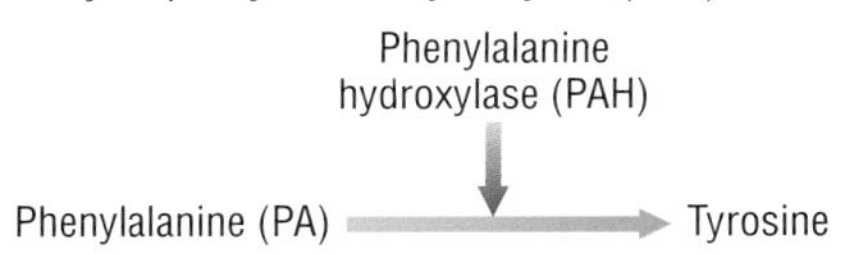

If PAH is not present, excess PA is converted to phenylpyruvic acid (PPA). Accumulation of PPA results in damage to developing brain tissue

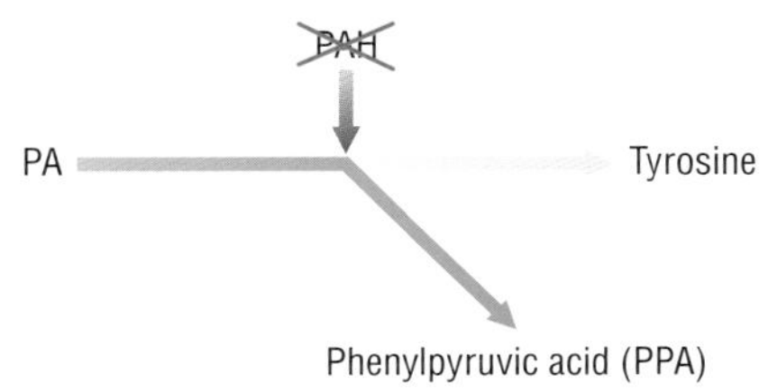

Figure 2 Phenylketonuria sequence and effects

Questions

1. ATP is an end-product inhibitor of one of the enzymes that catalyses an early reaction in the sequence of reactions involved in respiration. Explain the advantages to an organism of this inhibition.
2. Explain why end-product inhibition is usually inhibition of an enzyme at the start of a reaction sequence.
3. Suggest a reason why paralysis is one of the symptoms of multiple sclerosis.
4. Suggest why phenylalanine is not completely removed from the diet of an individual with phenylketonuria.
5. Why is diet restricted only in young people affected by phenylketonuria?
6. Tyrosine is normally changed to melanin. Suggest why individuals with phenylketonuria have very fair hair.
7. Phenylketonuria is rare – about 1 in 16 000 births per year. Suggest why all newborns are screened for phenylketonuria.

2.1 Biological molecules summary

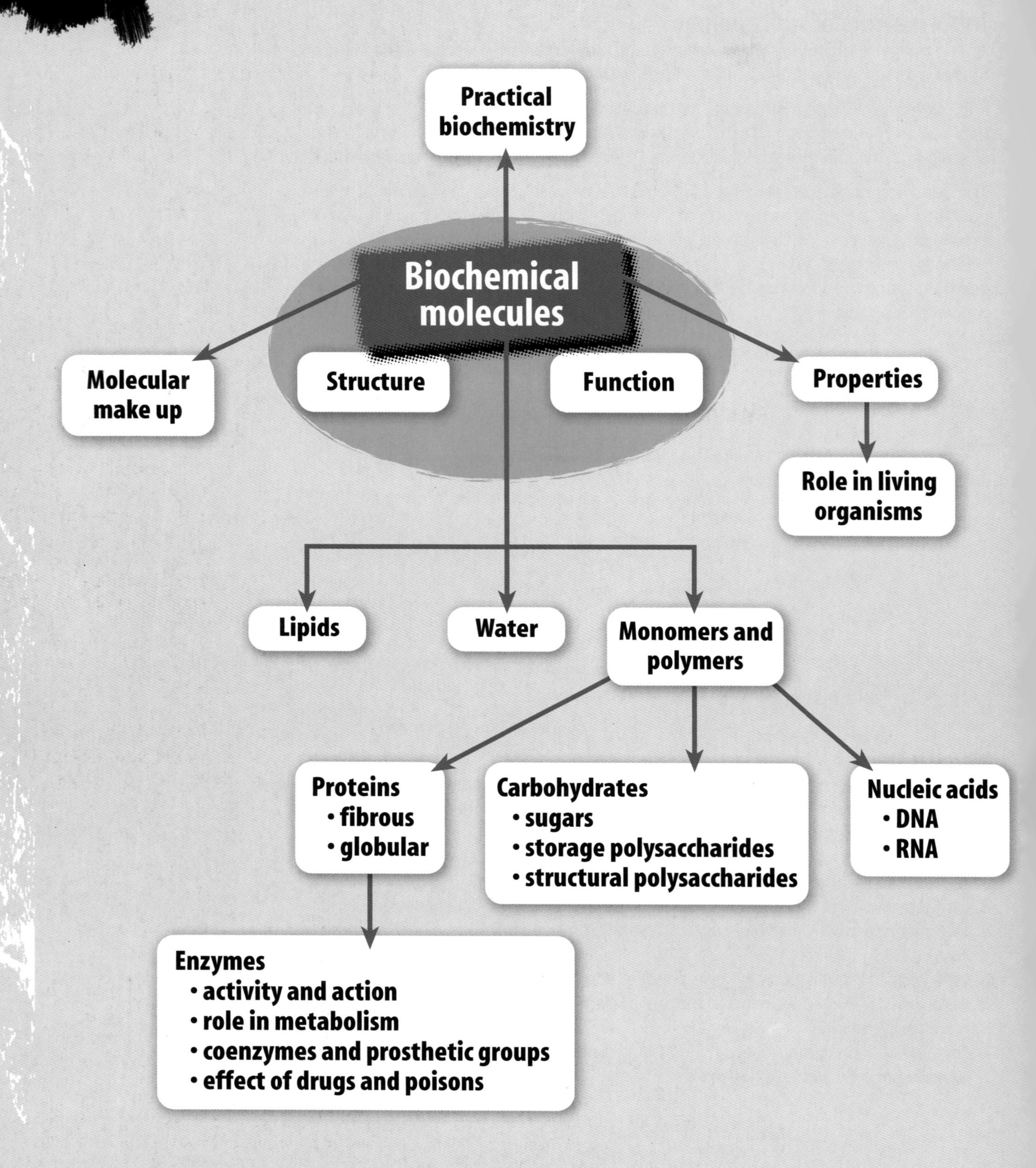

Practice questions

1 Name the sugar in DNA. [1]

2 State how many carbon atoms this sugar contains. [1]

3 State the name of the organic base that:
(a) is present in DNA but not in RNA; [1]
(b) is present in RNA but not in DNA. [1]

4 Complete the following passage by using the most suitable word(s) in each of the blank spaces.

Water is essential for life. It makes up a high proportion of the cytoplasm in a cell. Many different compounds can dissolve in it and it is therefore described as an excellent ....

Water remains in the .... state over a wide range of enviornmental temperatures.

As it cools below 4 °C it becomes less .... than warmer water. Ice floats on water, forming a layer that .... the water beneath with the result that large bodies of water rarely freeze entirely.

The .... bonds that form between water molecules are responsible for its high ...., which allows small insects such as pond skaters to move on its surface without sinking. [6]}

5 Explain why an enzyme which catalyses the conversion of starch to maltose is unable to catalyse the conversion of a protein into amino acids. [2]

6 If sucrose is tested using Benedict's test for reducing sugars, no change is observed. The bond between the glucose and fructose units must first be broken. The test for a reducing sugar can then be carried out.

Describe how the bond can be broken chemically before carrying out the Benedict's test. [2]

7 Complete the table to show the names of the types of molecule that are tested, the reagents used and the results obtained.

type of molecule tested	reagents used	positive result	negative result
protein			blue solution
	alcohol and water	white emulsion	clear liquid
starch			yellow solution

[5]

8 Explain how a **non-competitive** inhibitor affects the rate of an enzyme-catalysed reaction. [3]

9 In an investigation into the action of amylase, equal volumes of enzyme solution and starch solution were mixed. The quantity of maltose produced was measured during the course of two separate experiments, one carried out at 18 °C and the other at 23 °C. The results are shown in the graph.

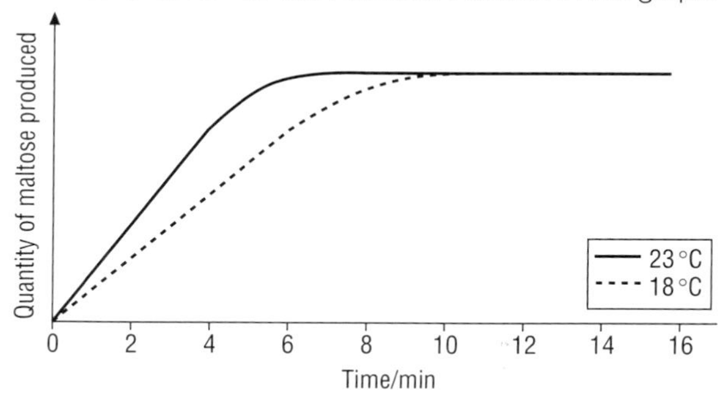

(a) Explain why the quantity of maltose produced eventually became constant. [1]
(b) Explain why, after 4 minutes, more maltose had been produced at 23 °C than at 18 °C. [2]
(c) Name the bond that is broken by the action of this enzyme. [1]

10 DNA replication is described as semi-conservative. Explain what is meant by the term *semi-conservative replication*. [3]

11 Glucose:
- is a carbohydrate
- is a hexose (six-carbon sugar)
- has the formula $C_6H_{12}O_6$
- has a six-membered ring structure

Figure 1 shows the molecular structures of two monosaccharide sugars, glucose and fructose.

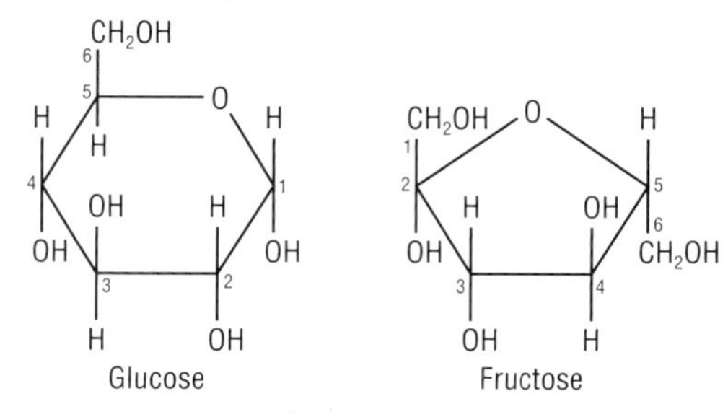

Figure 1

(a) State one way, visible in Figure 1, in which the structure of fructose is:
(i) similar to glucose? [1]
(ii) different from glucose? [1]
(b) Complete the diagram to show what happens when the glucose and fructose are joined together. [2]

2.1 Examination questions

1 **(a)** Describe a test which will indicate the presence of protein. [2]

Figure 1 shows a number of bonds or links, identified with the letters **A** to **E**, which are involved in the formation of proteins and polysaccharides. **C**, **D** and **E** are all covalent bonds or links.

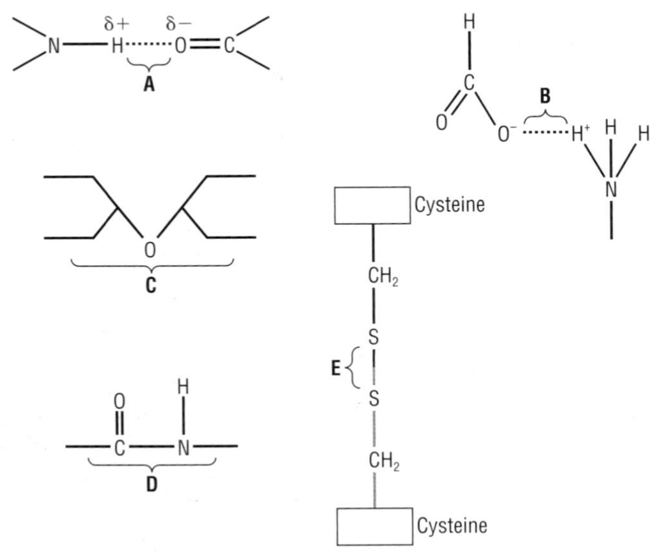

Figure 1

(b) **(i)** Name the bonds **A** and **B**. [2]

(ii) State the specific names given to bonds or links **C** to **E**. [3]

(c) With reference to Figure 1, use one of the letters **A** to **E** to indicate the bond or link that is used

(i) to join glucose molecules in the formation of a polysaccharide;

(ii) to join adjacent amino acids in the primary structure of a protein;

(iii) in the secondary structure of a protein;

(iv) only in the tertiary structure of a protein. [4]

[Total: 11]

(2801 Jan02)

2 A DNA molecule is made up of two polynucleotide strands which are twisted into a double helix, as shown in Figure 2.

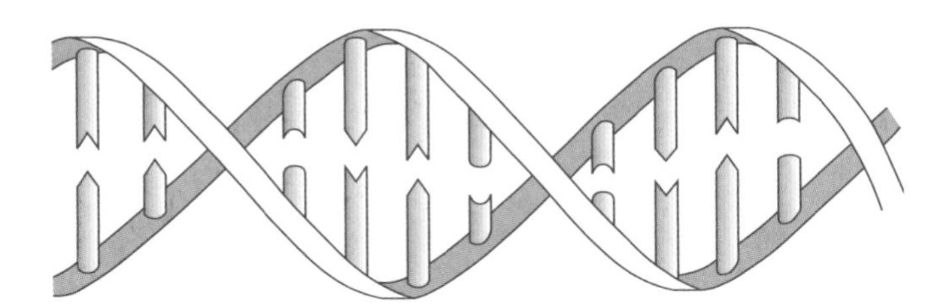

Figure 2

DNA is involved in transcription, which is part of protein synthesis, and replication.

Complete the table putting a tick (✓) in the column(s), where relevant, to indicate whether each statement refers to transcription, replication or both. The first one has been done for you. [5]

	transcription	replication
free nucleotides bond to the DNA strand	✓	
two new DNA molecules are produced		
only part of the DNA molecule containing the gene unwinds		
hydrogen bonds are broken between the two DNA strands		
cytosine nucleotides bond to guanine on the DNA strand		
uracil nucleotides bond to adenine on the DNA strand		

[Total: 5]

(2801 Jan04)

3 **(a)** DNA and RNA are nucleic acids.

State **two** ways in which the structure of DNA differs from that of RNA. [2]

(b) The DNA molecule is made of two chains of nucleotides, wound into a double helix.

(i) Describe the structure of a **DNA nucleotide**. You may draw a diagram if it will help your description. [3]

(ii) Describe how the two nucleotide chains in DNA are bonded together. [3]

(c) An enzyme, such as amylase, has a specific 3-dimensional shape.

Explain how DNA structure determines the specific shape of enzymes. [4]

[Total: 12]

(2801 Jan05)

Answers to examination questions will be found on the Exam Café CD.

4 **(a)** Figure 3 represents part of a collagen molecule.

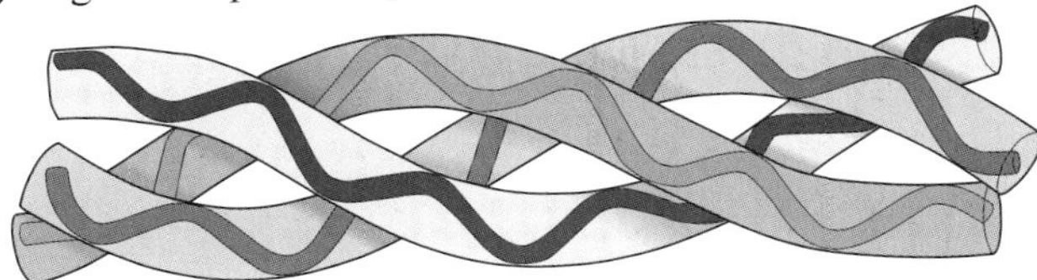

Figure 3

(i) Collagen is a protein made of three chains of amino acids, twisted together like a rope.
State the name given to a chain of amino acids. [1]

(ii) Name the amino acid that forms a high proportion of the collagen molecule. [1]

(iii) Collagen has tremendous strength, having about one quarter of the tensile strength of mild steel.
Using information given in Figure 2.1 to help you, explain how the structure of collagen contributes to its strength. [2]

(b) Complete the following passage by inserting the most appropriate terms in the spaces provided.
Cellulose and collagen are both fibrous molecules.
Cellulose, a carbohydrate, is the main component of the .... in plants.
Cellulose is made of chains of many .... glucose molecules which are joined by 1,4 .... bonds. Each glucose molecule is rotated .... relative to its neighbour, resulting in a .... chain. Adjacent chains are held to one another by .... bonds. [6]

[Total: 10]

(2801 Jun02)

5 A student was carrying out tests to determine which biological molecules were present in a food sample.

(a) **(i)** Describe a test that the student could carry out to discover whether this sample contained a lipid. [2]

(ii) State what would be seen if a lipid was present. [1]

(b) Describe how the **structure** of a phospholipid differs from that of a triglyceride.
You may use the space below for a diagram to help your answer. [3]

(c) **(i)** Describe a test that the student could carry out to discover whether the food sample contained protein. [1]

(ii) State what would be seen if protein was present. [1]

(d) Explain what is meant by the primary and secondary structure of a protein. [5]

[Total: 13]

(2801 Jan01)

6 **(a)** **(i)** State the components needed to synthesise a triglyceride. [2]

(ii) Name the chemical reaction by which these components are joined. [1]

(b) State one function of triglycerides in living organisms. [1]

Lipase is an enzyme that catalyses the hydrolysis of triglycerides. It is a soluble globular protein. The function of an enzyme depends upon the precise nature of its tertiary structure. Figure 4 represents the structure of an enzyme. The black strips represent the disulfide bonds which help to stabilise its tertiary structure.

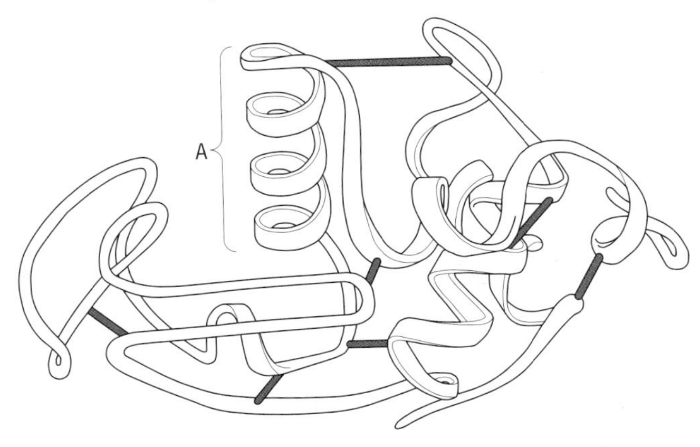

Figure 4

(c) **(i)** Describe the nature of the disulfide bonds that help to stabilise the tertiary structure of a protein such as lipase. [2]

(ii) Name **two other** types of bonding that help to stabilise tertiary structure. [2]

Region A on Figure 4 is a secondary structure.

(d) Describe the nature of region **A**. [2]

[Total: 10]

(2801 Jan03)

UNIT 2

Module 2 Food and health

Introduction

Food is essential to provide us with energy and the building blocks we need for growth. A balanced diet will provide all the necessary nutrients for good health. However, our diet and the range of foods we eat can have significant effects on our health – you can have too much of a good thing. If we consume too much energy, we will become obese. The amount and types of fat we eat are particularly important to the health of the circulatory system and heart.

We gain our nutrition by eating other living things or their products. In this module you will learn how humans have made use of other organisms to produce food. The range of organisms used includes examples from all five kingdoms. Modern food production increasingly uses fungi and bacteria. Humans have utilised the evolutionary process to breed ever-better varieties of domestic plants and animals, as well as fungi and bacteria. However, just as microorganisms have the ability to make our food, they have the ability to spoil it. You will also learn about how food can be spoilt and how we prevent that spoilage.

Certain diseases can be caused by bad habits such as smoking or over-eating. Infectious diseases are caused by pathogenic organisms. The human body is well adapted to defending itself against infectious diseases. We have evolved primary defences to stop pathogens entering our body, and a secondary response to kill them if they do gain entry. As we have learnt more about how to combat diseases, we have learnt that certain organisms can help us fight disease.

Our increasing effect on the environment threatens the existence of a wide range of living things. As we rely on them for our food and medicine, we must make attempts to conserve all living things.

Test yourself

1 What are the seven main components of a balanced diet?
2 What types of food give us most energy?
3 What foods are manufactured using bacteria or fungi?
4 Name two types of organism that cause disease.
5 Where do we get antibiotics from?
6 Which blood cells help us fight disease?
7 What do you understand by the term ‘conservation’?

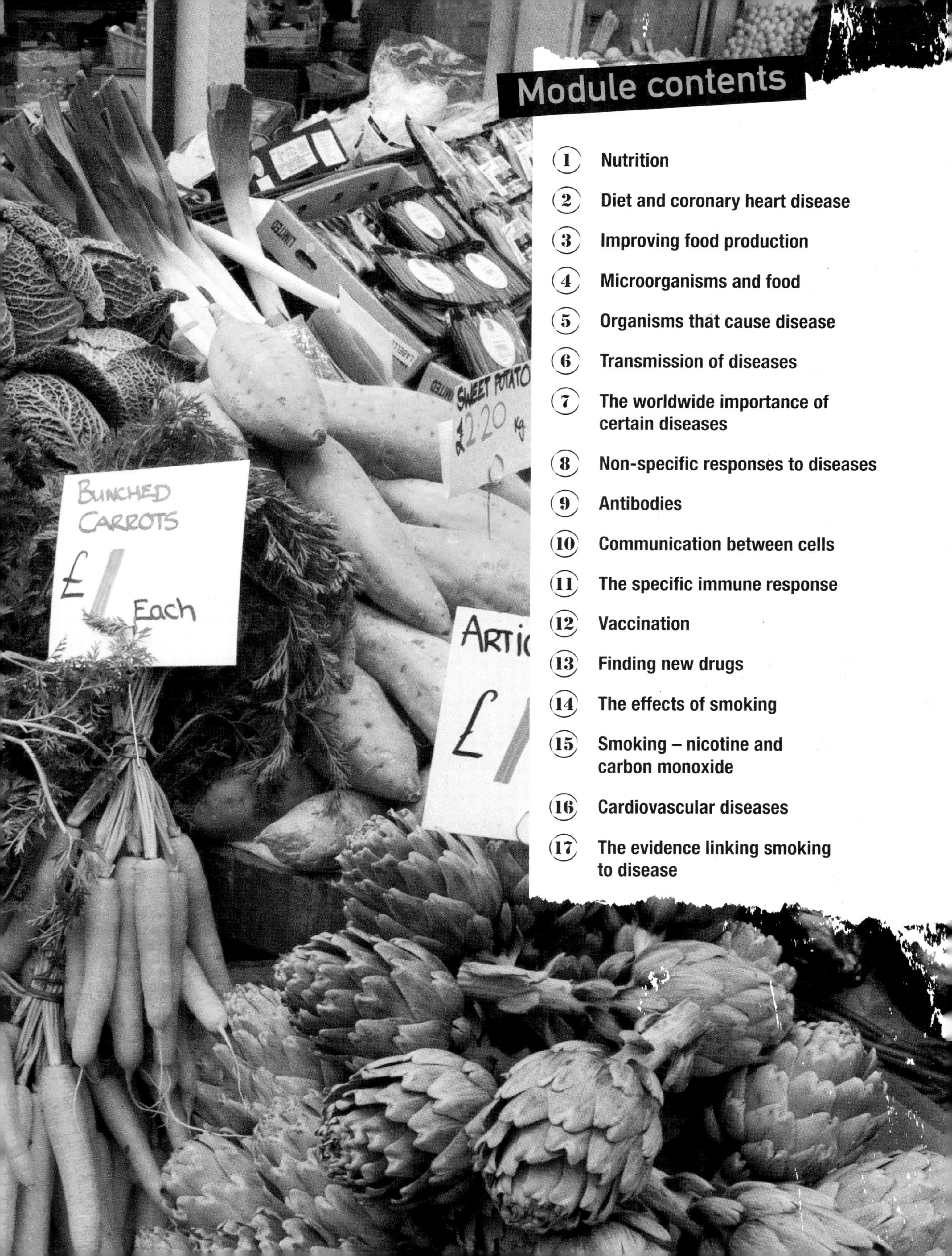

Module contents

2.2 (1) Nutrition

By the end of this spread, you should be able to . . .

* **Define the term 'balanced diet'.**
* **Explain how consumption of an unbalanced diet can lead to malnutrition, with reference to obesity.**

Key definitions

Nutrition is nourishment – the nutrients and energy needed for health and growth.

A **balanced diet** is one that contains all the nutrients required for health in appropriate proportions.

Obesity is when a person is 20% or more heavier than the recommended weight for their height.

Why do we eat?

Our nutrition comes from the food we eat. Good food is full of nutrients. There are different kinds of nutrients in different foods, and it is important to ensure that you eat a mixed, well balanced diet. Good nutrition will:

- provide better health
- ensure a stronger immune system
- mean that you become ill less often
- help you learn more effectively
- make you stronger
- make you more productive.

What's in a balanced diet?

A balanced diet must provide both the building blocks for growth and repair, and sufficient energy. There are seven components in a balanced diet.

- **Carbohydrates** – these are our main source of energy.
- **Proteins** – these are essential for growth and repair of muscle and other body tissues.
- **Fats** – these are essential as a source of energy and are also important in cell membranes, waterproofing, absorption of fat-soluble vitamins, and many other roles.
- **Vitamins** – these play many important roles in the chemical processes taking place inside cells. Some are water-soluble and some are fat-soluble.
- **Minerals** – these are the inorganic elements occurring in the body that are essential to its normal functions.
- **Water** – this is essential to body function. It is used in transporting substances around the body and as a main component of the body. About 60% of the human body is water.
- **Fibre or roughage** – this is the indigestible part of our food. It is essential for healthy functioning of the digestive system.

How much of each component do we need?

Everyone needs some of each of these components. The Government publishes guideline amounts for each component. You should note that the amount required by individuals will vary. Different people may need different amounts depending upon their age and level of activity (Table 1 gives an example).

Nutrient	Mass/g per day	Energy in food/kJ
Carbohydrates	250	4000
Fats	80	2960
Proteins	60	1020
Minerals	9.2	0
Fibre	12	0
Vitamins	traces	0
Water	variable	0

Table 1 Typical mass of each component of the diet required per day by a 17-year-old girl

One important aspect of nutrition is energy intake. It is impossible to give guideline advice on energy intake that will apply to everyone. This is because the amount of energy needed depends so much on the level of activity. In general, energy intake should come from the nutrients in the following proportions:

- 57% carbohydrates (bread, pasta, rice, sugar, sweets)
- 30% fats (dairy products, oil)
- 13% protein (eggs, meat, fish).

In an active person, the amount of carbohydrate, fat and protein eaten will need to be increased above the amounts in Table 1.

A good guide to whether your diet contains enough energy is change in mass or weight. If you gain weight, you are eating too much energy-containing food. If you lose weight, you do not have enough energy in your diet.

Obesity

Malnutrition is caused by an unbalanced diet. Most people think of malnutrition as deficiency. However, the biggest form of malnutrition in the developed world is obesity. Obesity is caused by consuming too much energy. The excess energy is deposited as fat in the **adipose** tissues. Obesity is the condition in which excessive fat deposition impairs health, and is usually defined as when a person has a **body mass index** (BMI) of 30 or over. This indicates a body weight of 20% or more above the weight recommended for your height. BMI (see Table 2) can be measured by the following formula:

$$\text{BMI} = \frac{\text{mass in kg}}{(\text{height in m})^2}$$

Key definition

BMI = mass in kg/(height in m)^2

BMI	Category
Less than 18.5	Underweight
Over 18.5–25	Desirable or healthy range
Over 25–30	Overweight
Over 30–35	Obese (Class I)
Over 35–40	Obese (Class II)
Over 40	Morbidly or severely obese (Class III)

Table 2 Body mass index

Obesity is not just a problem of excess fat. The location where the fat is deposited also has an effect. People with extra fat around the middle (apple-shaped) are known to be more at risk than people with excess fat around their hips and thighs (pear-shaped).

A growing problem

In the early 1980s, 6% of men and 8% of women in the UK were obese. Twenty years later, the figures show that 25% of men and 20% of women are obese. Of even more concern is that the number of obese children is also increasing. In the USA between 1995 and 2004, obesity in girls aged 11–15 rose from 15 to 26% and in boys from 14 to 24%.

Questions

1. List the nutrients in a balanced diet.
2. Why is it important to ensure that the level of energy consumed is consistent with the body's needs?

2.2  Diet and coronary heart disease

By the end of this spread, you should be able to . . .

* **Discuss the possible links between diet and coronary heart disease.**
* **Discuss the possible effects of a high blood cholesterol level on the heart and circulatory system, with reference to high-density lipoproteins (HDLs) and low-density lipoproteins (LDLs).**

Health risks caused by obesity

As we discussed on the previous pages, obesity is caused by consuming too much energy. Severe obesity can increase the risk of mortality (death). Up to 30 000 deaths a year in the UK are considered to be obesity-related. Obesity is thought to be the most important dietary factor in the following health problems:

- cancer
- cardiovascular disease
- type 2 diabetes.

Obesity is also linked to:

- gallstones
- osteoarthritis
- high blood pressure (hypertension).

Other components of the diet

The overall energy intake must be balanced with energy use to avoid becoming overweight or underweight. However, there are many other aspects of the diet that can affect your health. Here, we are concerned with components that may affect the risk of coronary heart disease (CHD). CHD is the result of deposition of fatty substances in the walls of the coronary arteries. This deposition is known as **atherosclerosis** and is described in spread 2.2.15. Deposition in the coronary artery walls narrows the size of the lumen. This restricts blood flow to the heart muscle, which may cause oxygen starvation.

Some components of the diet help to reduce the risk of CHD. Dietary fibre, moderate alcohol consumption and eating oily fish all appear to be beneficial. However, more is known about those components of the diet that increase the risk of CHD.

Salt

Excess salt in your diet will decrease the water potential of your blood. As a result more water is held in the blood and the blood pressure increases. This can lead to hypertension. Hypertension is a condition in which the blood pressure, and particularly the diastolic pressure (blood pressure when the heart is relaxing during its cycle), is maintained at a level that is too high. Hypertension can damage the inner lining of the arteries, which is one of the early steps in the process of atherosclerosis.

Fats (lipids)

Fats are an essential part of the diet. On any food label the nutritional information will tell you how much fat is present in the food and how much of that is 'saturates'. By 'saturates' the label means saturated fats. Animal fats tend to be saturated and plant oils tend to be unsaturated. In general it is recognised that saturated fats are more harmful than unsaturated fats. Polyunsaturated fats and monounsaturated fats, such as those found in olive oil, are particularly beneficial to health.

Cholesterol

Cholesterol is not a triglyceride but it has similar properties. It is found in many foods and is often associated with saturated fats in meat, eggs and dairy products. Cholesterol is also made in the liver from saturated fats. Too much cholesterol in the blood is harmful.

High blood cholesterol concentrations have been linked to 45–47% of deaths from coronary heart disease. The concentration of cholesterol in the blood should be maintained below 5.2 mmol dm^{-3}.

Why is cholesterol in the blood?

Cholesterol is essential to the normal functioning of the body. It is found in cell membranes and in the skin. It is also used to make steroid sex hormones and bile. Therefore cholesterol must be transported around the body. Like all the fats, cholesterol is not soluble in water. It must first be converted to a form that will mix with water. Cholesterol is transported, in the blood, in the form of lipoproteins. These are tiny balls of fat combined with protein. There are two types of lipoprotein:

- High-density lipoprotein (HDL)
- Low-density lipoprotein (LDL)

Both types of lipoprotein are released into the blood and can be taken up by cells that have the correct receptor sites.

High-density lipoprotein

HDLs are produced by the combination of unsaturated fats, cholesterol and protein. These tend to carry cholesterol from the body tissues back to the liver. The liver cells have receptor sites that allow the HDLs to bind to their cell surface membranes. In the liver the cholesterol is used in cell metabolism (to make bile) or is broken down. Therefore, high levels of HDLs are associated with reducing blood cholesterol levels. They reduce deposition in the artery walls by atherosclerosis and may even help to remove the fatty depositions of atherosclerosis. Since HDLs use unsaturated fats these fats are thought to be more beneficial to health than saturated fats.

Low-density lipoprotein

LDLs are produced by the combination of saturated fats, cholesterol and protein. These tend to carry cholesterol from the liver to the tissues. The tissue cells have receptor sites that allow LDLs to bind to their cell surface membranes. If too much saturated fat and cholesterol is consumed in the diet the concentration of LDLs in the blood will rise. A high blood concentration of LDLs causes deposition in the artery walls. Different fats affect the LDL receptors in a number of different ways.

- Saturated fats are thought to decrease the activity of the LDL receptors. Therefore as blood LDL concentration rises, less is removed from the blood. This results in higher concentrations of LDL in the blood and they are deposited in the artery walls.
- Polyunsaturated fats seem to increase the activity of the LDL receptors and so decrease the concentration of LDL in the blood.
- Monounsaturated fats also seem to help remove LDLs from the blood.

Diet and lipoproteins

It is important to remember that we do not eat lipoproteins. However, our diet has a significant effect upon the lipoprotein concentration in our blood. Overall it is best to keep to a low-fat diet which will maintain low lipoprotein concentrations. The ratio of HDL to LDL in the blood is important. Since LDLs are associated with greater deposition in the artery walls it is best to try to maintain a low proportion of LDLs. HDLs are associated with reduced deposition so it is best to maintain a high proportion of HDLs in the blood.

- Eating a lot of saturated (animal) fats will increase the concentration of LDLs in our blood.
- Eating a low-fat diet will reduce the overall concentration of lipoproteins.
- Eating a high proportion of unsaturated fats will increase the proportion of HDLs in the blood.
- Eating polyunsaturated fats helps to reduce the concentration of LDLs in the blood.
- Eating monounsaturated fats helps to reduce the concentration of LDLs in the blood.

Key definition

Lipoproteins are a combination of lipid, cholesterol and protein used to transport fats and cholesterol around the body.

Questions

1 Explain why it is unhealthy to add salt to every meal.

2 Explain why a high ratio of HDL to LDL in the blood is considered to be a healthy sign.

2.2 (3) Improving food production

By the end of this spread, you should be able to . . .

* **Explain that humans depend on plants for food, as they are the basis of all food chains (no details of food chains are required).**
* **Outline how selective breeding is used to produce plants with high yield, disease resistance and pest resistance.**
* **Outline how selective breeding is used to produce domestic animals with high productivity.**
* **Describe how the use of chemicals, including fertilisers and pesticides (plants) and antibiotics (animals), can boost food production.**

Plants as food

Plants can carry out **photosynthesis**. In the process of photosynthesis, plants convert the energy in light to a stored chemical form. They absorb carbon dioxide from the air and make **carbohydrates**. Most plants store energy as the carbohydrate **starch**. They also absorb minerals, such as nitrate, from the soil, and manufacture a range of other biological molecules. **Herbivores** are animals that make use of these biological molecules when they eat and digest plants.

Humans are **omnivores**. This means that we eat both plants and animals. We gain our nutrition both directly from plants and also indirectly by eating herbivorous animals. The human food chain tends to be short. By making the food chain more efficient, we can increase food production.

How can we make food production more efficient?

In plants it is possible to:

- improve the growth rate of crops
- increase the size of yield from each plant
- reduce losses of crops due to diseases and pests
- make harvesting easier by standardising plant size
- improve plant responses to fertilisers.

In animals it is possible to:

- improve the rate of growth
- increase productivity
- increase resistance to disease.

Key definitions

Selective breeding is where humans select the individual organisms that are allowed to breed according to chosen characteristics.

Fertilisers are minerals needed for plant growth, which are added to soil to improve its fertility.

A **pesticide** is a chemical that kills pests.

A **fungicide** is a chemical that kills fungi.

An **antibiotic** is a chemical that kills or prevents reproduction in bacteria.

Selective breeding

Charles Darwin was aware that people often bred animals with desirable traits and that, over time, such breeding exaggerated small differences. He recognised this as **artificial selection**. He described artificial selection as the intentional breeding of certain traits, or combinations of traits, in contrast to the process of **natural selection**. The purpose of artificial selection is to increase the benefit to humans.

There are three stages to selective breeding:

- isolation
- artificial selection
- inbreeding or line breeding.

In practice this means selecting a pair of animals or plants that display the characteristics you want. That pair is allowed to reproduce. The offspring produced are sorted carefully to select those with the best combination of characteristics. Only those offspring are allowed to reproduce. If this careful selection and controlled reproduction continues for many generations, the required characteristics become more exaggerated. In modern breeding programmes, detailed records are kept to help the selection process and to prove the ancestry of valuable individuals.

Humans have been selecting animals and plants for thousands of years. Ever since humans started agriculture, we have been selecting. We select which farm animals are allowed to breed, and which seeds to save for sowing the following year – instead of natural selection, *we* apply the **selection pressure**.

Recently, a new technique has been employed. This is known as marker-assisted selection. A section of **DNA** is used as a marker to recognise the desired characteristic. Once offspring have been produced from the selected parents, their DNA is checked for the marker. This allows selection at a very early stage.

Figure 1 Cows with different udder sizes

Some examples

- Farmers breed cattle for high milk yield or for meat production. Dairy cows can produce over 40 litres of milk a day.
- Farmed salmon have been selected to grow more quickly so that time-to-market has been cut by 30%. At the same time, the disease resistance and meat quality have been improved. The meat has less fat content, and is a better colour and texture.
- Chickens are bred for egg production or meat production. Egg-layers can produce over 300 eggs a year, while their unselected relatives will produce 20–30 a year.

Examples of marker-assisted selection

- Tomatoes have been bred with improved disease resistance. A wild tomato variety with good resistance to yellow leaf curl virus, which can devastate domesticated tomatoes, was found. The **allele** responsible for the resistance was identified and bred into a domestic variety.
- Apples have been bred with improved disease resistance, and work is continuing to develop varieties with improved flavour and texture.

Using chemicals

Food production can also be improved by the use of various chemicals. The response to these chemicals can be bred into the organisms.

Fertilisers replace minerals in the soil. These minerals may have been removed by previous crops. Fertilisers containing nitrate, potassium and phosphate are the most common. They increase the rate of growth and the overall size of the crops.

Pesticides are designed to kill organisms that cause diseases in crops. These diseases would reduce the yield or kill the crop. Many crops are sprayed with fungicides to reduce the effect of fungal growth in the leaves or roots. Animals can also be treated with pesticides – sheep are usually dipped to kill ticks that live on their skin under the wool.

Infected animals can be treated with antibiotics. These reduce the spread of disease among animals that are intensively farmed and in close proximity to each other. Such diseases could reduce the growth performance of the animals and may impair reproduction.

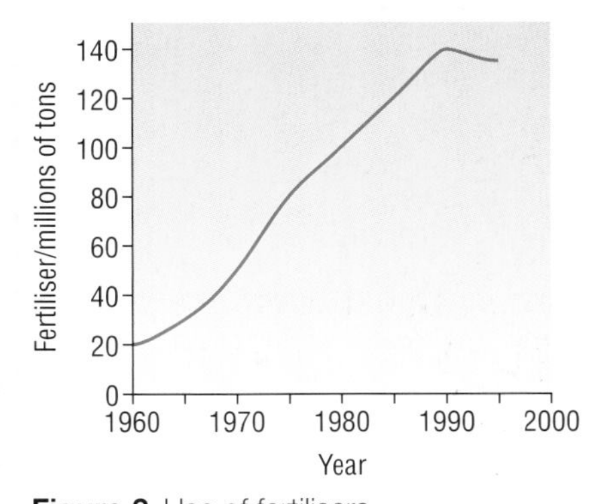

Figure 2 Use of fertilisers

Questions

1 Explain the difference between natural selection and artificial selection.

2 How does the application of a pesticide improve the yield of a crop such as wheat?

2.2 (4) Microorganisms and food

By the end of this spread, you should be able to . . .

- ✱ **Outline the methods that can be used to prevent food spoilage by microorganisms.**
- ✱ **Describe the advantages and disadvantages of using microorganisms to make food for human consumption.**

Microorganisms and food spoilage

Many microorganisms obtain their nutrition by digesting the organic matter around them. In doing so, they leave behind waste products. The organic matter they digest might well be our food. Once the effects of their activities are noticeable, we consider the food to be spoilt.

There are four main ways that microorganisms can spoil our food.

- Visible growth of microorganisms on food. This is most obvious when fungi grow on food. For example, colonies of the moulds *Mucor* and *Penicillium* often grow on bread. The mould has usually been growing for a few days before it becomes noticeable as either black (*Mucor*) or blue/green (*Penicillium*) mould.
- Microorganisms use an external digestion process. They release **enzymes** into the food and absorb the nutrients released by breakdown of the food molecules. When this happens, food often smells sweet as sugars are released from **carbohydrate** molecules. The food will eventually be reduced to a mush by the action of enzymes.
- The bacterium *Clostridium botulinum* produces a toxin called botulin. This causes botulism. If these bacteria are growing on food, the toxin will be present. It is one of the most toxic substances known – as little as 1 μg can kill a person.
- The presence of the microorganisms in food can cause infection. For example, *Salmonella* bacteria, sometimes present in poultry products, attack the lining of the stomach and digestive system.

Figure 1 Penicillium mould growing on bread

How do we prevent food spoilage?

We can prevent food spoilage by using food quickly, while it is still fresh. If food is to be kept for longer, it must be treated to prevent spoilage. This can be done by either killing any microbes already on the food, or preventing them from reproducing. The food must then be packaged to prevent further contamination with microbes.

Treatment methods to kill microorganisms or prevent their reproduction include:

- cooking – the heat denatures enzymes and other **proteins** and kills the microorganisms
- pasteurising – this involves heating to 72 °C for 15 seconds and then cooling rapidly to 4 °C, killing harmful microorganisms
- drying, salting and coating in sugar – these processes dehydrate any microorganisms as water leaves them by **osmosis**
- smoking – the food develops a hardened, dry outer surface, and smoke contains antibacterial chemicals
- pickling – this uses an acid pH to kill microorganisms by denaturing their enzymes and other proteins
- irradiation – ionising radiation kills the microorganisms by disrupting their **DNA** structure
- cooling and freezing – these do not kill microorganisms, but retard enzyme activity so their metabolism, growth and reproduction is very slow.

Methods to prevent further contamination include:

- canning – the food is heated and sealed in airtight cans
- vacuum wrapping – air is excluded so microbes cannot **respire** aerobically
- any plastic or paper packaging.

Using microorganisms to make food

Microorganisms have been used for many years in the manufacture of food. Many traditional foods are made with the help of microorganisms.

- Yoghurt is milk that has been affected by *Lactobacillus* bacteria. The *Lactobacillus* uses the lactose sugar in milk to make lactic acid, which causes the milk protein to thicken. The bacteria partially digest the milk, making it easily digestible by humans. Yoghurt can help to ensure that our digestive system contains the non-pathogenic bacteria it needs to aid digestion.
- Cheese is made from milk that has curdled. The solid portion of the milk (curds) is acted upon by *Lactobacillus* bacteria. The cheese can be given additional flavour by contamination with fungi such as *Penicillium* to produce 'blue' cheese.
- Bread is made to rise by yeast, which respires anaerobically to release carbon dioxide. Bubbles of the gas collect in the dough, causing the dough to rise.
- Alcohol is another product of the anaerobic respiration of yeast. Cereal grains containing the sugar **maltose** can be used to brew beer, as the yeast respires the sugar. Grapes contain the sugars fructose and glucose, which the yeast uses in making wine.

More recently, microorganisms have been used to manufacture protein that is used directly as food. This protein is known as single-cell protein (SCP). The best known protein made by microorganisms is Quorn™, a mycoprotein – a protein made by a fungus (in this case the edible fungus *Fusarium venenatum*). Quorn was first produced in the early 1980s. It has since been marketed as a meat substitute for vegetarians. It is also a healthy option for non-vegetarians, as it contains no animal fat or **cholesterol**.

Figure 2 Quorn™

There is huge potential in SCP production using such microorganisms as *Kluyveromyces*, *Scytalidium* and *Candida.* These fungi can produce protein with a similar **amino acid** profile to animal and plant protein. They can grow on almost any organic **substrate**, including waste materials such as paper and whey (curdled milk from which the curds have been removed).

Advantages of using microorganisms

- Production of protein can be many times faster than that of animal or plant protein.
- Production can be increased and decreased according to demand.
- There are no animal welfare issues.
- They provide a good source of protein for vegetarians.
- The protein contains no animal fat or cholesterol.
- SCP production could be combined with removal of waste products.

Disadvantages

- Many people may not want to eat fungal protein or food that has been grown on waste.
- Isolation of the protein – the microorganisms are grown in huge fermenters and need to be isolated from the material on which they grow.
- The protein has to be purified to ensure it is uncontaminated.
- Infection – the conditions needed for the useful microorganisms to grow are also ideal for pathogenic organisms. Care must be taken to ensure the culture is not infected with the wrong organisms (although meat, too, can become contaminated if hygiene is poor).
- Palatability – the protein does not have the taste or texture of traditional protein sources.

Questions

1 What is meant by food spoilage and how does it happen?
2 Describe how the processes of heating and cooling help to preserve foods.

2.2 (5) Organisms that cause disease

By the end of this spread, you should be able to . . .

* **Discuss what is meant by the terms health and disease.**
* **Define and discuss the meaning of the terms parasite and pathogen.**

Key definitions

Health is a state of mental, physical and social wellbeing, not just the absence of disease.

Disease is a departure from good health caused by a malfunction of the mind or body.

A **parasite** is an organism that lives in or on another living thing, causing harm to its host.

A **pathogen** is an organism that causes disease.

Health and disease

Health

Health can be defined as your physical, mental and social wellbeing. If you are in good health, you are:

- free from disease
- able to carry out all the normal physical and mental tasks expected in modern society
- well fed, with a balanced diet
- usually happy, with a positive outlook
- suitably housed with proper sanitation
- well integrated into society.

Disease

Disease can be defined simply as a departure from good health. But this is oversimplified. A disease is a malfunction of the body or mind, which causes symptoms. These symptoms may be physical, mental or social. There is a wide range of diseases that can be grouped into certain categories according to their cause. The diseases caused by living organisms are called infectious diseases. The symptoms are usually physical, but these diseases may have some effects on your mental and social health.

Parasites and pathogens

Parasites

Parasites are organisms that live in or on another living thing (the host). Parasites harm their host. They usually cause harm by taking their nutrition from the host. They may live all or part of their life in or on the host. Parasites that live on their host are called external parasites. An example is the head louse, which lives in people's hair. Parasites that live in their host are called internal parasites. Tapeworms that live in the digestive system are internal parasites.

Parasites may live almost unnoticed by their host. It is probably better for a parasite to be unnoticed so that the host does not try to remove it. If parasites become too numerous, they may become a huge burden to the host. Parasites may cause damage that allows other organisms to invade the host and cause secondary infections.

Figure 1 Head louse (×20)

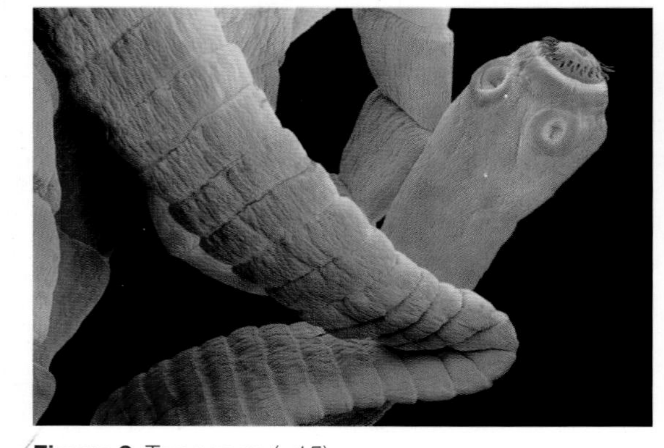

Figure 2 Tapeworm (×15)

Pathogens

The human body creates a good habitat in which microorganisms can live. As a result, there are numerous types of microorganism that live in or on our bodies. Many will cause us no harm – and may be beneficial. Pathogens are organisms that cause disease. They live by taking nutrition from their host, but also cause damage in the process. This damage can be considerable.

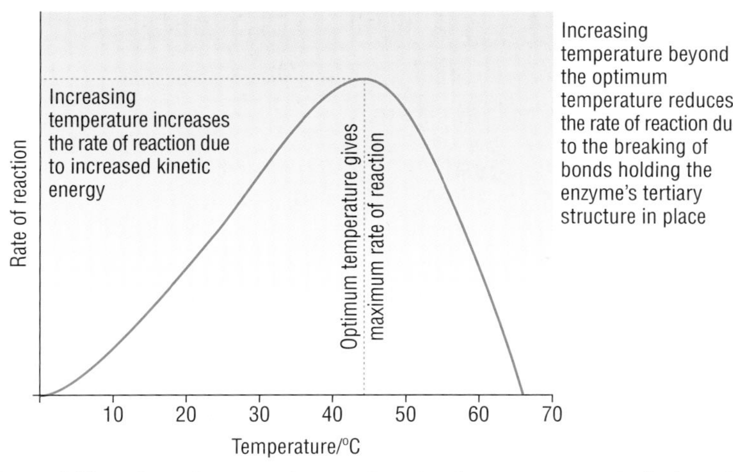

Figure 4 The effect of temperature on the rate of an enzyme-controlled reaction

Speeding up and slowing down – optimum temperature

- Increasing the temperature increases the rate of reaction of an enzyme-controlled reaction at first but, as the temperature increases further, the rate of reaction decreases. Eventually the enzyme will stop working.
- The temperature that gives the maximum rate of reaction is the enzyme's **optimum** temperature. It is a balance between increasing kinetic energy (which increases the number of collisions) and increasing the vibration of the enzyme molecule (which breaks the bonds that hold the tertiary structure in place).

Measuring the effect of temperature on enzyme action usually involves carrying out enzyme-controlled reactions at different temperatures, using a water bath controlled by a thermostat. We can measure the production of product or the disappearance of substrate in a variety of ways (see spreads 2.1.27 and 2.1.28).

Optimum temperatures vary between enzymes

Many enzymes have an optimum temperature somewhere between 40 and 50 °C. But there are some organisms for which such enzymes would be useless. The temperatures at which these organisms live means that they must have heat-resistant enzymes.

In modern genetic engineering (as you will discover in A2 biology), heat-resistant enzymes are used to catalyse reactions involved in making many copies of **DNA** segments. The process is known as the polymerase chain reaction (PCR) and is very similar to DNA replication.

Examiner tip

Don't assume that an enzyme described in an exam question has an optimum temperature of 37 °C. Remember that the optimum temperature is likely to be related either to the organism's environment or to the internal temperature it can maintain.

Questions

1. Explain why increased kinetic energy increases the rate of reaction in an enzyme-controlled reaction.
2. What types of bond would you expect to find in greater numbers holding the tertiary structure of heat-resistant enzymes, compared with more heat-sensitive enzymes? Explain your answer.
3. Suggest why the normal body temperature of mammals is slightly below the optimum temperature of most of the enzymes that occur in the organism.

2.1 (22) Enzymes at work – pH effects

By the end of this spread, you should be able to . . .

* **Describe and explain the effects of pH changes on enzyme activity.**

Hydrochloric acid, HCl, is called an acid because when it dissolves in water it splits up and one of the ions produced is H^+.

$$HCl \rightleftharpoons H^+ + Cl^-$$

The H^+ produced affects the pH of the solution. Acids like hydrochloric acid are known as 'proton donors'.

Figure 1 Proton donors

What is pH?

pH is a measure of the hydrogen ion (H^+) concentration, with values for pH ranging from 1 to 14. In this range, pH 7 is neutral, below pH 7 is referred to as acid, above pH 7 is alkaline (or basic).

The higher the concentration of H^+, the lower the pH value. This means that, in solution, acids contain a high concentration of H^+.

In chemical terms, an acid is defined as a 'proton donor' because hydrogen ions are also known as protons.

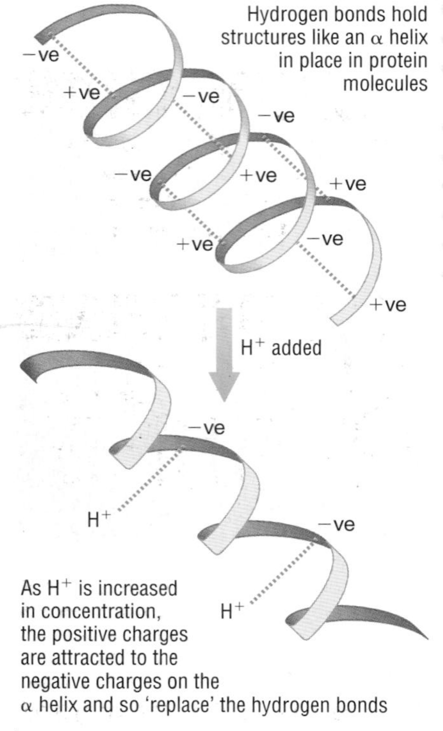

Figure 2 Tertiary structure bonds and interference with hydrogen ions

pH and bonds

A hydrogen ion carries a positive charge, so it will be attracted towards negatively charged ions, molecules or parts of molecules. As like charges repel, hydrogen ions will be repelled by positively charged ions, molecules or parts of molecules.

Large numbers of hydrogen bonds and ionic bonds are responsible for holding the **tertiary structure** of an **enzyme protein** in place. This, in turn, ensures that the **active site** of the enzyme is held in the right shape. These bonds are due to the attraction between oppositely charged groups on the **amino acids** that make up the enzyme protein.

Because of their charge, hydrogen ions can interfere with the **hydrogen bonds** and **ionic bonds** holding the tertiary structure in place. This means that increasing or decreasing the concentration of hydrogen ions around an enzyme can alter the tertiary structure of the enzyme molecule. Changes in pH can cause changes to the shape of the active site, and so change the rate of an enzyme-controlled reaction.

pH and active sites

The **induced-fit hypothesis** (spread 2.1.20) states that an important part of catalysis in the active site relies on charged groups on the R-groups of the amino acids that make up the active site. Increasing the concentration of hydrogen ions will alter the charges around the active site, as more hydrogen ions are attracted towards any negatively charged groups in the active site.

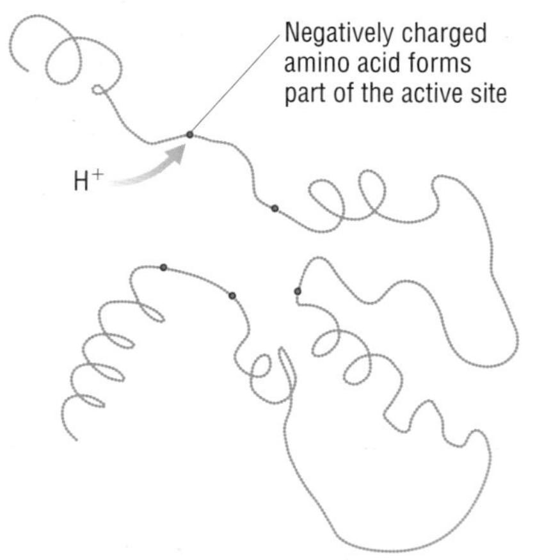

Figure 3 Active site charges and H^+ clustering

Organisms that cause infectious diseases

A wide range of organisms can cause disease. Bacteria, fungi, protoctists and viruses can all cause disease.

Bacteria

Bacteria belong to the kingdom **Prokaryotae** (see spread 2.3.6). Their cells are smaller than our cells, but they can reproduce rapidly. In the right conditions, some types of bacteria can reproduce every 20 minutes or so. Once in the human body, they multiply rapidly. Their presence can cause disease by damaging cells or by releasing waste products that are toxic to us. Cholera is caused by the bacterium *Vibrio cholerae*. Tuberculosis (TB) is caused by two species of bacterium, *Mycobacterium tuberculosis* and *Mycobacterium bovis*.

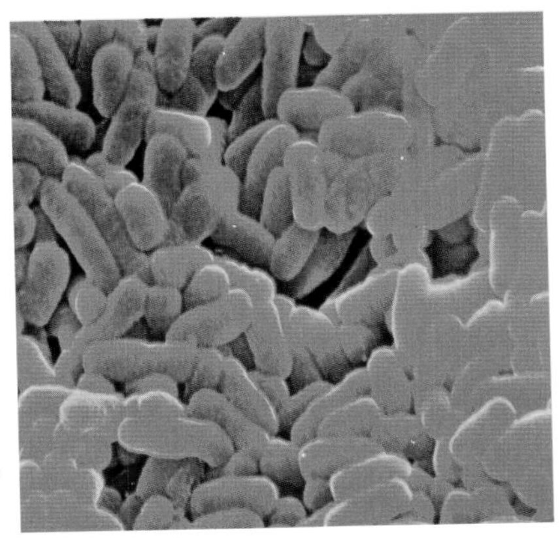

Figure 3 Tuberculosis bacteria (×5000)

Fungi

Fungi can also cause a variety of diseases. Athlete's foot and ringworm are caused by a fungus called *Tinea*. In fact there are a number of species that cause these diseases, including species from the genera *Microsporum, Trichophyton* and *Epidermophyton.* The fungus lives in the skin. When it sends out reproductive **hyphae**, these grow to the surface of skin to release spores. This causes redness and severe irritation.

Viruses

Viruses cause a lot of well-known diseases, including the common cold and influenza. HIV/AIDS is also caused by a virus, and so is the widespread plant disease tobacco mosaic virus (TMV), which affects many plants, including tomatoes. Viruses invade cells and take over the genetic machinery and other organelles of the cell. They then cause the cell to manufacture more copies of the virus. The host cell eventually bursts, releasing many new viruses.

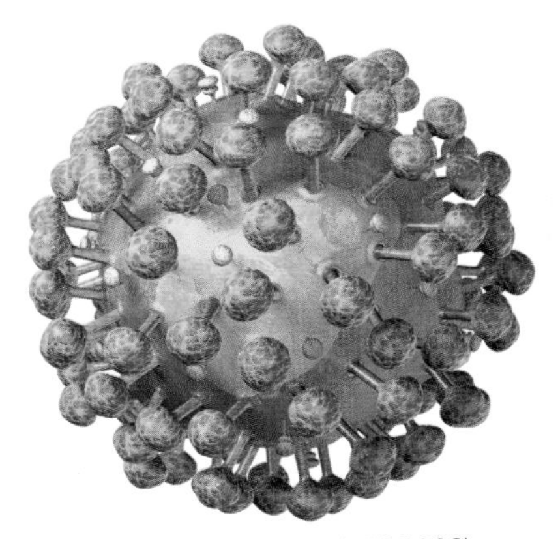

Figure 5 Model of HIV (×500 000)

Protoctista

There are a number of diseases caused by animal-like protoctista (protozoa). Amoeboid dysentery and malaria are two examples. These organisms usually cause harm by entering host cells and feeding on the contents as they grow. The malarial parasite *Plasmodium* has immature forms that feed on the contents of the red blood cells.

Examiner tip

If you are asked to name the causative agent of a disease, such as TB, always give the full name – *Mycobacterium tuberculosis*. However, in a long answer, it is acceptable to give the full name the first time that you use it, and then to abbreviate to *M. tuberculosis* after that.

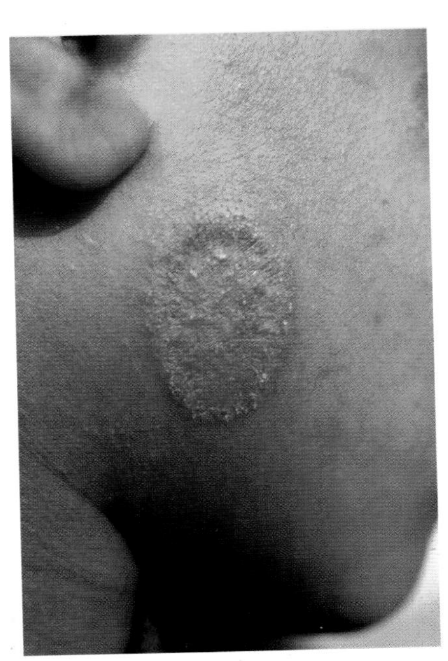

Figure 4 Ringworm (in fact a fungus)

Questions

1 Describe the differences between a parasite and a pathogen.
2 List two diseases caused by each of the following: viruses, bacteria, fungi and protoctista.

2.2 (6) Transmission of diseases

By the end of this spread, you should be able to . . .

* **Describe the causes and means of transmission of malaria, HIV/AIDS and TB.**

In order to cause a disease, a **pathogen** must be able to:
- travel from one host to another
- get into the host's tissues
- reproduce
- cause damage to the host's tissues.

Pathogenic organisms can be transmitted in a variety of ways. Once it reaches a new host, the pathogen will need to pass through any primary defences that the host has to prevent its entry. Once inside the host, the pathogen must overcome any secondary defences or immune responses. Most pathogenic organisms are well adapted to overcoming these obstacles.

The most common forms of transmission are:
- by means of a **vector**
- by physical contact
- by droplet infection.

Key definition

Transmission is the way in which a parasitic microorganism travels from one host to another.

Transmission of malaria

Malaria is caused by a **eukaryotic** organism from the genus *Plasmodium*. There are a number of different species. *Plasmodium falciparum* is the most widespread, but *P. vivax*, *P. ovale* and *P. malariae* also cause malaria.

Malaria is spread by a vector. The female *Anopheles* mosquito carries the *Plasmodium* from an infected person to an uninfected person. Female *Anopheles* mosquitoes feed on blood. They have mouthparts that are adapted as a fine tube or proboscis. This is used to penetrate a blood vessel and withdraw blood. Malarial **parasites** live in the red blood cells of the human host and feed on the **haemoglobin**.

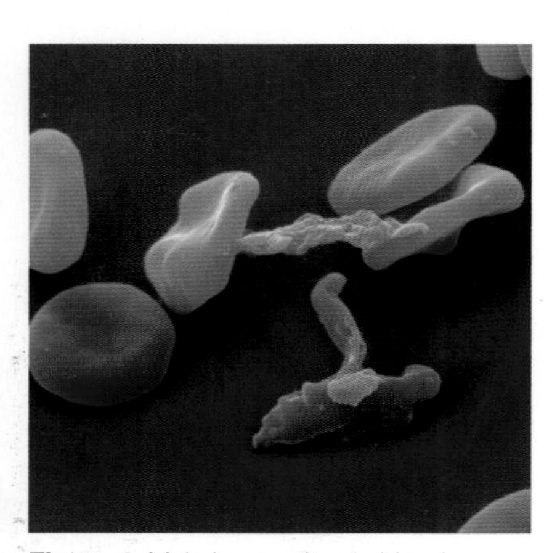

Figure 1 Malaria parasites in blood (×3000)

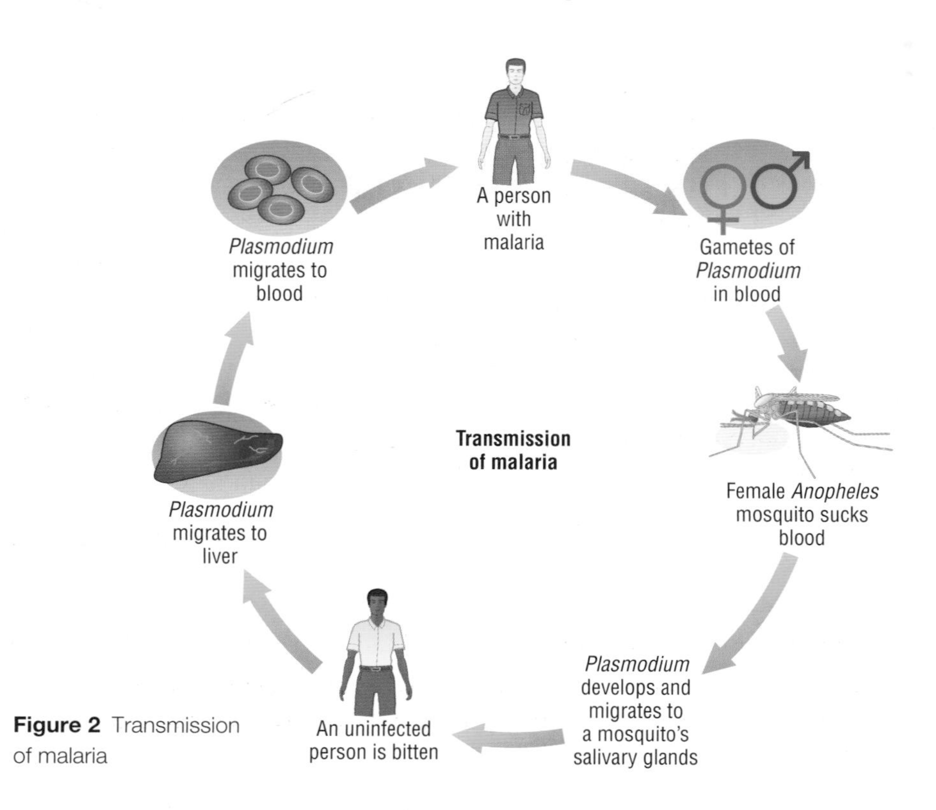

Figure 2 Transmission of malaria

The parasites are transmitted in the following cycle.

- If the host already has malaria, the mosquito will suck the parasite **gametes** into its stomach.
- The gametes fuse and the **zygotes** develop in the mosquito's stomach.
- Infective stages are formed and these move to the mosquito's salivary glands.
- When the mosquito bites another person, it injects a little saliva as an anticoagulant.
- This saliva contains the infective stage of the parasite.
- In the human host, the infective stages enter the liver, where they multiply before passing into the blood again.
- In the blood they enter red blood cells, where the gametes are produced.

The malarial parasite can also be transmitted by careless and unhygienic medical practices. Unscreened blood transfusions and use of unsterilised needles can transmit *Plasmodium*. It is also able to pass across the placenta into an unborn child.

Transmission of HIV/AIDS

HIV/AIDS is caused by the human immunodeficiency virus. The virus enters the body and may remain inactive. This is known as being HIV-positive. Once the virus becomes active it attacks and destroys T helper (T_h) cells in the immune system. These cells normally help to prevent infection. If they are destroyed, your ability to resist infection is reduced. You will be unable to defend yourself against any pathogen that enters your body and you may contract a range of **opportunistic infections**. It is the effect of these diseases that eventually kills a person with HIV. AIDS stands for acquired immune deficiency syndrome.

HIV can be transmitted in the following ways:

- exchange of body fluids such as blood-to-blood contact
- unprotected sexual intercourse
- unscreened blood transfusions
- use of unsterilised surgical equipment
- sharing hypodermic needles
- accidents such as 'needle-stick'
- across the placenta or during childbirth
- from mother to baby during breast feeding.

Transmission of tuberculosis

TB is caused by a bacterium. There are two species, *Mycobacterium tuberculosis* and *M. bovis*. These bacteria can affect many parts of the body, but TB is usually found in the lungs. TB is transmitted by droplet infection. It is thought that up to 30% of the world's population may be infected with TB. However, in many people it is inactive or controlled by their immune system.

Figure 3 Sneezing expels droplets

The old saying 'coughs and sneezes spread diseases' is very true of TB. The bacteria are contained in the tiny droplets of liquid that are released when an infected person coughs, sneezes, laughs or just talks. You can become infected when you inhale those droplets. Fortunately, it is not that easy to contract TB. It usually takes close contact with an infected person over a prolonged period to contract the disease. There are a number of conditions that make contraction and spread more likely:

- overcrowding – many people living and sleeping together in one house
- poor ventilation
- poor health – particularly if a person has HIV/AIDS, they are more likely to contract TB
- poor diet
- homelessness
- living or working with people who have migrated from areas where TB is more common.

TB can also be contracted from the milk or meat of cattle. This form of transmission is no longer a great issue in the developed countries, but it still remains a source of infection elsewhere.

Questions

1 What diseases can be spread by sneezing?

2 Explain why someone with AIDS is more likely to catch opportunistic diseases.

2.2 ⑦ The worldwide importance of certain diseases

By the end of this spread, you should be able to . . .

* **Discuss the global impact of malaria, HIV/AIDS and TB.**

Key definition

The World Health Organisation is a part of the United Nations. It is the specialised agency for health. WHO's objective is the attainment by all peoples of the highest possible level of health.

Why are diseases important?

The World Health Organisation (WHO) maintains that good health is a human right. Poor health causes a lot of suffering. Ill health also has an economic cost as a result of the need to provide medical services, and due to the loss of productivity. People who are ill can't work. In many parts of the world, people do not have access to the basic requirements for good health. In many less economically developed countries, the following aspects may contribute to the poor health of the people. There may be:

- poverty
- lack of proper shelter
- lack of purified water
- poor nutrition
- poor hygiene
- lack of investment by the government
- poor or inadequate health services
- inadequate education about the causes of disease and how they are transmitted
- civil unrest or warfare
- inadequate transport facilities that prevent people reaching medical assistance.

Malaria

Malaria kills about 3 million people each year. About 300 million people are affected by malaria worldwide, and the number is increasing. Malaria is limited to the areas in which the vector mosquito, *Anopheles*, can survive. This is currently the tropical regions. Of all people with malaria, 90% live in sub-Saharan Africa. With the onset of global warming, the *Anopheles* mosquito may soon be able to survive further north, even in parts of Europe.

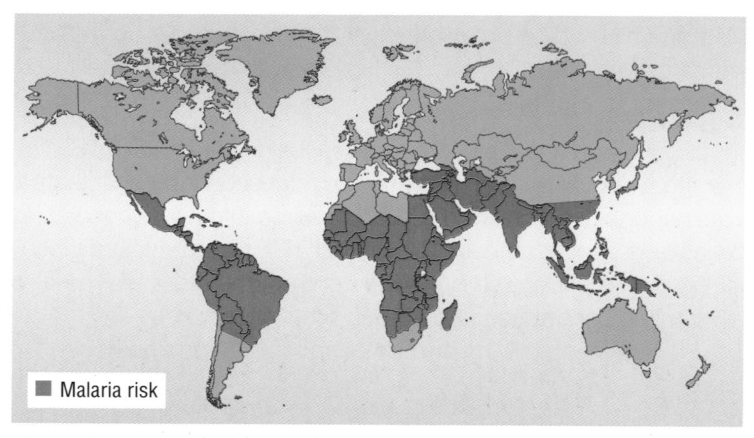

Figure 1 Distribution of malaria

HIV/AIDS

HIV/AIDS is a worldwide disease. It is still spreading in **pandemic** proportions. There were approximately 45 million people living with HIV/AIDS at the end of 2005. More than half of these were in the region of sub-Saharan Africa.

Every year about 5 million people are newly infected. By the end of 2005, nearly 30 million people had died from HIV/AIDS-related diseases. By 2006–07 HIV/AIDS was spreading rapidly in China, Russia and other Eastern European countries. It is thought the number of people with HIV/AIDS in China will soon exceed the number in any other country.

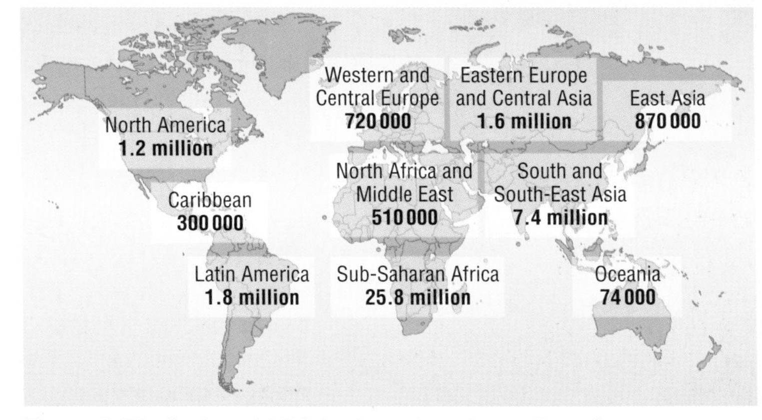

Figure 2 Distribution of AIDS (estimated numbers infected)

Tuberculosis

TB is another worldwide disease. WHO declared TB to be a public health emergency in 1993. Since then the number of cases rose steadily each year. Approximately 1% of the world's population is newly infected each year, and 10–15% of those go on to develop the disease. The percentage of the world's population struck by TB peaked in 2004 and held steady in 2005. In 2005 about 8.8 million new cases of TB were recorded and approximately 1.6 million people died of TB. Current estimates suggest that up to 30% of the world's population may be infected with *Mycobacterium*. It is therefore still a great threat. TB is particularly common in South-east Asia and in sub-Saharan Africa. It is also rising in Eastern Europe. There is an increasing threat from new strains of *Mycobacterium* that are resistant to most of the drugs available to treat it.

Health organisations

Health organisations such as WHO have limited resources with which to combat worldwide disease. So they have to prioritise their efforts. They use **epidemiology** to study the spread of disease and the factors affecting that spread. Through epidemiology, it is possible to:

- identify the cause of a disease
- identify **risk factors** associated with a disease
- determine the **incidence** of a disease (the number of new cases in a population per year)
- determine the **prevalence** of a disease (the number of people with the disease at a given time)
- determine the **mortality** (the number of people who die from the disease per year)
- determine the **morbidity** (the number of people with the disease as a proportion of the population)
- study how quickly it is spreading
- identify a disease as **endemic** (always present in the population), **epidemic** (spreading rapidly to a lot of people over a large area) or **pandemic** (a worldwide epidemic)
- identify countries at risk
- identify which parts of a population are at risk
- check how well control programmes are working.

With this sort of information, health organisations and public health authorities can plan to use their resources most effectively. This may mean:

- targeting education programmes at the people who are most at risk, to inform them of the risks and how to avoid them
- targeting advertisements to raise awareness
- targeting screening programmes to identify individuals at risk
- providing specialised healthcare in certain areas
- providing vaccination programmes for the major diseases
- targeting research to find cures for the major diseases.

Examiner tip

When describing rates of disease always use any numbers or figures you can, and include units.

Questions

1 Explain why people in the less economically developed countries are more likely to suffer from infectious diseases.

2 Explain the difference between prevalence and incidence.

8 Non-specific responses to diseases

By the end of this spread, you should be able to . . .

* Define the term **immune response**.
* Describe the primary lines of defence against **pathogens** and **parasites** (including skin and mucous membranes) and outline their importance (no details of skin structure are required).
* Describe the structure and mode of action of **phagocytes**.

Key definitions

The **primary defences** are those that attempt to prevent pathogens from entering the body.

The **immune response** is the specific response to a pathogen, which involves the action of lymphocytes and the production of antibodies.

Primary defences

Pathogenic organisms need to enter the body of their host before they can cause harm. **Evolution** has **selected** hosts **adapted** to defend themselves against such invasions. The mechanisms that have evolved to prevent entry of pathogenic organisms are called **primary defences**.

The skin

The body is covered by the skin. This is the main primary defence. The outer layer of the skin is called the **epidermis**. The epidermis consists of layers of cells. Most of these cells are called **keratinocytes**. These cells are produced by **mitosis** at the base of the epidermis. They then migrate out to the surface of the skin. As they migrate, they dry out and the cytoplasm is replaced by the protein **keratin**. This process is called keratinisation. It takes about 30 days. By the time the cells reach the surface, they are no longer alive. Eventually the dead cells slough off. The keratinised layer of dead cells acts as an effective barrier to pathogens.

Mucous membranes

Certain substances, such as oxygen and the nutrients in our food, must enter our blood. This leaves the body exposed to infection, as the barrier between the blood and our environment is reduced. In areas such as the airways, lungs and the digestive system, there is great potential for infection. We take air and food in from our environment, and these materials could be harbouring many microorganisms.

These areas are protected by mucous membranes. The epithelial layer contains **mucus**-secreting cells called **goblet cells**. In the airways, the mucus lines the passages and traps any pathogens that may be in the air. The **epithelium** also has ciliated cells. The **cilia** are tiny, hair-like organelles that can move. They move in a coordinated fashion to waft the layer of mucus along. They move the mucus up to the top of the trachea, where it can enter the oesophagus. It is swallowed and passes down the digestive system. Most pathogens in the digestive system are killed by the acidity of the stomach, which can be pH 1–2. This denatures the pathogen's **enzymes**.

Mucous membranes are also found in the gut, genital areas, anus, ears and nose.

Other primary defences

- The eyes are protected by **antibodies** in the tear fluid.
- The ear canal is lined by wax, which traps pathogens.
- The vagina is protected by maintaining relatively acidic conditions.

Phagocytes – a secondary defence

Many trapped pathogens are not killed by the conditions in the body. These pathogens must be killed before they reproduce and cause the symptoms of a disease. This is the job of the numerous non-specific **phagocytes**.

There are two types of phagocyte.

- The most common phagocytes are **neutrophils**. You can recognise these cells by their multilobed nucleus. Neutrophils are manufactured in the bone marrow. They travel in the blood and often squeeze out of the blood into tissue fluid. They may also be found on epithelial surfaces such as the lungs. Neutrophils are short-lived, but they will be released in large numbers as a result of an infection.
- **Macrophages** are larger cells manufactured in the bone marrow. They travel in the blood as **monocytes**. They tend to settle in the body organs, particularly in the lymph nodes. Here they develop into macrophages. Macrophages also play an important role in the specific responses to invading pathogens.

How phagocytes work

Phagocytes engulf and destroy pathogenic cells. When a pathogen invades the body, it is recognised as foreign by the chemical markers on its outer membrane. These markers are called **antigens**. Antigens are specific to the organism. Our own cells have antigens, but these are recognised as our own and do not produce a response.

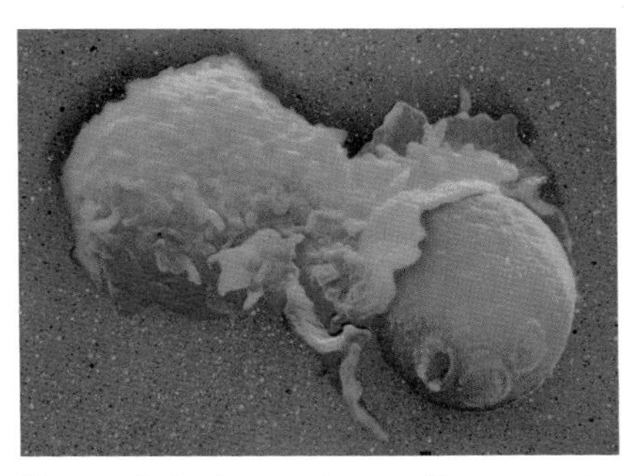

Figure 1 A phagocyte engulfing a yeast (*Candida*) (×1500)

Proteins in the blood, called **antibodies**, attach to the foreign antigens. Phagocytes have membrane-bound proteins that act as **receptors**. The receptor binds to the antibodies already attached to the pathogen. This process may be assisted by other proteins called opsonins.

Once the phagocyte is bound to the pathogen, it will envelop the pathogen by folding its membrane inwards. The pathogen is trapped inside a vacuole called a **phagosome**. **Lysosomes** fuse with the phagosome and release **enzymes** into it. These enzymes are called lysins, and these digest the bacterium. The end products are harmless nutrients that can be absorbed into the cytoplasm. The neutrophils are short-lived and will die soon after digesting a few pathogens. They may collect in an area of infection to form pus.

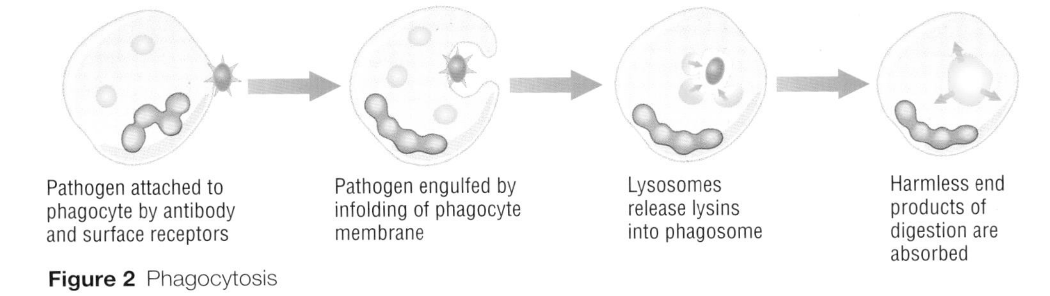

Figure 2 Phagocytosis

Examiner tip

Don't say that phagocytes 'eat' microorganisms. They engulf and ingest them.

The role of macrophages

Infected cells release chemicals such as histamine, which attracts neutrophils to the area. Histamine also causes a response that makes the capillaries more leaky. As a result, more fluid leaves the capillaries in the area of infection. This causes swelling and redness, but it also means that more tissue fluid passes into the **lymphatic system**. This leads the pathogens towards the macrophages waiting in the lymph nodes.

The macrophages play an important role in initiating the specific response to a disease. This is known as the **immune response**. The immune response is the activation of **lymphocytes** in the blood to help fight the disease.

Questions

1 Describe how a neutrophil performs phagocytosis.
2 Explain what is meant by a non-specific response.

2.2 (9) Antibodies

By the end of this spread, you should be able to . . .

* **Define the terms antigen and antibody.**
* **Describe, with the aid of diagrams, the structure of antibodies.**
* **Outline the mode of action of antibodies, with reference to the neutralisation and agglutination of pathogens.**
* **Compare and contrast the primary and secondary immune responses.**

Key definitions

Antigens are molecules that stimulate an immune response.

Antibodies are protein molecules that can identify and neutralise antigens.

Specificity: an antibody is specific to a particular antigen because of the shape of the variable region. Each type of antibody has a differently shaped variable region.

Examiner tip

Never say that the variable region of an antibody changes shape to fit the antigen. A separate type of antibody is made for each type of antigen.

Antigens

Antigens are molecules that can stimulate an **immune response**. Almost any molecule could act as an antigen. Antigens are usually large molecules that have a specific shape. A foreign antigen will be detected by the immune system and will stimulate the production of **antibodies**. These antibodies will be specific to the antigen. As the antigen is specific to the organism, we can think of the antibody being specific to the **pathogen**. Antigens are usually a **protein** or **glycoprotein** in or on the **plasma membrane** (cell surface membrane). Our own antigens are recognised by our immune system and do not stimulate any response.

Antibodies

Antibodies are molecules produced by the **lymphocytes** in the immune system. They are released in response to an infection. Antibodies are large proteins and are also known as immunoglobulins. They have a specific shape that is **complementary** to that of a particular antigen. Therefore antibodies are specific to particular antigens. Our immune system must manufacture one type of antibody for every antigen that is detected. Antibodies attach to antigens and render them harmless.

The structure of an antibody

Antibody molecules are Y-shaped and have two distinct regions. The structure of an antibody molecule includes the following features.

- Four **polypeptide** chains held together by disulfide bridges.
- A constant region, which is the same in all antibodies. This enables the antibody to attach to **phagocytic** cells and helps in the process of phagocytosis.
- A variable region, which has a specific shape and differs from one type of antibody to another. This is the result of its **amino acid** sequence. It ensures that the antibody can attach only to the correct antigen. The shape of the variable region is complementary to that shape of the antigen, and can bind to that antigen.
- Hinge regions, which allow a certain degree of flexibility. These allow the branches of the Y-shaped molecule to move further apart in order to allow attachment to more than one antigen.

Figure 1 An antibody molecule

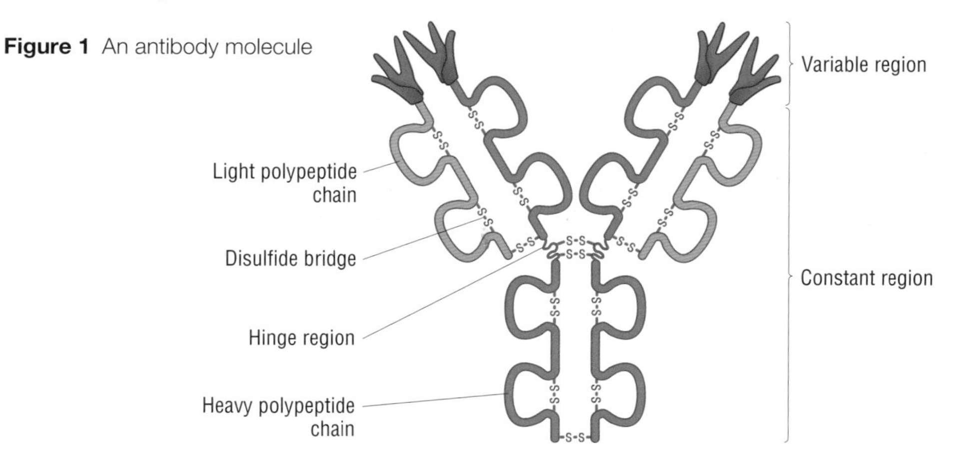

How antibodies work

Most antibodies work by attaching to the antigens on a pathogen. The antigen is a molecule on the cell surface membrane of the pathogen. The pathogen may have another use for this molecule. For example, it may be a binding site, which would be used to bind to the host cell. If the antibody blocks this binding site, the pathogen cannot bind to its host cells. This is called neutralisation.

Some antibodies are larger than the Y-shaped molecule described above. They resemble several Y-shaped molecules attached together, and have many specific variable regions. Each variable region can act as a binding site to bind to an antigen on a pathogen. If the antibody has a number of binding sites, it may be able to attach to a number of pathogens at the same time. This is called agglutination. When the pathogens are stuck together, they cannot enter host cells.

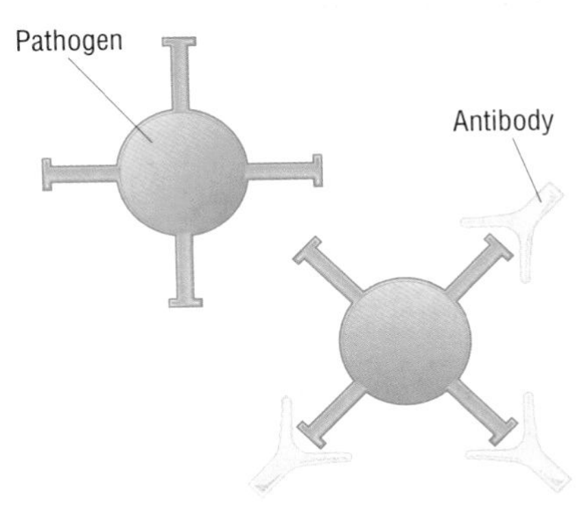

(a) Neutralisation – antibodies covering the pathogen binding sites prevent the pathogen from binding to a host cell and entering the cell

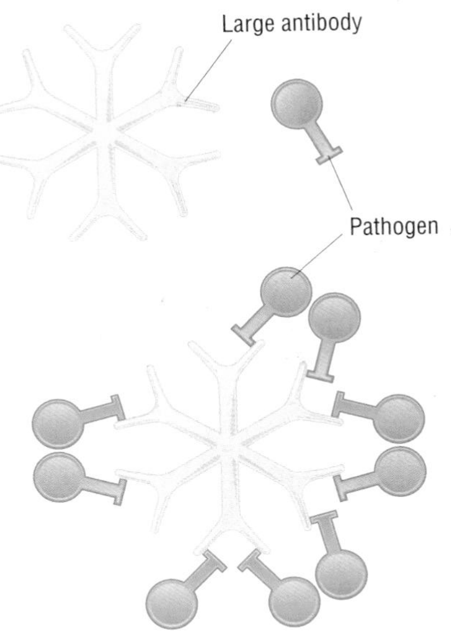

(b) Agglutination – a large antibody can bind many pathogens together. The group of pathogens is too large to enter a host cell

Figure 2 Neutralisation and agglutination

Producing antibodies

Antibodies are produced in response to infection. When an infecting agent is first detected, the immune system starts to produce antibodies. But it takes a few days before the number of antibodies in the blood rises to a level that can combat the infection successfully. This is known as the primary **immune response**. Once the pathogens have been dealt with, the number of antibodies in the blood drops rapidly.

Antibodies do not stay in the blood. If the body is infected a second time by the same pathogen, the antibodies must be made again. But the immune system can swing into action more quickly. This time the production of antibodies starts sooner and is much more rapid. So the concentration of antibodies rises sooner and reaches a higher concentration. This is known as the secondary immune response.

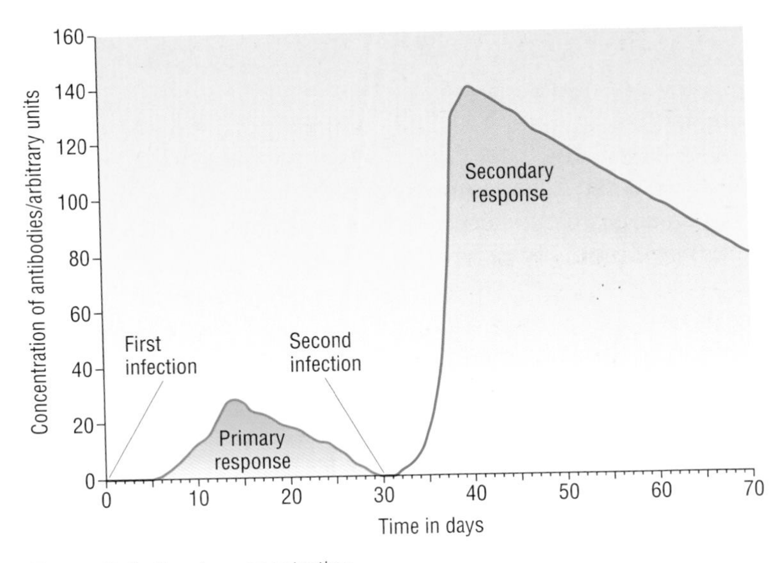

Figure 3 Antibody concentration

Examiner tip

Don't confuse pathogens with antigens. The pathogen is the infecting agent. It has antigens on its surface. The antibody is made in response to the antigens and fits onto the antigens. So don't call a pathogen an antigen.

Questions

1 Explain the difference between antigens and antibodies.
2 Describe how the structure of an antibody is related to its function.

2.2 (10) Communication between cells

By the end of this spread, you should be able to . . .

* **Describe the structure and mode of action of T lymphocytes and B lymphocytes, including the significance of cell signalling.**

Key definition

Cell signalling is the communication between cells that allows effective coordination of a response.

Cell signalling

The **immune response** is a specific response to the detection of **pathogens** in the body. It involves a coordinated response between a wide range of cells. In order to work together effectively, these cells need to communicate. This is known as **cell signalling**. Figure 1 shows the sequence of events in the immune response. You can see how many cell types are involved and why it is so important that they can communicate effectively.

This communication is achieved through cell surface molecules and through the release of hormone-like chemicals called **cytokines**. In order to detect a signal, the target cell must have a cell surface **receptor**. B lymphocytes (B cells) and T lymphocytes (T cells) have receptors that are **complementary** in shape to the foreign **antigen**. The antigen may be an isolated **protein**, it may be attached to a **pathogen**, or it may be on the surface of a host cell. When the antigen is detected, the **lymphocyte** is activated or stimulated. Chemical signals are also detected by their target cells using specialised cell surface receptors.

What sort of information is communicated?

Identification

The first signalling is actually done by the pathogen. The pathogen carries antigens on its cell surface. These act as flags or markers that say 'I am foreign'. These are detected by our body cells.

Sending distress signals

When a body cell is infected by a pathogen, it is usually damaged in some way. The internal cell organelles such as **lysosomes** will attempt to fight the invader. As a result, a number of pathogen cells are damaged. Parts of the pathogen often end up attached to the host **plasma membrane**. These have two effects:

- they can act as a distress signal and can be detected by cells from the immune system
- they also act as markers to indicate that the host cell is infected – T killer cells can recognise that the cell is infected and must be destroyed.

Antigen presentation

This process has been taken a step further by the **macrophages** in the lymph nodes. Macrophages act like **phagocytes** to engulf and digest the pathogen. But they do not fully digest it. They separate out the antigens and incorporate them into a cell surface molecule. This is exposed on the surface of the macrophage, which becomes known as an **antigen-presenting cell**. Its function is to find the lymphocytes that can neutralise that particular antigen.

Instructions

There are a range of cytokines released by cells. These chemical signals act as instructions to their target cells. They generally act over short distances at very low concentration. They act by binding to specific membrane-bound receptors on the target cell. This will cause the release of second messengers inside the cell, which alter its behaviour through gene expression.

Communication using cytokines includes the following:

- macrophages release **monokines** that attract **neutrophils** (by **chemotaxis** – the movement of cells towards a particular chemical)

- macrophages release monokines that can stimulate B cells to differentiate and release **antibodies**
- T cells, B cells and macrophages release **interleukins**, which can stimulate proliferation and differentiation of B and T cells
- Many cells can release **interferon**, which can inhibit virus replication and stimulate the activity of T killer cells.

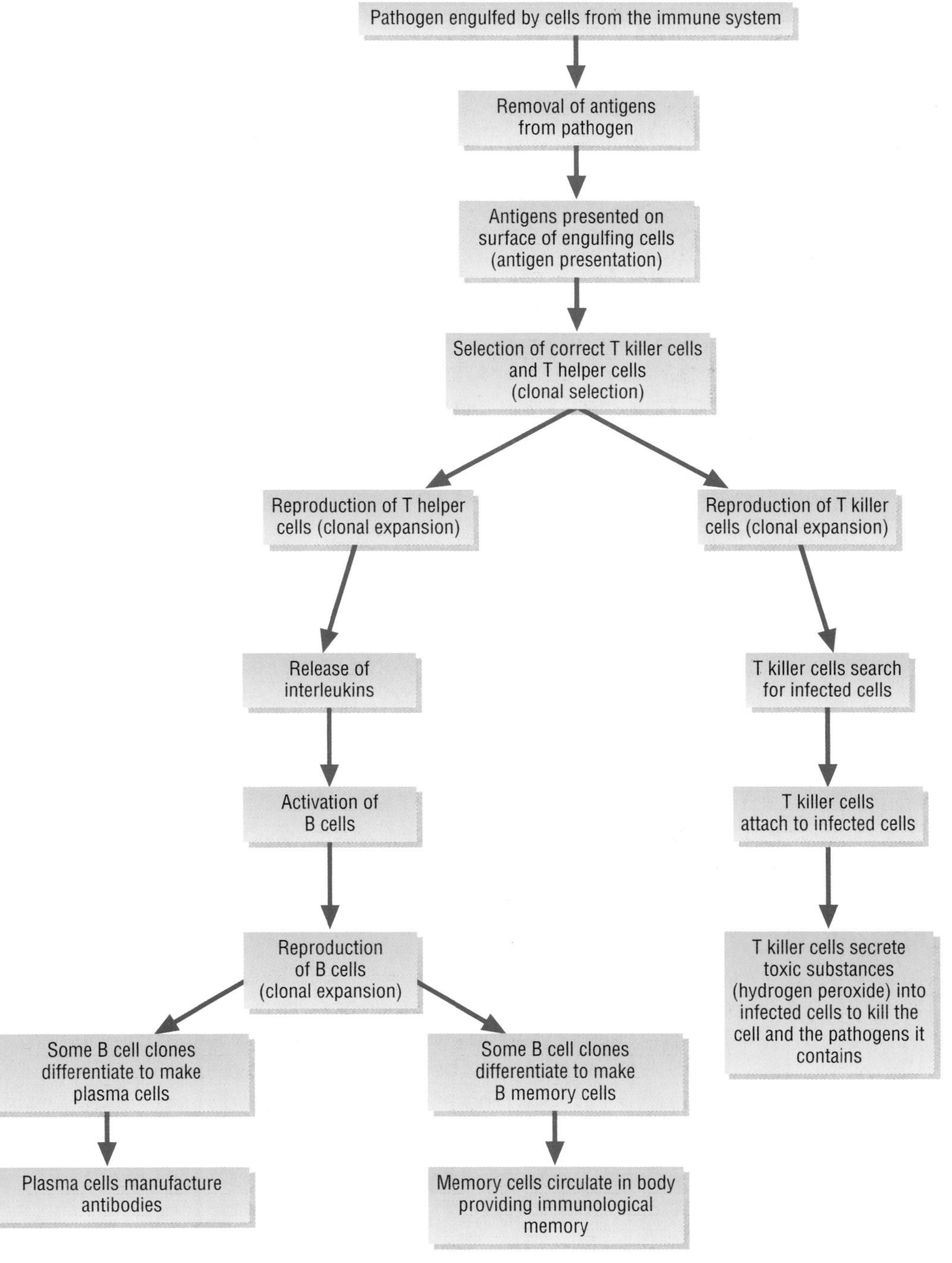

Figure 1 Communication between cells in the specific immune response

Examiner tip

There are many types of **receptors** in and on cells. Remember to state that these are bound to the plasma membrane (cell surface membrane).

Questions

1 Explain why cell surface receptors must be specific.
2 Explain why the immune system does not attack our own cells.

2.2 (11) The specific immune response

By the end of this spread, you should be able to . . .

* **Describe the structure and mode of action of T lymphocytes and B lymphocytes, including the role of memory cells.**

Key definitions

Memory cells are cells that circulate in the blood after an immune response. They speed up the response to a subsequent attack by the same pathogen.

Lymphocytes are white blood cells that circulate around the body in the blood and lymph. B cells mature in the bone marrow, while T cells originate in the bone marrow but mature in the thymus gland.

The immune response

The **immune response** is a specific response to the detection of **pathogens** in the body. It involves B **lymphocytes** (B cells) and T lymphocytes (T cells). These are white blood cells with a large nucleus and specialised **receptors** on their **plasma membranes** (cell surface membranes). The immune response produces **antibodies**. It is the antibodies that actually neutralise the foreign antigens. The immune response also provides long-term protection from the disease. It produces **immunological memory** through the release of **memory cells** which circulate in the body for a number of years.

Starting the response

An invading pathogen has foreign **antigens**. In order to trigger the immune response, these must be detected by specific T lymphocytes and B lymphocytes. These lymphocytes carry the correct **receptor** molecules on their membranes. The receptor molecules are cell surface **proteins** that have a shape that is **complementary** to the shape of the antigen. Once the correct T or B lymphocytes detect the antigens, the immune response can begin.

However, there may be only one or a very few of the correct T and B lymphocytes in the body. It may take some time for them to find the antigens. The presentation of foreign antigens by a number of cells increases the chances that the correct B and T lymphocytes will locate the antigens. Cells that are attacked by the pathogen will display antigens on their surface. **Macrophages** in the lymph system can become antigen-presenting cells. Thus many copies of the pathogen's antigens are displayed.

The selection of the correct B and T lymphocytes is known as **clonal selection**. Before they can become effective in fighting the pathogen, these lymphocytes must increase in numbers. This is called **clonal expansion**. The lymphocytes divide by **mitosis** a number of times – they are cloned.

Differentiation

The B and T lymphocytes do not manufacture the antibodies directly. Once selected, the T lymphocytes develop or differentiate into three types of cell:

- T helper (T_h) cells, which release **cytokines** (chemical messengers) that stimulate the B cells to develop and stimulate **phagocytosis** by the **phagocytes**
- T killer (T_k) cells, which attack and kill infected body cells
- T memory (T_m) cells.

The B lymphocytes develop into two types of cell:

- effector or **plasma cells**, which flow around in the blood, manufacturing and releasing the antibodies
- B memory cells that remain in the body for a number of years and act as the immunological memory.

Examiner tip

It is the *plasma cells* that manufacture antibodies, not the lymphocytes.

This all takes time

Each process takes time. The correct lymphocytes must be selected. The cells must divide to increase in number. They must differentiate into plasma cells. The plasma cells must manufacture the antibodies. All this means that it may be a few days before the number of antibodies in the blood starts to rise.

The immune response leaves memory cells in the body. If there is a second invasion by the same pathogen, these can stimulate the production of plasma cells and antibodies much more quickly.

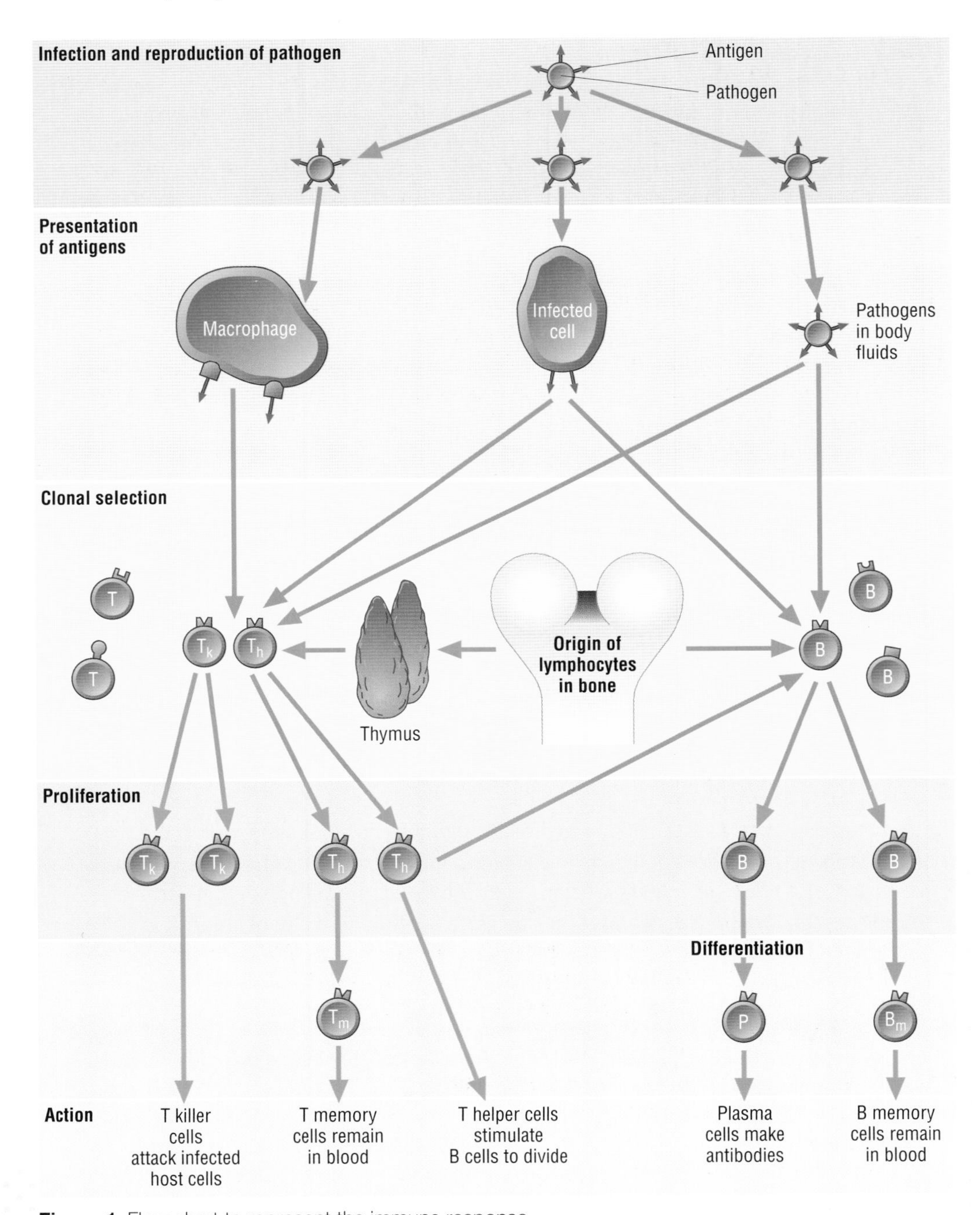

Figure 1 Flow chart to represent the immune response

Questions

1 What is the difference between T helper and T killer cells?

2 What is the role of the memory cells?

2.2 (12) Vaccination

By the end of this spread, you should be able to . . .

* **Explain how vaccination can control disease.**
* **Discuss the responses of governments and other organisations to the threat of new strains of influenza each year.**
* **Compare and contrast active, passive, natural and artificial immunity.**

Key definition

A **vaccination** is a deliberate exposure to antigenic material, which activates the immune system to make an immune response and provide immunity.

What is a vaccination?

Vaccination provides immunity to specific diseases. A person who has been vaccinated has artificial **active immunity**. This is created by deliberate exposure to **antigenic** material that has been rendered harmless. The immune system treats the antigenic material (see spreads 2.2.10 and 2.2.11) as a real disease. As a result, the immune system manufactures **antibodies** and **memory cells**. The memory cells provide the long-term immunity.

The antigenic material used in vaccines can take a variety of forms.

- Whole, live microorganisms – usually ones that are not as harmful as those that cause the real disease. But they must have very similar antigens so that the antibodies produced will be effective against the real pathogen (e.g. the smallpox vaccine).
- A harmless or attenuated version of the **pathogenic** organism (e.g. measles and TB vaccines).
- A dead pathogen (e.g. typhoid and cholera vaccines).
- A preparation of the antigens from a pathogen (e.g. hepatitis B vaccine).
- Some harmless toxin (called a toxoid) (e.g. tetanus vaccine).

Vaccination can be achieved by injection, or the vaccine can be taken orally.

Vaccination and control of diseases

Vaccination can be used to control disease by providing immunity to all those who are at risk. It is necessary to have some form of reporting procedure so that new cases of the disease can be dealt with quickly. There are two ways to use vaccination.

Herd vaccination

Herd vaccination is using a vaccine to provide immunity to all or almost all of the population at risk. Once enough people are immune, the disease can no longer spread. In order to be effective, it is essential to vaccinate almost all the population. To eradicate smallpox, it was necessary to vaccinate 80–85% of the population. It is estimated that at least 95% of the population would need to be immunised in order to prevent the spread of measles.

In the UK there is a vaccination programme to immunise young children against the following diseases: TB, diphtheria, tetanus, whooping cough, polio, meningitis, measles, mumps and rubella.

Ring vaccination

Ring vaccination is used when a new case of a disease is reported. Ring vaccination involves vaccinating all the people in the immediate vicinity of the new case(s). This may mean vaccinating the people in the surrounding houses, or even in the whole village or town. Ring vaccination is used in many parts of the world to control the spread of livestock disease.

Possible future threats

Many pathogenic organisms can form a new strain by **mutation**. This may mean that the existing vaccines will have little or no effect on the new strain. Diseases caused by viruses are a particular threat.

Influenza

Influenza is a killer disease caused by a virus. It affects the respiratory system. People over 65 years of age and those with other respiratory tract conditions are particularly at risk. Occasionally a new strain of flu virus arises that is particularly virulent. This may cause an **epidemic**. For example, in 1918 a flu epidemic killed at least 40 million people worldwide (some estimates are as high as 100 million). More recently in 1968/69 about one million people were killed by Hong Kong flu, also known as the H1N1 strain of the virus. Such a large-scale outbreak of a disease is called a **pandemic**.

In an attempt to avoid another pandemic, people at risk may be immunised. In the UK there is a vaccination programme to immunise all those aged over 65 and those who are at risk for any other reason. In 2006/07 about 74% of people over 65 were vaccinated along with about 42% of younger people in 'at-risk' groups. The strains of flu used in this immunisation programme change each year. Research is undertaken to determine which of the strains of flu are most likely to spread that year.

Active and passive immunity

Active immunity is the immunity that is achieved by activation of the immune system. **Lymphocytes** in the body manufacture antibodies and release them into the blood. This form of immunity can usually last for many years or even for a lifetime.

In contrast, **passive immunity** is provided by antibodies that have not been manufactured by stimulating the recipient's immune system. Antibodies may be provided by a mother across the placenta or via breast milk (but not by bottle-feeding). Antibodies can also be provided by intravenous injection. This form of immunity is often short-lived.

Natural and artificial immunity

Natural immunity is gained in the normal course of living processes. It may be gained as a result of an infection that stimulates an **immune response**.

Artificial immunity is gained by deliberate exposure to antibodies or antigens.

Natural immunity can be active or passive. Artificial immunity can also be active or passive. Some examples of natural, artificial, passive and active immunity are shown in Table 1.

	Natural	Artificial
Passive	Antibodies provided via the placenta or via breast milk. This makes the baby immune to diseases that the mother is immune to. It is very useful in the first year of the baby's life, when its immune system is developing.	Immunity provided by injection of antibodies made by another individual (e.g. tetanus injections).
Active	Immunity provided by antibodies made in the immune system as a result of infection. A person suffers from the disease once and is then immune (e.g. immunity to chicken pox).	Immunity provided by antibodies made in the immune system as a result of vaccination. A person is injected with a weakened, dead or similar pathogen, or with antigens, and this activates his/her immune system (e.g. immunity to TB and influenza).

Table 1 Types of immunity

Questions

1. Why is bird flu considered to be such a threat?
2. Explain how herd vaccination could be used to protect us from bird flu.

13 Finding new drugs

By the end of this spread, you should be able to . . .

* **Outline possible new sources of medicines, with reference to microorganisms (e.g. antibiotics) and plants (e.g. from tropical rainforest species) and the need to maintain biodiversity.**

The need for new drugs

There are currently over 6000 different kinds of medicine available in the UK. But new drugs are needed because:

- new diseases are emerging
- there are still many diseases for which there are no effective treatments
- some **antibiotic** treatments are becoming less effective – the microorganisms that cause disease continue to **evolve**. As soon as we start to use a new drug, it acts as a **selection pressure**. Any strains of microorganism that are less susceptible to, or are resistant to, the drug will be at a selective advantage. These resistant organisms are more likely to survive and reproduce, and the next generation will be more resistant.

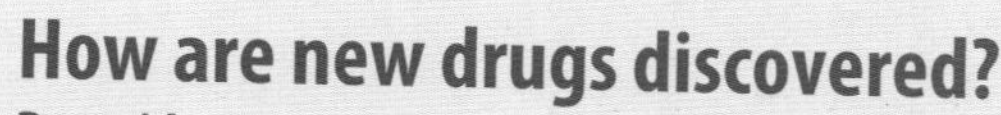

How are new drugs discovered?

By accident

The accidental discovery of penicillin by Alexander Fleming is well documented. Most antibiotics currently in use are made by the bacterium *Streptomyces*. The antibiotics neomycin, chloramphenicol and streptomycin are all made from *Streptomyces*. But streptomycin is now rarely used, as most bacteria are resistant to it. The important antifungal drug Nystatin is derived from *Streptomyces noursei*.

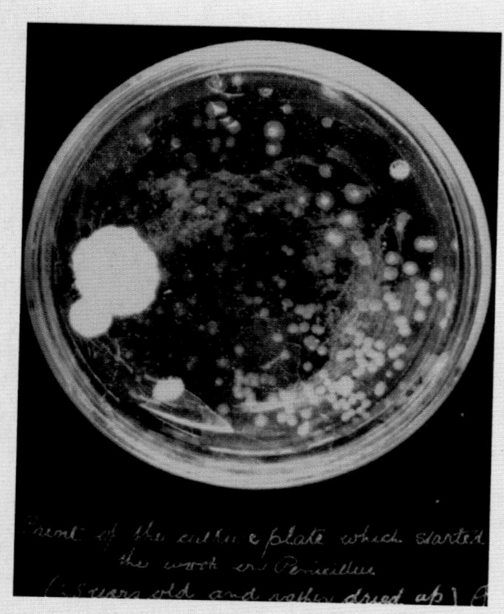

Figure 1 Fleming's bacterial culture

Traditional medicine

Many drugs have been used for centuries. The World Health Organization calculates that 80% of the world's population relies on traditional medicines. In India some 7000 different plants are used for their medicinal properties, and in China they use about 5000 different plants. In Europe, too, some of our modern drugs have their origins in traditional medicine.

Anaesthetics

The sap of unripe poppies was used in Neolithic times in parts of southern Europe and Egypt. In the twelfth century opium from poppies was used as an anaesthetic and by the nineteenth century morphine and opium were being used. These opiate drugs reduce nervous action in the central nervous system. If the nerves cannot carry impulses, no pain is felt.

Observation of wildlife

There are many examples of animals self-medicating. For example:

- monkeys, bears and other animals rub citrus oils on their coat as insecticides and antiseptics to prevent insect bites and infection
- chimpanzees swallow leaves folded in a particular way to remove parasites from their digestive tract
- elephants roam miles to find clay to counteract dietary toxins
- birds line their nests with medicinal leaves to protect chicks from blood-sucking mites.

Modern research

Scientists have used traditional plant medicines and animal behaviour as a starting point in their search for new drugs. Research into the plants used allows them to isolate the active ingredient.

The example of aspirin

Hippocrates (a physician in ancient Greece) used an extract from willow bark to relieve pain and fever. Throughout the Middle Ages a similar extract was used in Britain. It was not until 1828 that Johann Buchner extracted the active ingredient salicin. This could be used to relieve pain, but caused stomach bleeding as a side effect. In 1897 a way was found to reduce the side effects by adding an acetyl group to salicin. By 1971 further research had revealed that salicin works by inhibiting enzymes involved in the synthesis of **prostaglandins**. Prostaglandins are hormone-like substances that have a variety of roles in cell communication. Further research has revealed other drugs that have the same effect.

The hunt is on

Natural medicines

In recent decades, discovery of natural drugs has concentrated on tropical plants. Owing to their great diversity, there are hopes that there may be many new medicinal drugs to discover. But it is important to remember that there may be many potential uses of wild and cultivated plants in the UK. New chemical fingerprinting technology is enabling scientists to screen natural chemicals more effectively for their activity as potential medicines.

Further research

Most antibiotics developed over the past 50 years come from the bacterium *Streptomyces*. Biologists hope to learn how the bacterium produces antibiotics by finding out what its **genes** do. They can then use this information to improve current production methods.

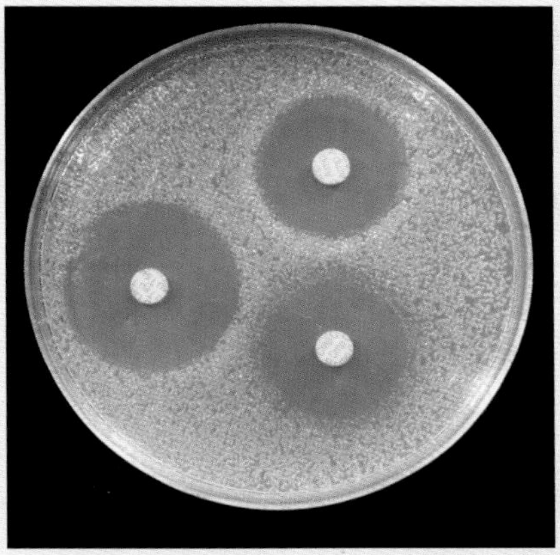

Pharmaceutical companies have been conducting research into the way that microorganisms cause disease. Many make use of **receptors** on **plasma membranes** (cell surface membranes). If the receptor site can be blocked by a drug, then the disease-causing **pathogen** cannot gain access to the cell. For example, the HIV virus uses a receptor called the CD4 receptor. The CD4 receptor can be isolated and sequenced. Once the **amino acid** sequence is known, molecular modelling can be used to determine the shape of the receptor. The next step is to find a drug that could block that receptor without causing major side effects.

This technology can be taken a step further by sequencing the genes that code for the CD4 receptor or for other receptors. Once this technology is developed, it may be possible to compare the **DNA** of a person with that of a plant or microorganism. This could identify potential medicinal drugs from the DNA. This is known as genomics. It also has potential uses in **vaccines** – by sequencing the genes of microorganisms, we should find a range of candidates from which vaccines can be made.

Figure 2 The effect of antibiotic on the growth of bacteria

Questions

1. Explain how our ancestors might have discovered that poppy seeds contain a chemical that has anaesthetic properties.
2. How did Fleming discover penicillin?

(14) The effects of smoking

By the end of this spread, you should be able to . . .

* **Describe the effects of smoking on the mammalian gas exchange system, with reference to the symptoms of chronic bronchitis, emphysema (chronic obstructive pulmonary disease) and lung cancer.**

Cigarette smoke

Cigarette smoke contains over 4000 different chemicals. Many of these are harmful. The harmful substances in cigarette smoke include:

- tar, which is a mixture of chemicals including **carcinogens**
- carbon monoxide
- nicotine.

What harm does tar cause?

Short-term effects

Tar is a combination of chemicals, which settles on the lining of the airways and **alveoli**. This increases the diffusion distance for oxygen entering the blood and for carbon dioxide leaving the blood.

The presence of many chemicals in the tar lying on the surface of the airway may cause an allergic reaction. This causes the smooth muscles in the walls of the airways to contract. The lumen of the airway gets smaller and this restricts the flow of air to the alveoli.

The tar paralyses or destroys the **cilia** on the surface of the airway, so they are unable to move the layer of **mucus** away and up to the back of the mouth. The tar also stimulates the **goblet cells** and mucus-secreting glands to enlarge and release more mucus. This mucus collects in all the airways.

Examiner tip

Never suggest that the *blood vessels* get blocked by tar. Tar does not enter the blood.

Bacteria and viruses that become trapped in the mucus are not removed. They can multiply in the mucus, and eventually a combination of mucus and bacteria may block the **bronchioles**.

The presence of bacteria and viruses means that the lungs are more susceptible to infection. Smokers are more likely to catch diseases such as influenza or pneumonia.

Longer-term effects

Smoker's cough is an attempt to shift the bacteria-laden mucus that collects in the lungs. It results from the irritation of the airways by the mucus and bacteria, as well as from the need to clear the airways in order to get air down into the alveoli.

Unfortunately, a constant cough produces its own effects. The delicate lining of the airways and alveoli can become damaged. The lining will eventually be replaced by scar tissue, which is thicker and less flexible. Also, the layer of smooth muscle in the wall of the bronchioles thickens. This reduces the **lumen** of the airway, and the flow of air is permanently restricted.

Frequent infections brought about by the presence of bacteria and viruses in the mucus will inflame the lining of the airways. This damages the lining and, in particular, the layer of **epithelium**. Also, it attracts white blood cells. These are brought in to deal with **pathogenic** microorganisms. The white blood cells have to make their way out of the blood and into the airways. In order to do this, they release **enzymes**. These enzymes digest parts of the lining of the lungs in order to pass through into the air spaces. In particular, the enzyme elastase is used. This damages the elastic tissue in the lining of the lungs. The greatest effect is in the small bronchioles and alveoli. Loss of elastic tissue can reduce the elasticity of the alveolus wall. As we exhale, the alveolus walls do not recoil to push the air out. The bronchioles collapse, trapping air in the alveoli. This can cause the alveoli to burst as pressure in the lungs increases.

Lung cancer

Cigarette smoke contains a large number of carcinogenic compounds – those that cause cancer. Benzopyrene is one of the most harmful. These compounds are in the tar that lies on the delicate surface of the lungs. Carcinogenic compounds enter the cells of the lung tissue. They enter the nucleus of these cells, and have a direct effect upon the genetic material. Any change to the genetic material is called a **mutation**. If the mutation affects the **genes** that control cell division, then uncontrolled cell division may take place. This is cancer. Cancers often start at the entrance to the **bronchi**, as the smoke hits this fork in the airway and deposits tar. Lung cancer often takes 20–30 years to develop, and a cancer may have been growing for years before it is discovered.

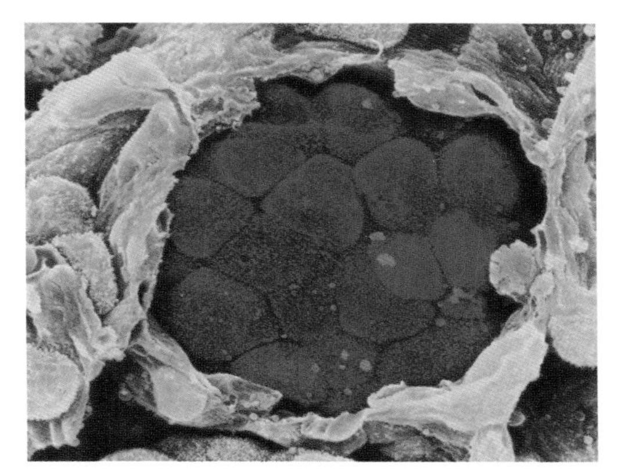

Figure 1 Scanning electron micrograph of a tumour (×1000)

Key definitions

Carcinogens are chemicals that cause cancer.

Chronic obstructive pulmonary disease (COPD) is a combination of diseases including chronic bronchitis and emphysema.

What diseases are associated with smoking?

Chronic bronchitis

Chronic bronchitis is the inflammation of the lining of the airways. This is accompanied by damage to the cilia and the overproduction of mucus, so that mucus collects in the lungs. The symptoms are irritation in the lungs, continual coughing, and coughing up mucus that is often filled with bacteria and white blood cells. It leads to an increased risk of lung infection.

Emphysema

Emphysema is the loss of elasticity in the alveoli, which causes the alveoli to burst. The lungs have a reduced surface area as larger air spaces are formed. This means that there is less surface area for gaseous exchange. A person with emphysema will often be short of breath, especially when exerting themselves. The loss of elasticity in the alveoli makes it harder to exhale. In severe cases, the breathing will become shallow and more rapid. The blood is less well oxygenated and fatigue occurs.

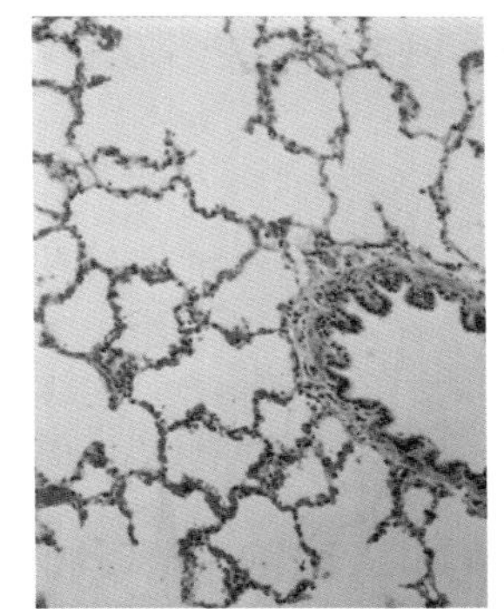

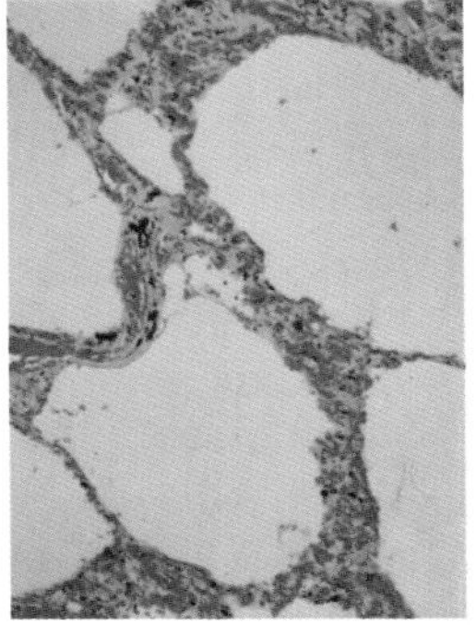

Figure 2 A section of a healthy lung compared to one from a person with emphysema (×50)

Chronic obstructive pulmonary disease

COPD is a combination of diseases that include chronic bronchitis, emphysema and asthma. The symptoms are a combination of the symptoms described above.

Lung cancer

Lung cancer can be recognised by continual coughing and a shortness of breath. There may be pain in the chest and blood coughed up in the sputum. This blood is often the first sign that alerts a person to the possibility of lung cancer.

Questions

1. Name the three main substances in tobacco smoke that can be harmful.
2. Explain how the deposition of tar can lead to smoker's cough.
3. Explain why smokers are more likely than non-smokers to suffer from infections in the lungs.

Smoking – nicotine and carbon monoxide

By the end of this spread, you should be able to . . .

* **Describe the effects of nicotine and carbon monoxide in tobacco smoke on the cardiovascular system with reference to the course of events leading to atherosclerosis, coronary heart disease and strokes.**

Key definitions

Atherosclerosis is the deposition of fatty substances in the walls of the arteries.

Cardiovascular disease is a disease of the heart or circulatory system.

Nicotine and carbon monoxide

These two chemicals, found in cigarette smoke, enter the lungs and pass through the lung surface into the blood. In the blood they cause changes to the circulation. These changes lead to increased risk of cardiovascular diseases including:

- **atherosclerosis**
- coronary heart disease (CHD)
- stroke.

Nicotine

Nicotine is the chemical in cigarette smoke that causes addiction. It has a variety of effects on the body. The body becomes used to these effects and the smoker no longer feels well unless he or she has nicotine in their blood.

- Nicotine mimics the action of transmitter substances at the synapses between nerves. This makes the nervous system more sensitive and the smoker feels more alert.
- Nicotine causes the release of the hormone adrenaline. This hormone has a variety of effects that prepare the body for activity. These include increasing the heart rate and breathing rate, and causing constriction of the arterioles. This raises the blood pressure in the arterioles.
- Nicotine also causes constriction of the arterioles leading to the extremities of the body. This reduces blood flow and oxygen delivery to the extremities. In extreme cases it can lead to the need for amputation.
- Nicotine also affects the **platelets** to make them sticky. This increases the risk that a blood clot or **thrombus** may form.

Carbon monoxide

- Carbon monoxide enters the red blood cells and combines with **haemoglobin**. It combines much more readily than oxygen, and forms the stable compound **carboxyhaemoglobin**. This reduces the oxygen-carrying capacity of the blood. Smokers feel this when they exercise. The body will detect the lower level of oxygen, and it may cause the heart rate to rise.
- Carbon monoxide can also damage the lining of the arteries.

Examiner tip

Don't confuse *atherosclerosis* with *arteriosclerosis*. These words look similar but mean different things. When describing atherosclerosis, be sure to describe how the deposition occurs under the endothelium of arteries, not on the surface. It should be described as *in* the walls of the arteries, not *on* the walls.

Problems caused by changes to the blood system

The changes described above are part of a chain of events that can lead to serious diseases such as coronary heart disease (CHD). CHD is a multifactoral disease – this means there is no single factor that causes it. A number of factors contribute to the risk of a person having CHD – these are called risk factors.

Atherosclerosis

Carbon monoxide can damage the inner lining (**endothelium**) of the arteries. If the person also has high blood pressure (this increases with ageing and other factors, such as too much salt in the diet), this will add to such damage. The damage is repaired by the action of white blood cells (**phagocytes**). These encourage the growth of **smooth**

muscle and the deposition of fatty substances. The deposits include cholesterol from low-density lipoproteins – tiny balls of fat and protein that are used to transport **cholesterol** around in the blood. High blood pressure also increases the deposition of cholesterol. The deposits, or **atheromas**, may also include fibres, dead blood cells and platelets. The process of deposition is called atherosclerosis.

This build-up of atheromas occurs under the endothelium, in the wall of the artery. It may grow enough to break through the inner lining of the artery. The atheroma eventually forms a **plaque**, which sticks out into the **lumen** of the artery. This leaves the artery wall rougher and less flexible than in a healthy artery. It also reduces the size of the lumen of the artery, which reduces blood flow.

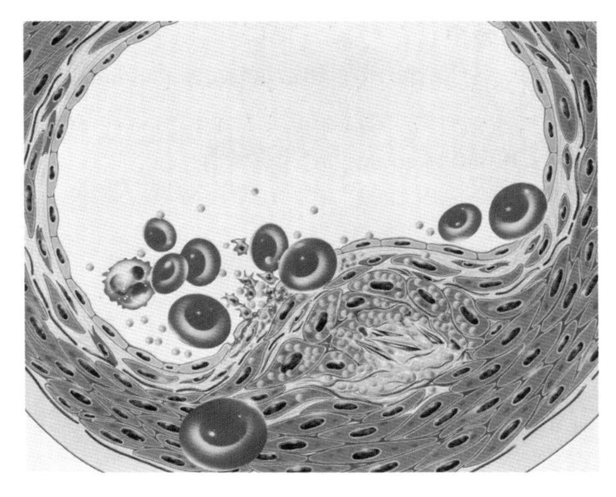

Figure 1 Atherosclerosis

Thrombosis

Blood flowing past the plaque cannot flow smoothly. This increases the chance that it will clot. The stickiness of the platelets, caused by nicotine, increases this chance significantly. If the delicate membrane that covers the plaque is damaged, red blood cells also stick to the exposed fatty deposits. A blood clot is known as a thrombus. A clot in an artery may stop the blood flow in that artery. Sometimes a clot, or part of a clot, may break free and be carried around in the blood until it reaches a narrow artery. Then it will lodge and stop blood flowing through that artery.

Coronary heart disease

The arteries that carry blood to the heart muscles are called the **coronary arteries**. They branch off the aorta close to the heart and carry blood at high pressure. Therefore they are prone to damage and atherosclerosis. When the lumen of a coronary artery is narrowed by plaques, this reduces the blood flow to the heart muscles, which receive less oxygen for respiration. This can lead to CHD, which takes three forms:

- angina – a severe pain in the chest, which may extend down the left arm or up into the neck
- heart attack or myocardial infarction – the death of a part of the heart muscle, usually caused by a clot in the coronary artery blocking the flow of blood to the heart muscle
- heart failure – when the heart cannot sustain its pumping action; this can be due to the blockage of a major coronary artery, but there are other types and causes of heart failure.

Stroke

Stroke is the death of a part of the brain tissue. It is caused by the loss of blood flow to that part of the brain. This can happen very suddenly. There are two possible causes:

- a blood clot (thrombus) floating around in the blood blocks a small artery leading to part of the brain
- an artery leading to the brain bursts (haemorrhage).

Examiner tips

Don't confuse blood flow into the heart, which occurs in *veins*, with blood flow to the heart muscle. This occurs in the coronary *arteries* and carries oxygen to the cardiac muscle.

Don't say 'the artery becomes narrower'; say 'the *lumen* of the artery becomes narrower'.

Questions

1 List the effects of nicotine on the body.
2 Explain why the blood of a smoker is more likely to clot than that of a non-smoker.
3 Describe how atherosclerosis occurs and explain how it can lead to angina.

Cardiovascular diseases

By the end of this spread, you should be able to . . .

* **Describe the effects of nicotine and carbon monoxide in tobacco smoke on the cardiovascular system.**
* **Discuss the possible links between diet and coronary heart disease.**

Key definitions

Coronary heart disease (CHD) is a disease of the heart caused by malfunction of the coronary arteries.

Stroke is the death of part of the brain due to a lack of blood flow to that part of the brain and subsequent oxygen deficiency.

Cardiovascular diseases

Cardiovascular diseases are those diseases that affect the heart and circulatory system. These include:

- **atherosclerosis**
- coronary heart disease (CHD)
- stroke
- **arteriosclerosis**.

Cardiovascular disease is one of the greatest causes of premature death in the more economically advanced countries around the world. A person with cardiovascular disease is often unaware that they have it until the symptoms become very obvious.

Coronary heart disease and stroke often result from **atherosclerosis**. The events that lead to atherosclerosis have been described in spread 2.2.15. Atherosclerosis may start in adolescence. The accumulation of fatty deposits may take many decades before they have a serious effect on blood flow. Once blood flow is significantly reduced, there may be noticeable symptoms of cardiovascular disease.

Cardiovascular disease can be disabling and will eventually cause death. Treating cardiovascular disease can be very expensive, as it may involve long treatments with drugs to reduce **blood pressure** and blood **cholesterol** levels, or it may involve surgery. For this reason, health-promotion centres aim to prevent or reduce atherosclerosis. If atherosclerosis can be reduced by reducing the risk factors, then the need for treatment later in life will be reduced. This emphasis on prevention needs to start from childhood.

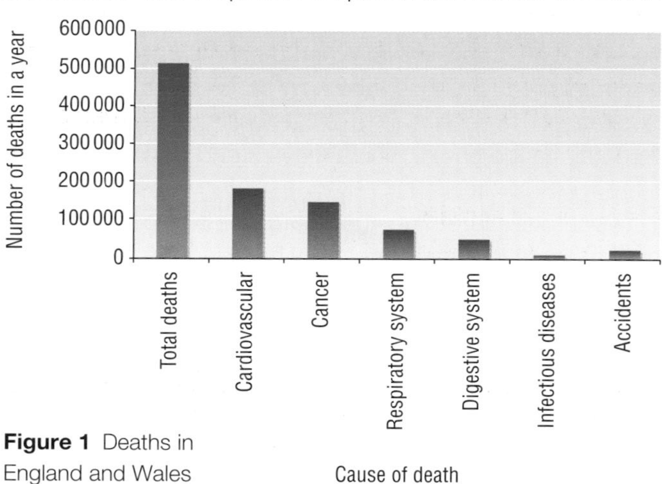

Figure 1 Deaths in England and Wales

Symptoms of cardiovascular disease

High blood pressure and hypertension are usually the first signs. These result from narrowing of the artery **lumen** caused by deposition of **atheroma**. The narrower lumen increases the friction between the blood and the artery wall. The heart pumps against this increased friction, raising the blood pressure. The atheroma also makes the artery walls less elastic. This means the artery walls cannot dilate and recoil as easily. The process of **arteriosclerosis** makes this worse. Arteriosclerosis is the hardening of the artery walls, which makes them even less flexible. This is caused by deposition of minerals, such as calcium, in the walls and particularly in the atheromas.

Coronary heart disease

A person with CHD may find it more difficult to exercise, and may feel out of breath after only a small amount of exertion. This is due to atherosclerosis in the coronary arteries. The atheroma narrows the lumen of the artery, reducing blood flow to the cardiac muscle. When exercising, the heart needs to increase its output. The heart muscle does not receive sufficient oxygen for aerobic **respiration** and is put under strain in an effort to pump more blood. As the cardiac muscle does not have enough oxygen, it does not pump as strongly as required. This results in insufficient quantities of blood flowing to the body, muscles and lungs.

In advanced cases, angina may occur, which is a severe pain in the chest felt during exercise. The pain subsides quickly after exercise. A heart attack (myocardial infarction) may also occur. This is often felt as a severe and disabling pain in the chest and arm. It may feel as if a tight band has been placed around the chest. If not treated quickly, it may lead to heart failure (the heart cannot pump enough blood to meet the metabolic needs of the body). But sometimes a heart attack starts slowly and the patient just feels general discomfort in the chest area, which may be mistaken for indigestion. (Heart failure also has other causes, such as valve damage.)

Stroke

The symptoms of stroke are always sudden, because it results from part of the brain receiving insufficient oxygen:

- sudden numbness or weakness of the face, arm or leg, especially on one side of the body
- sudden confusion and difficulty in speaking or understanding
- sudden difficulty with seeing in one or both eyes
- sudden trouble with walking – dizziness, loss of balance or lack of coordination
- sudden, severe headache with no known cause.

What factors increase the risk of CHD?

Remember that CHD is a multifactoral disease – no one factor can be said to cause it. But there are a number of factors that increase the risk of developing CHD:

- age – the risk increases as you get older
- sex – men are much more likely to die of CHD under age 50 than are women
- cigarette smoking
- obesity
- high blood pressure (hypertension)
- high blood cholesterol concentration
- physical inactivity/sedentary lifestyle
- diet – a high level of animal fats (called saturated fats)
- a high salt intake
- an absence of healthy fats (called polyunsaturated fats) such as omega-3
- an absence of antioxidants such as vitamins A, C and E
- genetic factors/family history of cardiovascular disease
- diabetes
- stress.

CHD is less of a problem in less economically advanced countries. This is because the inhabitants of these countries lead a very different lifestyle, and have a lower life expectancy. They are more likely to die from some other disease, such as malaria or HIV/AIDS, before they have lived long enough to develop CHD. They often have a more active lifestyle and their diet contains fewer harmful elements, so they are less likely to be obese. Because of the economic situation, they are less likely to be heavy smokers. Smoking is increasing in these countries, and so is the incidence of CHD – but it is masked by the huge numbers of people dying from infectious diseases.

Key definition

A **risk factor** is a factor that increases the chance or risk that a person will develop a disease.

Questions

1. Describe the symptoms of a stroke.
2. Explain what is meant by a multifactoral disease.
3. Explain how each of the risk factors may contribute to the chances of a person developing CHD.

2.2 (17) The evidence linking smoking to disease

By the end of this spread, you should be able to . . .

* **Evaluate the epidemiological and experimental evidence linking cigarette smoking to disease and early death.**

Smoking and disease

Smoking did not become very widespread in the western world until cigarettes became widely available. This was shortly after the start of the twentieth century. Many men took up the habit during the First World War. The number of women who smoked increased during the 1940s and 1950s. In the later decades of the twentieth century, it became apparent to the medical profession that certain diseases had become more common. These included lung cancer (but also cancer of the mouth and throat), chronic bronchitis, emphysema and cardiovascular disease.

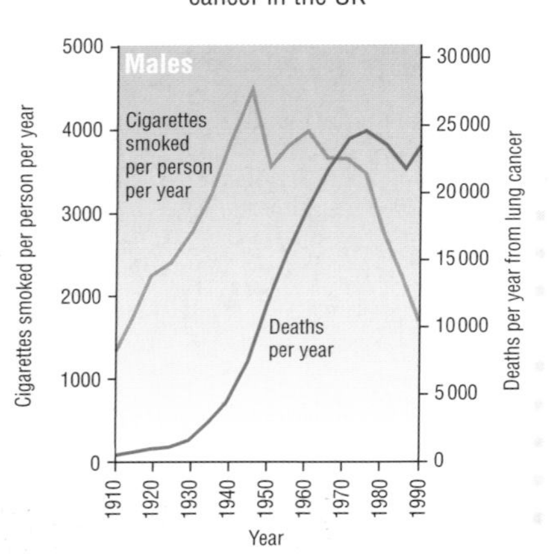

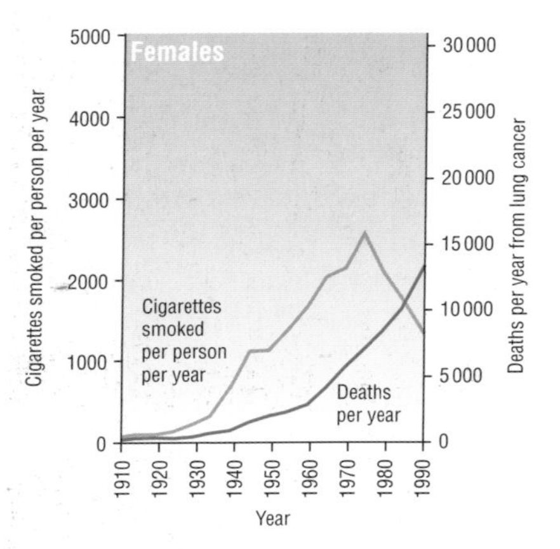

Figure 1 The rise in cigarette smoking and lung cancer in the UK

Epidemiology

The link between smoking cigarettes and lung cancer appears obvious when we look at the facts and figures. Such a comparison of data is called **epidemiology**. It is used by health professionals and organisations such as the World Health Organisation. Such organisations can use epidemiology to identify links between diseases and certain contributory factors (risk factors). It may help to identify:

- which countries may be at greater risk
- which age range of the population may be at greater risk
- which sex may be at greater risk
- which lifestyle factors may increase or decrease the risk.

Key definition

Epidemiology is the study of the distribution of a disease in populations, and the factors that influence its spread.

The information gained can be used to:

- help countries or organisations target further spending
- help target research at particular risk factors to find a cause or a cure
- help target screening procedures and find those at risk early
- help target advice and education at the parts of the population most at risk
- predict where a disease might become more prevalent in the future
- target geographical areas at risk, using preventative measures (e.g. vaccination) to prevent the disease spreading
- check how well campaigns and preventative measures are working.

Epidemiological evidence linking smoking and disease

When we look at the epidemiological evidence, it is clear that increased cigarette smoking increases the risks of early death. Epidemiologists have discovered the following links:

- a regular smoker is three times more likely to die prematurely than a non-smoker
- 50% of regular smokers are likely to die of a smoking-related disease
- the more cigarettes a person smokes per day, the more likely he/she is to die prematurely, and the younger he/she is likely to die.

Links to lung cancer

- A smoker is 18 times more likely than a non-smoker to develop lung cancer.
- 25% of smokers die of lung cancer.
- A heavy smoker (more than 25 cigarettes a day) is 25 times more likely than a non-smoker to die of lung cancer.
- The chance of developing lung cancer reduces as soon as a person stops smoking.

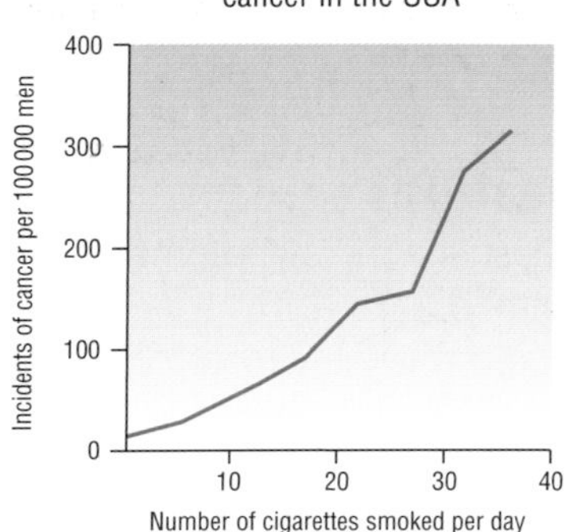

Figure 2 Cigarette consumption and lung cancer in the USA

Links to other lung diseases

- Chronic obstructive pulmonary disease is rare in non-smokers.
- 98% of people with emphysema are smokers.
- 20% of smokers have emphysema.

Links to cardiovascular diseases

It is not easy to link smoking with cardiovascular disease using epidemiological evidence. This is because there are so many risk factors contributing to cardiovascular disease. Some recent research suggests that the link between smoking and cardiovascular disease is less direct and less clear than previously thought. However, it is known that the substances released in cigarette smoke can influence the circulatory system in a way that is likely to enhance atherosclerosis and other circulatory diseases.

Examiner tip

When describing data from a graph or table, or describing trends, always use some figures from the data to back up your points.

Experimental evidence

The use of animals in experiments is an emotive subject. It can generate much heated debate, and the use of animals in such research is now very closely monitored and controlled. In the 1960s experiments were carried out using dogs to investigate the direct effects of smoking. Many dogs were attached to devices that made them inhale smoke from cigarettes.

Some dogs were forced to breathe unfiltered smoke. These dogs developed changes in their lungs that were similar to those of chronic obstructive pulmonary disease. They also developed early signs of lung cancer.

Another group of dogs had to breathe smoke from filtered cigarettes. These dogs remained healthier, but further analysis of their lung tissue showed that some cells were developing changes that lead to cancer.

This research suggested that the filter removed some of the substances in smoke that are particularly harmful. As a result, further research was carried out on the tar that was collected by the filter. Chemical analysis of the tar revealed that it contained substances known to cause cancer (carcinogens) such as benzopyrene. Another study collected tar from cigarette smoke and tested it on the bare skin of mice. These mice developed cancer in the skin cells covered by the tar.

Figure 3 Poorer health and a shorter life

Smoking – the future

The prevalence of smoking among adults in the western world has been in decline over recent years. However, this trend has not been seen in younger people. If anything, the opposite is true, with more young people taking up the habit.

Also, more people in the less economically developed countries are smoking. This is of great concern to organisations such as the World Health Organisation. The effects of smoking on your health can take 20 years or more to develop. So the effects of this increase in smoking are unlikely to be detected for some time to come.

Questions

1. Describe the experimental evidence linking cigarette smoke to lung cancer.
2. Explain why it is not easy to prove a link between smoking and coronary heart disease.
3. Describe the ways in which a health authority might use the information provided by epidemiologists.

Food and health summary

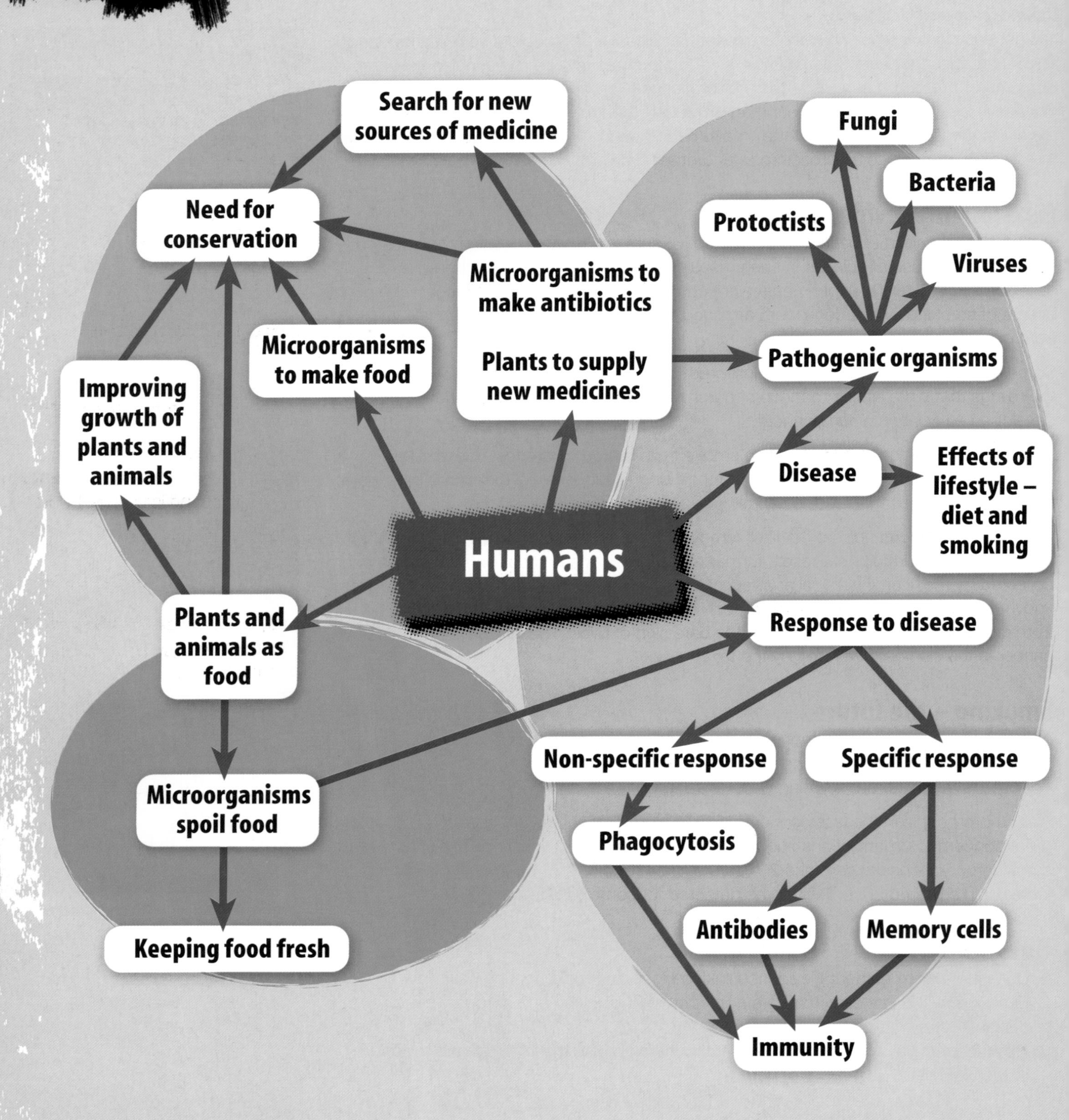

Practice questions

1. (a) Name the organism that causes AIDS. [1]
   (b) How is AIDS transmitted? [3]

2. How do farmers use pesticides and fertilisers to improve their crops? [3]

3. (a) What is Quorn™? [2]
   (b) Name one microorganism that is capable of making edible food for humans. [1]

4. (a) How do microorganisms cause food to spoil? [3]
   (b) Review the range of methods used to keep food fresh. [4]

5. Describe the primary defences of our body against disease. [3]

6. (a) Name two types of phagocyte. [2]
   (b) Describe the process of phagocytosis. [3]

7. (a) Explain how influenza is able to cause epidemics. [4]
   (b) Suggest why scientists are studying avian flu very closely. [3]

8. (a) Suggest three advantages of using microorganisms to produce food. [3]
   (b) Suggest three disadvantages of using microorganisms to produce food. [3]

9. Explain the difference between antigens and antibodies. [2]

10. (a) Explain what is meant by clonal selection during the immune response. [3]
    (b) How are antibodies made specific to the antigen? [2]
    (c) Apart from specificity, describe how the structure of an antibody molecule is suited to its function. [3]

11. (a) Suggest why TB is difficult to cure. [2]
    (b) Using examples, explain the potential role in medicine of molecules that are able to block membrane receptor sites. [4]

12. With reference to the use of plants as sources of medicine and microorgansims as sources of antibiotics, explain why it is important to conserve global biodiversity. [6]

13. (a) Coronary heart disease (CHD) is known as a multifactorial disease. Explain what is meant by 'multifactorial'. [2]
    (b) List three dietary factors that increase the risk of developing CHD. [3]
    (c) Describe how the lumen of the coronary artery becomes narrowed by atherosclerosis. [4]

14. (a) Describe the effects of tar from cigarette smoke on the lungs. [4]
    (b) Describe how nicotine contributes to the risk of developing CHD. [3]
    (c) What evidence is there that smoking increases the risk of causing cancer? [3]

15. (a) What is the difference between passive immunity and active immunity? [2]
    (b) Explain what is meant by an 'attenuated' pathogen. [1]
    (c) Describe how an attenuated pathogen could be used to create artificial active immunity. [3]

2.2 Examination questions

1 Figure 1.1 is a diagram showing the structure of a typical antibody.

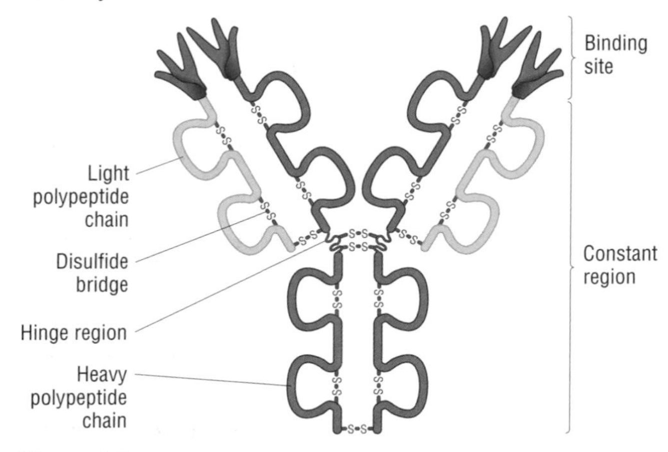

Figure 1.1

(a) Name the type of cell that produces antibodies. [1]

(b) (i) State **one** function for each of the component parts listed below:
binding site
disulfide bridge
constant region
hinge region. [4]

(ii) Explain why the part of the antibody molecule incorporating the binding site is often called the variable region. [2]

Figure 1.2 shows the concentration of antibodies in the blood following a first infection by a pathogen on day 0.

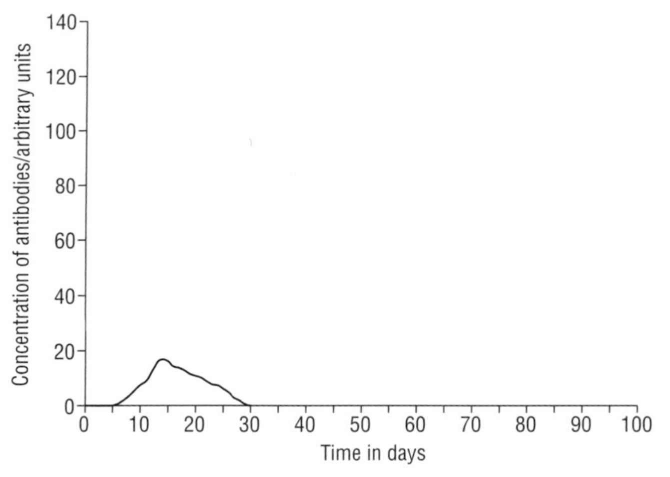

Figure 1.2

(c) (i) Explain why there is a delay between the **first** infection by the pathogen and the appearance of antibodies in the blood. [2]

(ii) On Figure 1.2, draw a curve to show the expected concentration of antibodies in the blood following a **second** infection by the same pathogen at day 30. [2]

(d) Antibiotics can be used to combat bacterial infections. However, resistance to antibiotics is common among bacteria. For example, the so-called 'superbug' MRSA (methicillin-resistant *Staphylococcus aureus*) is resistant to many antibiotics.

(i) State how a population of bacteria develops resistance to an antibiotic. [1]

(ii) Suggest **two** measures, **apart from use of antibiotics**, that could be taken in a hospital to prevent possible infection with MRSA. [2]

[Total: 14]

(Jan 07 2802)

2 Over the past few years there has been much public concern over the diet of people in the UK and its effects on their weight and health. The body mass index (BMI) is a calculation used by doctors to indicate whether a person is underweight or overweight. People with a BMI higher than 30 are classed as obese.

(a) State two causes of obesity. [2]

As part of a long-term survey into the health of the nation, a random sample of the English population was selected every 5 years. The BMI of each member of the sample was calculated and the percentage of people fitting into each mass category was recorded. The results are shown in Table 2.1.

BMI category	year				
	1980	1985	1990	1995	2000
Underweight	12	8	6	6	5
Acceptable	53	51	46	41	36
Overweight	28	32	34	36	39
Obese	7	9	14	17	20

Table 2.1

(b) Describe the trends shown in Table 2.1. [4]

Table 2.2 shows the daily intake of certain components in three diets, **A**, **B** and **C**, from men in the UK.

Diet **A**
- a normal balanced diet for a typical man

Diet **B**
- a weight-reducing low-fat diet
- restricted to avoid fats
- includes any fruit, vegetables and proteins
- energy intake is monitored carefully

Diet **C**
- a weight-reducing low-carbohydrate diet
- restricted to avoid carbohydrates
- excludes fruit, which contain sugars
- includes any non-starchy vegetables, proteins and fats
- energy intake is not counted and may exceed 10 000 kJ on some days

diet	energy /kJ	fats /g	carbohydrates /g	proteins /g	combined minerals /g
A normal balanced diet	9720	87	275	88	12
B weight-reducing low-fat diet	6000	34	200	76	12
C weight-reducing low-carbohydrate diet	8000	124	20	165	18

Table 2.2

In any unbalanced diet, it is possible that there may be a deficiency of certain nutrients.

(c) Suggest **one** nutrient that may be deficient in diet **B** and **one** in diet **C**. [2]

(d) **(i)** **Explain** which diet, **B** or **C**, is likely to cause more rapid weight loss. [2]

(ii) State the relationship between energy intake and energy use that would allow a person to lose weight. [1]

[Total: 11]

(Jan 07 and June 05 2802)

UNIT 2 Module 3 Biodiversity and evolution

Introduction

Biodiversity is the variety of life in all its forms. We consider diversity at a number of levels:

- the genetic diversity within species that causes variation
- the differences between species
- the differences between the habitats in which living things survive.

This module incorporates some fieldwork skills. These are used to measure biodiversity in your local habitats.

Just how many species are there? One million? 10 million? 100 million? How can we be sure? One thing is for certain – there are so many that we need to place them in groups, or classify them, in order to help us understand their diversity.

Variation is the key to biodiversity. It is through variation that species can change over time and evolution occurs. Older species can give rise to modern species. In Victorian times, some members of the scientific community believed that evolution occurred, but did not understand how. When Charles Darwin thought up a mechanism for evolution, he fully understood how important it was. He spent almost a decade writing a book before publishing his ideas. Darwin's book was called *On the Origin of Species by Means of Natural Selection*. When it was published, it caused uproar. People would not believe that we, the superior species, and apes share a common ancestor!

Our understanding of natural selection has allowed us to make use of selection in driving the evolution of certain species. We can artificially select features and steer the evolution of species in ways that suit our needs. However, we have also come to realise the importance of maintaining biodiversity for the future. Conservation of species, habitats and genetic diversity are high on the agenda of modern biologists.

Test yourself

1. Can you name the five kingdoms?
2. Into which kingdom do humans fit?
3. How many species of animals are there?
4. What are the differences between a mammal and a lizard?
5. State three ways in which any two people will be similar, and three in which they will be different.
6. Who was the naturalist who described the process of natural selection?
7. What is meant by artificial selection?

Module contents

2.3 (1) Biodiversity

By the end of this spread, you should be able to . . .

* **Define the terms species, habitat and biodiversity.**
* **Explain how biodiversity may be considered at different levels – habitat, species and genetic.**
* **Discuss current estimates of global biodiversity.**

Key definitions

A **species** is a group of individual organisms very similar in appearance, anatomy, physiology, biochemistry and genetics, whose members are able to interbreed freely to produce fertile offspring.

A **habitat** is the place where an organism lives.

Biodiversity is the variety of life – the range of living organisms to be found.

Some definitions

Species

There is a huge number of different types of organism living in the world today. A **species** is just one of these types of organism. The term refers to the fundamental unit of biological classification. A species consists of individual organisms that are very similar in appearance, anatomy, physiology, biochemistry and genetics. As a result, individuals in a species can interbreed freely to produce fertile offspring (offspring that can breed to give rise to more offspring).

Habitat

A **habitat** is the place where individuals in a species live. It is a specific locality with a specific set of conditions and organisms living there. Organisms are often very well-adapted to their habitat, especially if the conditions are extreme. Examples of habitats include oak woodland, coniferous woodland, freshwater ponds and rock pools on a rocky shore.

Biodiversity

Biodiversity includes all the different plant, animal, fungus and microorganism species worldwide, the **genes** they contain, and the **ecosystems** of which they form a part.

Biodiversity is not just about the number of different species. It also represents the degree of nature's variety. Biodiversity takes into account the number of individuals and in how many places they can be found.

As an example, a wild meadow might have 25 species of grasses and herbs per square metre. Compare that with a garden lawn, or a managed cow pasture. Here there may be 25 plant species, but one or two grass species dominating, with just odd individuals of the other species dotted about. The wild meadow is much more diverse.

(a)

(b)

Figure 1 **a** Wild meadow and **b** lawn

Biodiversity is about the structural and functional variety in the living world. We can consider it at a number of levels.

- The range of habitats in which different species live. Even in your school grounds or in a local park, there may be a variety of habitats. Carefully manicured lawns, ponds, dark corners between buildings or a small patch of trees are all different habitats. Each habitat will be occupied by a range of organisms.
- The differences between species. All species are different from each other. These could be obvious structural differences, such as the difference between a tree and an ant. They could also be functional differences, such as the differences between bacteria that cause decay and those that help us to digest food.
- Genetic variation between individuals belonging to the same species. This is the variation found within any species that ensures we do not all look alike.

How many species are there?

In June 2001 two organisations, Species 2000 and the Integrated Taxonomic Information System, decided to work together to create the Catalogue of Life. The Catalogue of Life is planned to become a comprehensive catalogue of all known species on Earth. They aim to cover all known species by 2011. By 2007 the annual checklist contained 1 008 965 species. This is probably just over half of the world's known species, which may be nearer to 1 800 000. Even so, some scientists believe this may be only 10% of the actual total number of species.

Table 1 shows the number of known species in the UK and in the world. These figures are for terrestrial and freshwater habitats. They do not include marine species.

Group	UK species	World species
Bacteria	Unknown	4000+
Protoctista	40 000+	80 000+
Fungi	15 000+	70 000+
Plants	4080	293 000+
Invertebrates	30 000+	1 280 000+
Vertebrates	488	33 700+
Total	89 000+	1 730 000+

Source: UK Biodiversity Action Plan.

Table 1 Number of known species

These figures are estimates for the number of known species. We cannot be sure how accurate they are because:

- we cannot be sure we have found all the species on Earth
- new species are being found all the time
- evolution and **speciation** are continuing
- many species are endangered and some are becoming extinct.

It is important to remember that biodiversity is not just a count of how many species exist. These estimates of the number of species on Earth do not take any account of the numbers of individuals in each species. Nor do they give any indication of the variation between different species, or within a species.

Questions

1 Explain what is meant by a habitat.
2 Make a list of all the habitats you can see in your school or college grounds.
3 Explain why we share so many of our genes with plants.

2.3

How Science Works

2 Sampling plants

By the end of this spread, you should be able to . . .

* **Explain the importance of sampling in measuring the biodiversity of a habitat.**
* **Describe how random samples of plants can be taken when measuring biodiversity.**

Random
1. Apply co-ordinates to area.
2. Use random number generator to get co-ordinates.
3. Place quadrat on coordinates min 30.

Sampling

In order to measure the **biodiversity** of a **habitat**, you need to observe all the species present, identify them, and count how many individuals of each species there are. Ideally, you should do this for all the plants, animals, fungi, bacteria and other single-celled organisms living in the habitat. Obviously this is not practical, as the numbers of fungi, bacteria and single-celled organisms will be impossible to count. One estimate suggests that there may be billions of single-celled organisms per square metre of soil and possibly hundreds of thousands of mites per square metre!

Instead, you can sample a habitat. This means you select a small portion of the habitat and study that carefully. Then you can multiply up the numbers of individuals of each species found, in order to estimate the number in the whole habitat.

Randomising samples

If you go to sample a habitat such as a meadow, you will first look around to see what species you can spot. If the meadow contains a few large or colourful plant species, you may be tempted to make sure you include them in your sample. You should certainly list these plants, but if you deliberately include them in a **quantitative** sample this may bias the numbers. The results of the survey may suggest that those plants are far more common than is correct.

So you must randomly choose the position of your sample sites inside the habitat. You do this by estimating the size of the habitat and deciding where to take samples before you study any area in detail. There are three simple ways to do this – you can:

- take samples at regular distances across the habitat
- use random numbers, generated by a computer or a random number table, to plot coordinates within the habitat
- select coordinates from a map of the area and use a portable global positioning satellite system to find the exact position inside the habitat.

How many samples should you take?

The number of samples will depend on the size of the habitat and the time available. You may also need to take into account the diversity of the habitat being studied. The number of samples should be sufficient to give an accurate measure of the number of species in the habitat and their relative abundance.

In a field such as a football pitch, you may need to record 10 samples, each being one square metre. But in a wild-flower meadow of similar size it may be worth recording twice as many samples, because some of the species may occur in widely separated patches. If you are comparing the two areas as part of the same study, you should take the same number of samples in each case.

Recording your results

Before starting your survey, you should prepare a table in which to record your results. This table should allow space for all the species you may find and space to record the data for each sample site. An example of part of a table has been provided on the left.

Sample	Species	
	Meadow grass	Clover
1		
2		
3		
4		
5		
6		
7		
8		
9		
10		

Sampling for plants

You can count large plants, such as trees, individually, but many smaller herbs and grasses may be too numerous to count. For these plants, it is more usual to measure the percentage of ground cover. But beware – the random sampling technique can miss plants that occur only infrequently. It is therefore important to carry out a visual survey of the habitat to spot odd individuals that are not found as part of the random sampling technique. You should record these as present but with no **abundance** (frequency of occurrence) data – this is known as **qualitative** data. You cannot use it for statistical analysis.

Using random quadrats

A **quadrat** is a square frame used to define the size of the sample area. A quadrat may be any size, but it is often 50 cm or 1 m on each side. You place the quadrat at random in the habitat. To do that, you need to imagine the area as a large grid and place the quadrats at the coordinates you generated as pairs of random numbers. First, you need to identify the plants found within the frame. Then you have to measure their abundance. This can be done in a number of ways.

- An abundance scale is not precisely quantitative, but it does give estimates of abundance. The commonest scale is known as the ACFOR scale (Table 1). You observe the contents of the quadrat and apply an abundance score to each species.
- It may be possible to estimate the percentage cover, although most students tend to underestimate. Some quadrats have a grid of string that divides the quadrat into a number (usually 100) of smaller squares. This grid can help your estimates.
- You can measure percentage cover using a point frame. This is a frame holding a number of long needles or pointers. You lower the frame into the quadrat and record any plant touching the needles. If the frame has 10 needles, and is used 10 times in each quadrat, you will have 100 readings. So each plant recorded as touching the needle will have 1% cover. As one needle may touch several plants, it is possible to find you have 300–400% cover. Don't forget to record bare ground.

A	Abundant
C	Common
F	Frequent
O	Obvious
R	Rare

Table 1 The ACFOR scale

Figure 1 Using a quadrat and point frame

Using a transect

A transect is a line taken across the habitat. You stretch a long rope or tape measure across the habitat and take samples along the line.

In a large habitat, you might use a line transect. In this case you would record the plants touching the line at set intervals along it. You may also decide to use a quadrat at set intervals along the line. This is called an interrupted belt transect. This will provide quantitative data at intervals across the habitat.

Alternatively, you may use a continuous belt transect. In this case you place a quadrat beside the line, and move it along the line so you can study a band or belt in detail. You should study each quadrat as described above. This will provide quantitative data in a band or belt across the habitat.

Figure 2 Using a line transect

Questions

1. Explain why it is essential to take samples of a habitat.
2. Explain why samples must be random.
3. Describe how you would compare the vegetation in a meadow with that in a lawn.

2.3 How Science Works ③ Sampling animals

By the end of this spread, you should be able to . . .

* **Explain the importance of sampling in measuring the biodiversity of habitat.**
* **Describe how random samples of animals can be taken when measuring biodiversity.**

The trouble with animals

Animals move. Any attempt to sample the animals in a habitat will disturb the habitat. Many animals – particularly larger ones – will detect your presence and hide away. This presents a problem when sampling a habitat. To overcome this problem, you need to catch or trap the animals and then estimate the numbers from your trapped sample.

You should not trap larger animals. You can note the presence of many larger animals by careful observation. It may be possible to hide and wait quietly for any animals to emerge from hiding or to pass through the habitat. Alternatively, you may be able to see signs that the animals have left behind. For example, many animals leave easily identified droppings – owls deposit pellets of undigested food, rabbits have obvious burrows, and deer damage the bark of trees in a particular way.

Catching animals

There are many ways of catching animals. The technique you use will depend on the habitat and the type of animals you are hoping to catch.

Sweep netting

This technique involves walking through the habitat with a stout net. You sweep the net through the vegetation in wide arcs. Any small animals, such as insects, will be caught in the net. Then you can empty the contents of the net onto a white sheet to identify them.
You need to be careful, as many of the animals may crawl or fly away as soon as you release them from the net. You can use a device called a pooter to collect the animals before they fly away. This type of sampling is suitable for low vegetation that is not too woody. You can use a similar technique to take samples in water.

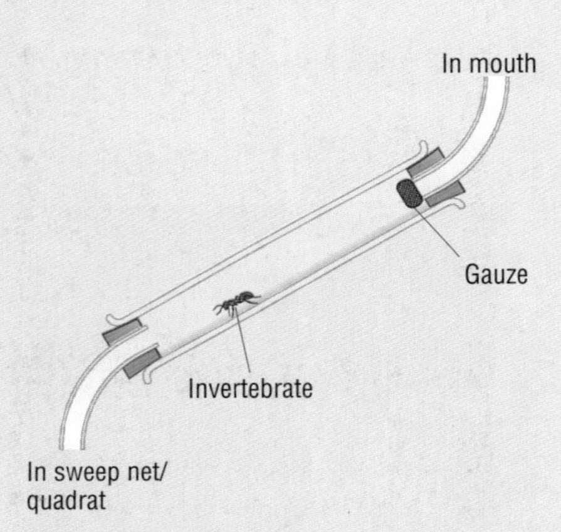

Figure 2 A pooter

Figure 1 Sweep netting

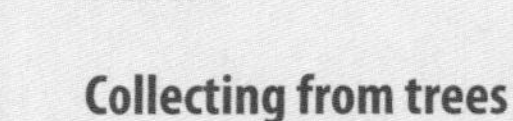

Collecting from trees

Sweeping a net through tree branches is unlikely to work well – here it is better to spread a white sheet out under a branch, and knock the branch with a stout stick. The vibrations caused dislodge any small animals, which then drop onto the sheet. Again, you will need to be quick to identify and count the animals before they crawl or fly away.

Figure 3 Collecting from a tree

A pitfall trap

This is a trap set in the soil to catch small animals. It consists of a small container buried in the soil so that its rim is just below the surface. Any animals moving through the plants or leaf litter on the soil surface will fall into the container. The trap should contain a little water or scrunched paper to stop the animals crawling out again. In rainy weather, the trap should be sheltered from the rain so that it does not fill up.

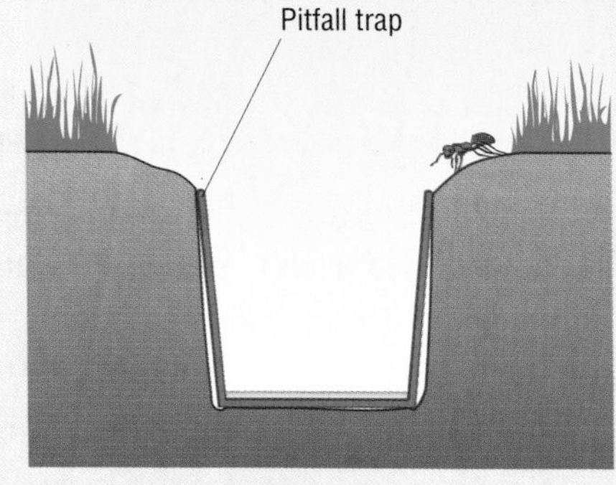

Figure 4 A pitfall trap

A Tullgren funnel

A Tullgren funnel is a device for collecting small animals from leaf litter. You place the leaf litter in a funnel. A light above the litter drives the animals downwards as the litter dries out and warms up. They fall through the mesh screen to be collected in a jar underneath the funnel.

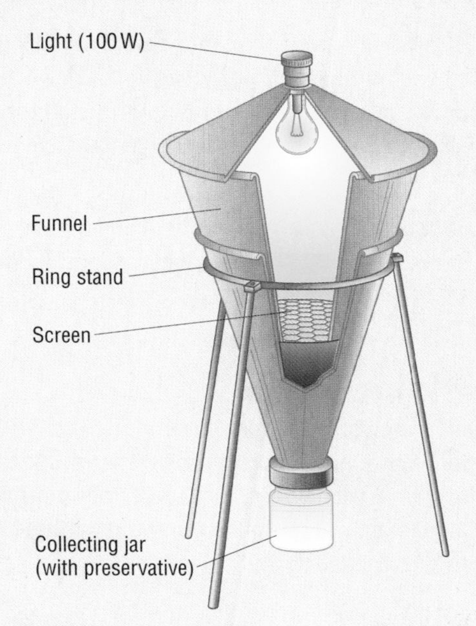

Figure 5 Tullgren funnel

A light trap

You can use a light trap to collect flying insects at night. It consists of an ultraviolet light that attracts the insects. Under the light is a collecting vessel containing alcohol. Moths and other insects attracted to the light eventually fall into the alcohol.

The need to sample

Sampling can cause disturbance to a habitat. This may be a temporary disturbance due to your presence, or it may be more long term. Your sampling may damage the habitat in some way – for example, you will cause damage when you dig a pitfall trap or trample the vegetation.

In order to study a habitat, it is essential to take samples as discussed on the previous pages. But why do we need to study the habitat?

- Human activities affect the environment in a number of ways.
- Unless we study how our activities affect our environment, we cannot assess the effect we have. Environmental impact assessment (EIA) is a very important part of the planning process.
- EIA is used to estimate the effects of a planned development on the environment (see spread 2.3.19).
- The importance of maintaining habitats and reducing the damage we do to them is discussed in spreads 2.3.15 and 2.3.16.

Questions

1 Describe how you would collect animals from leaf litter.
2 List the ways that your sampling might damage the habitat.
3 Consider how building a new park-and-ride facility might affect the local habitat.

2.3 How Science Works

By the end of this spread, you should be able to . . .

* **Describe how to measure species richness and species evenness in a habitat.**
* **Use Simpson's Index of Diversity (*D*) to calculate the biodiversity of a habitat, using the formula $D = 1-(\Sigma(n/N)^2)$.**
* **Outline the significance of both high and low values of Simpson's Index of Diversity (*D*).**

Species observed	Percentage cover	
	Field A	Field B
Cocksfoot grass	57	38
Timothy grass	32	16
Meadow buttercup	3	14
White clover	3	22
Creeping thistle	1	5
Dandelion	4	5
Total	100	100

Table 1 Two surveys

Species richness is the number of species present in a habitat

Species eveness number of individuals in a species

Species richness and evenness

When measuring **biodiversity**, we have to consider **species** richness – the number of species found in a **habitat**. The more species are present, the richer the habitat. However, richness is not sufficiently quantitative to be a measure of **diversity** on its own. It does not take into account the number of individuals in each species. For this, we need to estimate **evenness**.

Evenness is a measure of the relative numbers or **abundance** of individuals in each species. A habitat in which there are even numbers of individuals in each species is likely to be more diverse than one in which individuals of one species outnumber all the others. As discussed in spread 2.3.2, we can measure abundance in plants as percentage cover, rather than as numbers of individuals.

Table 1 shows an example of two simple surveys. Fields A and B have equal richness as they both contain six species. However, field B has greater evenness. Therefore field B would be considered more diverse.

Estimating species richness

You can use a **qualitative** survey to estimate species richness. This means that you need to make observations within the habitat and record all the different species you see. You should start by taking some samples, as described in spread 2.3.2. Your method of sampling will depend on the habitat. You should also take time to walk around the habitat and make further observations in case your samples missed some species.

Estimating species evenness

Estimating species evenness is more difficult. For this you need to carry out a **quantitative** survey.

Surveying the frequency of plants

First use the sampling techniques described in spread 2.3.2 to take your samples. Then count the number of plants of each species per unit area, or measure the percentage cover of each species. With large plants it is better to count the number of individuals per unit area. With smaller plants, such as grasses and herbs, it is better to measure the percentage cover. You can use similar techniques in both terrestrial and aquatic habitats. You may need to take extra precautions when sampling aquatic habitats!

Measuring the density of animals in a habitat

Measuring the density of animals in a habitat means calculating how many animals of each species there are per unit area of the habitat.

- For larger animals, such as badgers or deer, you will need to observe carefully and count the individuals present. However, this is not possible for most animals, as they are too small.
- For smaller animals, you will need to take samples of the animals present. As mentioned in spread 2.3.3, sampling animals involves trapping them. But you cannot be certain you have trapped all the animals in the population. You can calculate population size using the mark-and-recapture technique.

- First you need to capture a sample of animals, then mark each individual in some way that causes it no harm. The number captured will be C_1. Then release them and leave the traps for another period of time. The number captured on this second occasion will be C_2. The number of already marked animals captured on the second occasion is C_3.
- Then you can calculate the total population using the formula:

  total population = $(C_1 \times C_2)/C_3$

- Mark-and-recapture will not work for the numerous tiny animals living in soil. Here the only way to estimate population size is to take a sample of soil and sift through it to find all the individuals and count them.
- Sampling in water is a similar process. You can use a net to sample in the body of the water and to sift through the mud at the bottom. Then you can estimate population size and density.

Simpson's diversity index

Simpson's diversity index is a measure of the diversity of a habitat. It takes into account both species richness and species evenness. It is calculated by the formula:

$$D = 1 - [\Sigma(n/N)^2]$$

where n is the number of individuals of a particular species (or the percentage cover for plants) and N is the total number of all individuals of all species (or the percentage cover for plants).

Table 2 shows how to apply Simpson's diversity index to the results for fields A and B above.

	Field A			Field B		
	n	n/N	$(n/N)^2$	n	n/N	$(n/N)^2$
Cocksfoot grass	57	0.57	0.3249	38	0.38	0.1444
Timothy grass	32	0.32	0.1024	16	0.16	0.0256
Meadow buttercup	3	0.03	0.0009	14	0.14	0.0196
White clover	3	0.03	0.0009	22	0.22	0.0484
Creeping thistle	1	0.01	0.0001	5	0.05	0.0025
Dandelion	4	0.04	0.0016	5	0.05	0.0025
Sum (Σ)	–	–	0.4308	–	–	0.243
1 – Σ	–	–	0.5692	–	–	0.757

Table 2 Simpson's diversity index

Using Simpson's diversity index

A high value of Simpson's index indicates a diverse habitat. Such a habitat provides a place for many different species and many organisms to live. A small change to the environment may affect one species. If this species is only a small part of the habitat, the total number of individuals affected is a small proportion of the total number present. Therefore the effect on the whole habitat is small. The habitat tends to be stable and able to withstand change.

A low value for diversity suggests a habitat dominated by a few species. In this case, a small change to the environment that affects one of those species could damage or destroy the whole habitat. Such a small change could be a disease or predator, or even something that humans have done nearby.

Questions

1 Explain the difference between species richness and species evenness.

2 Suggest what precautions you may need to take when measuring populations of aquatic animals or plants.

3 Explain why a habitat with high diversity tends to be more stable than one with low diversity.

(5) Classification and taxonomy

By the end of this spread, you should be able to . . .

* **Define the terms *classification*, *phylogeny* and *taxonomy*.**
* **Explain the relationship between classification and phylogeny.**

Key definitions

Biological classification is the process of sorting living things into groups. Natural classification does this by grouping things according to how closely related they are. Natural classification reflects evolutionary relationships.

Taxonomy is the study of the principles of classification.

Phylogeny is the study of the evolutionary relationships between organisms.

Classification

Biological classification is the process of placing living things into groups. This involves detailed study of the individuals in a **species**. We can then place the species in a category with other species that show a number of similar characteristics.

Some classifications are artificial and are made for our convenience. For example, in a wild flower guide the plants are grouped according to the colour of the flower. This enables you to turn quickly to the section of the guide containing all the flowers of that colour.

Taxonomy

Taxonomy is the study of the principles behind classification. This means the study of the differences between species. These differences can be used to classify species. Species are usually grouped according to their physical similarities. All species that look very similar are placed together. Species that look quite different are placed in separate groups.

Natural classification

The basic unit of natural classification is the species. A species is a group of individual organisms that are very similar in appearance, anatomy, physiology, biochemistry and genetics. Individual members of a species will show **variation**. For example, all the varieties of dog, from a chihuahua to a German shepherd, are members of the same species. Underneath the obvious visible differences, all dogs are very similar.

As all members of the species are very similar, we can consider them to be closely related. In the same way, different species that are very similar can be considered to be closely related. Two closely related species will be placed in a group together. Then closely related groups will be placed together in a larger group. In this way the whole of the living world can be organised into a series of ranked groups – a hierarchy.

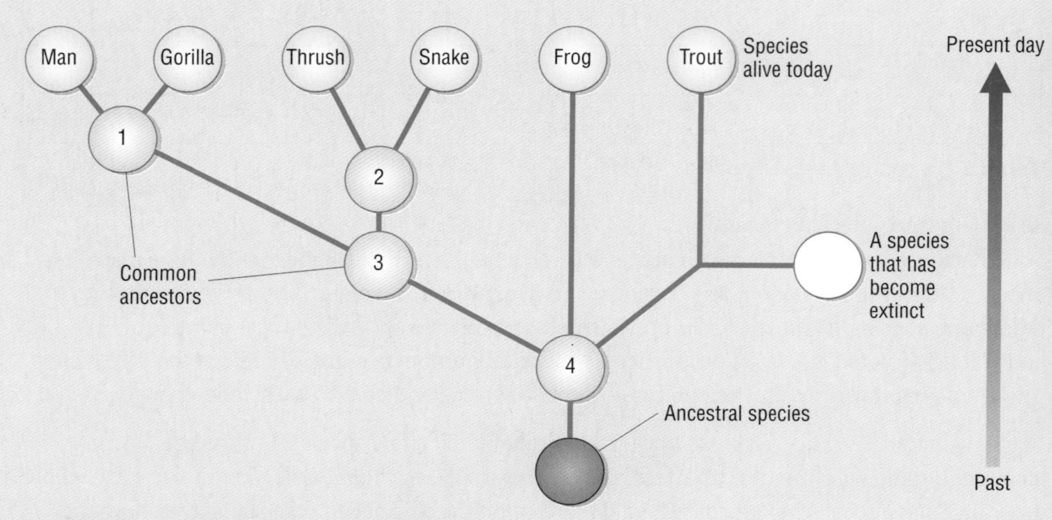

Figure 1 The evolutionary tree of vertebrates

Increasingly, modern classification has come to reflect the **evolutionary distance** between species. We can think of all living things as belonging to an evolutionary tree. Any two species living today have had a common ancestor at some time in the past. The time at which the two species started to evolve separately is a branch point on the tree. The common ancestor appears on the tree at that branch point. The more recent the common ancestor, the more closely related the two species are – they have a short evolutionary distance.

Phylogeny

Phylogeny is the study of how closely different species are related. It reflects the evolutionary relationships of the species. We can see this in the evolutionary tree. The more closely two species are related, the closer they appear together on the evolutionary tree. Also the more recently in the past they shared a common ancestor. This is used as a basis for natural classification.

Using the evolutionary tree in Figure 1, we can see certain evolutionary relationships that indicate how closely related the species are.

- Humans and gorillas share many features and are closely related.
- We have a common ancestor in the recent past (species 1).
- We can call humans and gorillas **monophyletic** – because they belong to the same phylogenetic group.
- Humans and gorillas can be placed (classified) in the same taxonomic group.
- The thrush is more closely related to the snake than to the mammals.
- We can see this because the common ancestor (species 2) shared by the thrush and the snake is more recent than the common ancestor shared by the thrush and the mammals (species 3).
- Therefore the thrush must be placed in a different group from the mammals.
- Similarly, the snake is more closely related to the thrush than to the frog or the trout.
- All species 1–6 can be considered to be monophyletic as they all evolved from the same species (species 4).

It should be noted that the common ancestors do not survive today. We cannot say we evolved from the apes, or from the gorillas, or even from modern-day fish. We evolved from an ancestor that lived at some time in the past. It happens that the gorillas also evolved from that same ancestor.

Questions

1. What is a common ancestor?
2. Explain the difference between taxonomy and classification.
3. Why do we study how closely related we are to other organisms?

2.3 6 The five kingdoms

By the end of this spread, you should be able to . . .

* **Outline the characteristic features of the following five kingdoms: Prokaryotae, Protoctista, Fungi, Plantae, Animalia.**

The five kingdoms

Traditionally, all living things have been grouped into a number of kingdoms. For many years the generally accepted number of kingdoms was two. All living things were grouped into either plants or animals. As more living things were discovered and studied closely, it became clear that not all living things could fit easily into one of these categories. For some years now, the accepted number of kingdoms has been five.

Key definitions

Prokaryotes have no nucleus.

Protoctists include all the organisms that don't fit into the other four kingdoms. Many are single-celled, but some are multicellular.

Fungi are organisms that are mostly saprophytic. They consist of a mycelium with walls made from chitin.

Plants are multicellular organisms that gain their nutrition from photosynthesis.

Animals are heterotrophic multicellular eukaryotes.

Prokaryotes

The **prokaryotes** are the organisms belonging to the kingdom Prokaryota (which used to be called the Monera). These organisms include the bacteria and the cyanobacteria (these used to be called blue-green algae). Prokaryote means 'before nucleus'. The main feature of all the prokaryotes is given in their name – they evolved before the nucleus became the place to keep **DNA**. So all prokaryotes have no nucleus. The prokaryotes:

- have no nucleus
- have a loop of naked DNA (DNA that is not associated with histore proteins) that is *not* arranged in linear **chromosomes**
- have no membrane-bound **organelles**
- have smaller **ribosomes** than in other groups
- carry out **respiration** not in mitochondria, but on special membrane systems (mesosomes)
- have cells smaller than those of **eukaryotes**
- may be free-living or parasitic; some cause diseases.

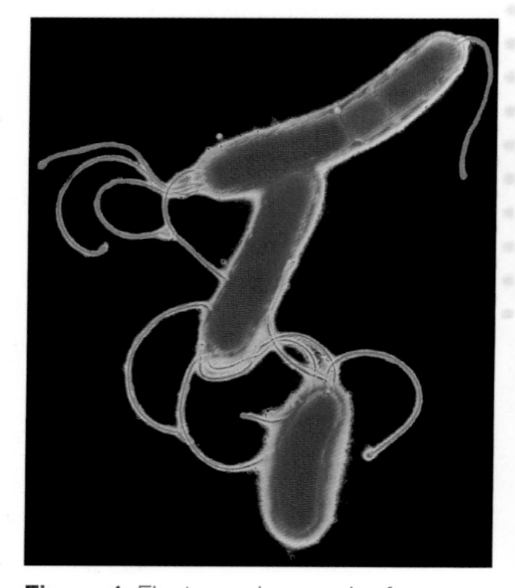

Figure 1 Electron micrograph of *Helicobacter pylori* bacteria

Protoctists

All members of the kingdom Protoctista are eukaryotes. This group contains organisms that are single-celled. It also contains some multicellular organisms, e.g. algae. (The smaller members of this kingdom used to be known as protists.) The protoctists:

- are eukaryotes
- are mostly single-celled
- show a wide variety of forms
- show various plant-like or animal-like features
- are mostly free-living
- have autotrophic or heterotrophic nutrition – some photosynthesise, some ingest prey, some feed using extracellular enzymes (like fungi do), and some are parasites.

The only thing that all Protoctists have in common is that they do not qualify to belong to any of the other four kingdoms!

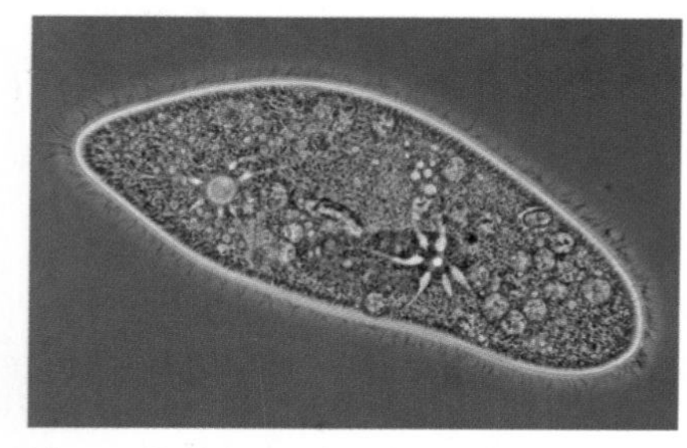

Figure 2 *Paramecium* – a protoctist

Fungi

The members of the kingdom Fungi are a group of organisms in which the body consists of a **mycelium** – a network of numerous strands called **hyphae**. The cytoplasm is surrounded by a wall of a **polysaccharide** called chitin. The cytoplasm is usually not divided into cells. It has many nuclei (it is multinucleate). A few fungi are cellular. The fungi:

- are eukaryotes
- have a mycelium, which consists of hyphae
- have walls made of chitin
- have cytoplasm that is multinucleate
- are mostly free-living and saprophytic – this means that they cause decay of organic matter.

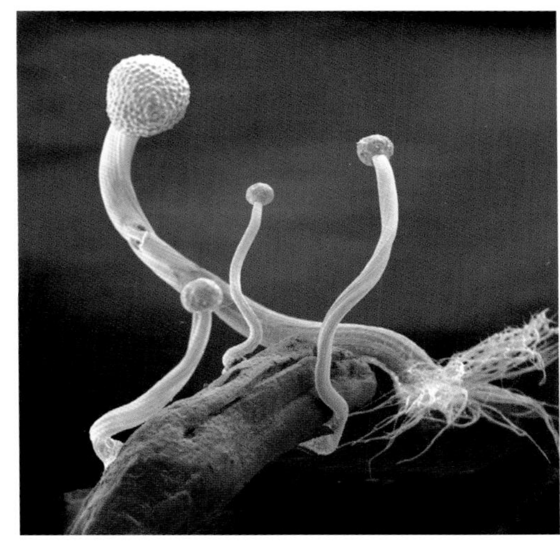

Figure 3 Scanning electron micrograph of ergot fungus

Plants

The green plants (kingdom Plantae) are all multicellular organisms. They gain their nutrition from **photosynthesis**. This means they are **autotrophs**. The plants:

- are eukaryotes
- are multicellular
- have cells surrounded by a cellulose cell wall
- produce multicellular embryos from fertilised eggs
- have autotrophic nutrition.

Figure 4 An oak tree – an example of a plant

Animals

The animals (kingdom Animalia) are multicellular organisms that gain their nutrition by digesting and absorbing organic matter – they are **heterotrophic**. The animals:

- are eukaryotes
- are multicellular
- have heterotrophic nutrition
- have fertilised eggs that develop into a ball of cells called a blastula
- are usually able to move around.

Figure 5 Impala – an example of animals

Questions

1 Describe the differences between fungi and plants.

2 Produce a table to compare the prokaryotes with the protoctists.

7 Classifying living things

By the end of this spread, you should be able to . . .

* **Describe the classification of species into the taxonomic hierarchy of domain, kingdom, phylum, class, order, family, genus and species.**

Why we classify living things

To put things in order, or to sort them out, seems to be a basic human need. To categorise all living things is a huge task. Current estimates suggest that there are nearly 2 million different **species** of living organisms. Each species must be studied in detail, then placed in a group of similar organisms. This is done:

- for our convenience
- to make the study of living things more manageable
- to make it easier to identify organisms
- to help us see the relationships between species.

Understanding that humans are animals, and that we have close relatives in the animal kingdom, is important in helping us understand our own evolution.

Carl Linnaeus devised the system of classification that is still in use today. Around 250 years ago, Linnaeus classified about 70 000 organisms. He was more concerned with how these organisms could be classified methodically than with simply collecting and describing them.

He studied each one closely and organised them according to their visible features. Two organisms with many similar visible features were grouped closer together. Linnaeus decided to put his organisms into a series of ranked categories. These categories are called taxonomic groups or taxa (singular taxon). His original classification contained five taxa: kingdom, class, order, genus and species.

As more organisms have been found and described, this original system of classification has had to be modified. We now have more kingdoms, and these are grouped into even larger categories called domains (see spread 2.3.9). Below the rank of kingdom we now have phylum, which contains the classes, and genera are now grouped into families.

Classification

Aristotle was the first person to attempt to classify all known living organisms, in the fourth century B.C. His classification extended to categorising organisms as plants or animals.

The current system of classification

The current system of classification uses eight taxa:

- domain
- kingdom
- phylum
- class
- order
- family
- genus
- species.

The species is the basic unit of classification. All members of a species show some variations, but all are essentially the same. As you rise through the ranks of classification (taxa), the individuals grouped together show more and more diversity. The number of similarities, and therefore the level of relatedness, get less and less. There will be one or more species in each genus (plural: genera). The species in a genus are all closely related and show many similar features. There may be a number of genera in each family, and a number of families in each order. This pattern continues up to the domains. All living things are classified as belonging to one of the three domains: Bacteria (Eubacteria), Archaea (Archaebacteria) and Eukaryotae (see spread 2.3.9).

Classifying a species

At the higher levels of this ranked system, the differences between individual species can be very great. It is therefore quite easy to place a species in its domain or kingdom. It is also relatively easy to place a species in its phylum. The differences between phyla are quite major. For example, two phyla in the animal kingdom are the Chordata and the Arthropoda. The Chordata have a nervous system with a central bundle of nerves running along their back, usually protected by a series of bones called the vertebral column. Most members of the phylum Chordata are called vertebrates. In contrast, the Arthropoda have a hard exoskeleton (skeleton on the outside of the body rather than the inside) and jointed limbs.

Within a phylum, the species must be placed in a class. This becomes a little more difficult as the differences between the classes in one phylum may not be very great. A longer description of the species may be needed. For example, two classes in the phylum Arthropoda are the insects and the arachnids (spiders). All members of the class Insecta can easily be recognised as having a number of features in common:

- three body parts
- six legs
- usually two pairs of wings.

Figure 1 A fruit fly – an example of an insect (×20)

All members of the class Arachnida can be recognised as having a number of features in common, which are different from those of the Insecta:

- two body parts
- eight legs.

As you descend to the lower taxonomic groups, it becomes increasingly difficult to separate closely related species and to place a species accurately. A more and more detailed description of the species is needed.

Figure 2 A spider – an example of an Arachnid (×5)

Some examples of classification

Table 1 lists some examples of classification.

Taxonomic rank/name	Named taxonomic group			
Domain	Eukaryotae	Eukaryotae	Eukaryotae	Bacteria (Eubacteria)
Kingdom	Animalia	Animalia	Animalia	Prokaryotae
Phylum	Chordata	Chordata	Arthropoda	Proteobacteria
Class	Mammalia	Mammalia	Insecta	Proteobacteria
Order	Primate	Primate	Diptera	Enterobacteriales
Family	Hominidae	Hominidae	Drosophilidae	Enterobacteriaceae
Genus	*Homo*	*Gorilla*	*Drosophila*	*Escherichia*
Species	*sapiens*	*gorilla*	*melanogaster*	*coli*
Common name	Human	Gorilla	Fruit fly	*E. coli*

Table 1 Examples of classification

Questions

1 Explain why a spider is not an insect.

2 Using information from spreads 2.3.1 and 2.3.5–7, define the term species.

(8) Naming living things

By the end of this spread, you should be able to . . .

* **Outline the binomial system of nomenclature and the use of scientific (Latin) names for species.**
* **Use a dichotomous key to identify a group of at least six plants, animals or microorganisms.**

Key definitions

The **binomial system** uses two names to identify each species: the genus name and the species name.

A **dichotomous key** uses a series of questions with two alternative answers to help you identify a specimen.

The binomial system

Carl Linnaeus devised a system of naming living organisms that uses two names. Before Linnaeus, **species** were identified by a common name, or a long and detailed description. Some scientists even used long descriptive names in Latin. Using a common name does not work well because:

- the same organism may have a completely different common name in different parts of one country
- different common names are used in different countries
- translation of languages or dialects may give different names
- the same common name may be used for different species in other parts of the world.

Linnaeus used Latin as a universal language. This means that whenever a species is named, it is given a universal name. Every scientist in every country will use the same name. This avoids the potential confusion caused by using common names.

Binomial means 'two names': in this case, the genus name and the species name. The binomial classification system uses the name *Homo sapiens* for humans. *Homo* refers to the genus to which humans belong. The genus name is always given a capital first letter. The species name is *sapiens*. Thus humans are *Homo sapiens*, which can be abbreviated to *H. sapiens*. The binomial Latin name is always written in a style that makes it stand out. In printed text this is in italics, in handwritten text it is underlined.

Identifying living organisms

The importance of identifying organisms

Before any large development can take place, we need to check that it will not harm the environment too much. To do this, we would need to carry out an environmental impact assessment (EIA). When carrying out an EIA, scientists need to know what species are present in the area. If there is a species that is very rare, then losing that habitat would have a major impact on that species and on the environment.

For example, any pond that contains great crested newts (*Triturus tristatus*) must not be harmed. The great crested newt is protected under the Wildlife and Countryside Act 1981 and the Habitat Regulations Act 1994. It is illegal to catch, possess or handle great crested newts without a licence. It is also illegal to cause them harm or death, or to disturb their habitat in any way.

Using a dichotomous key

A dichotomous key is a way of identifying and naming a specimen you have found. The key provides a series of questions. Each question has two answers – usually 'yes' or 'no'. The answer to each question leads you to another question. Eventually the answers will lead you to the name of the specimen. A good dichotomous key will have one question fewer than the number of species it can identify.

The following six winter twigs can be identified using a key with just five questions.

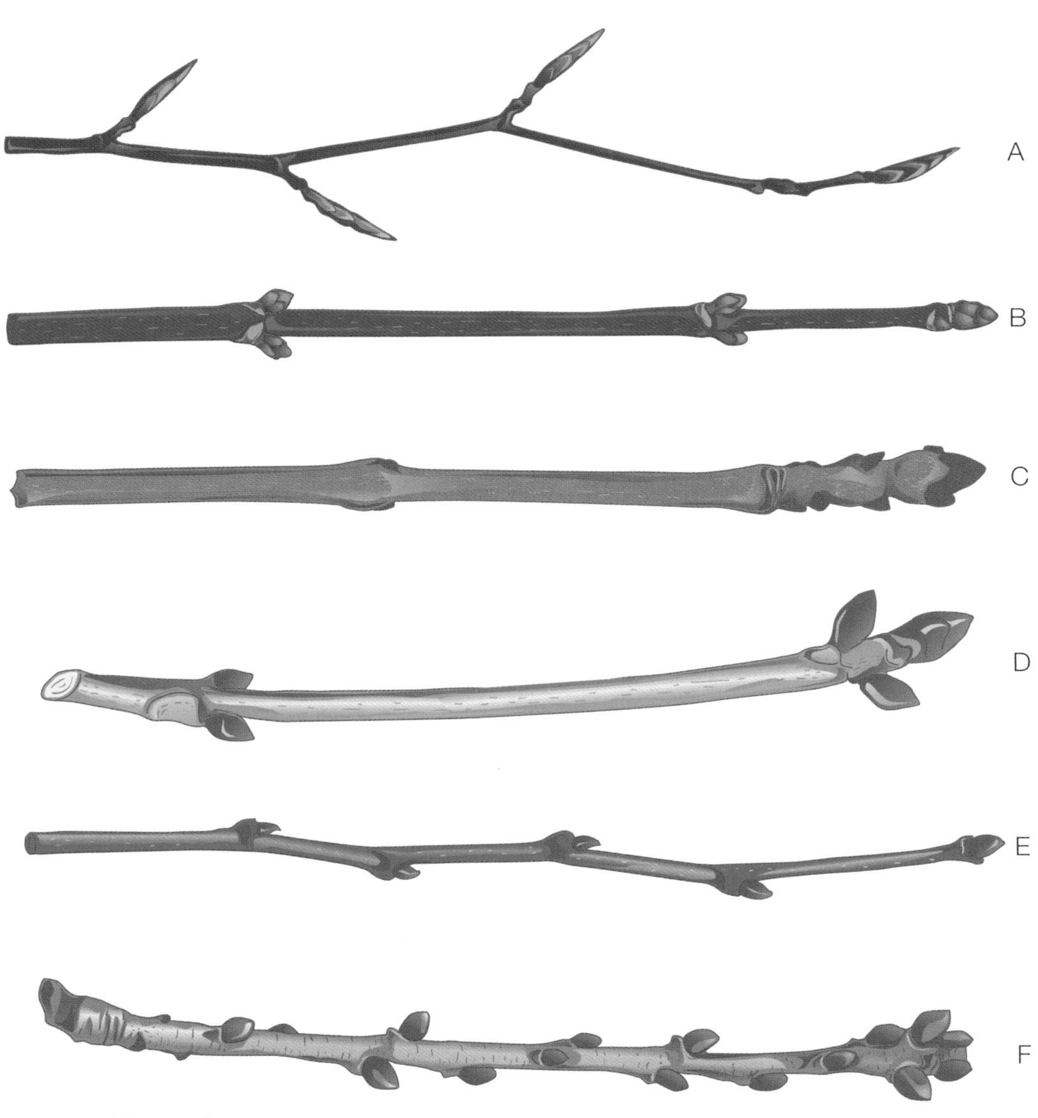

Figure 1 Winter twigs

Key:

Question 1 Are the buds paired?	yes – go to question 2 no – go to question 4
Question 2 Are the buds black?	yes – **ash** (*Fraxinus excelsior*) no – go to question 3
Question 3 Are the buds green?	yes – **sycamore** (*Acer pseudoplatanus*) no – **horse chestnut** (*Aesculus hippocastanum*)
Question 4 Are the buds distinctly pointed and longer than they are wide?	yes – **beech** (*Fagus sylvatica*) no – go to question 5
Question 5 Are the buds bunched together at the tip?	yes – **English oak** (*Quercus robur*) no – **sweet chestnut** (*Castanea sativa*)

Answers:

A – beech
B – sycamore
C – ash
D – horse chestnut
E – sweet chestnut
F – English oak

Questions

1 Explain why using common names for living organisms can cause confusion.
2 Why do we need to identify living things?

2.3 (9) Modern classification

By the end of this spread, you should be able to . . .

* **Discuss the fact that classification systems were originally based on observable features, but more recent approaches draw on a wider range of evidence to clarify relationships between organisms, including molecular evidence.**
* **Compare and contrast the five-kingdom and three-domain classification systems.**

Early classification systems

The early classification systems of Linnaeus and other scientists (see spread 2.3.7) were based on observable features. This means they were limited to those features of organisms that you can see. By the seventeenth century, scientists had microscopes to help. Later electron microscopes revealed details inside cells. Even as late as the second half of the twentieth century, scientists had little more information to work on.

Let us consider our earlier definition of a **species**:
'A species is a group of individual organisms that are very similar in appearance, anatomy, physiology, biochemistry and genetics.'

Early classification systems were based only on appearance and anatomy. For many species, this provided enough information to allow accurate classification. But it is easy to make mistakes. Aristotle classified all living things as either plant or animal. He further subdivided the animals into three groups – those that:

- live and move in water
- live and move on land
- move through the air.

This was based on the similarities he observed – some animals have fins, some have legs and some have wings. Unfortunately, this grouped fish with turtles, birds with insects, and mammals with frogs.

Such early classifications have been adapted and made more accurate as more research is done and more information becomes available.

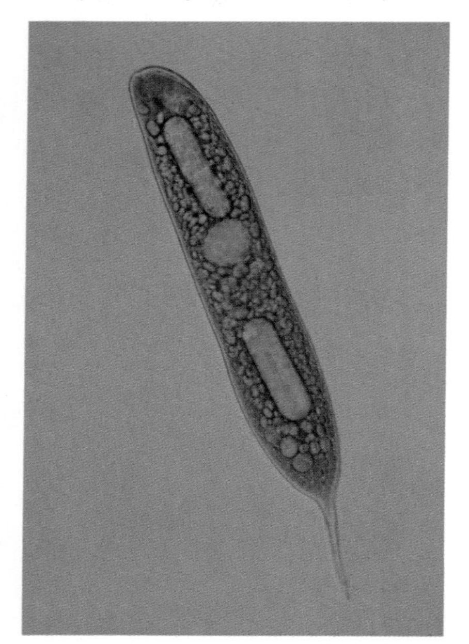

Figure 1 Euglena – a single-celled organism that can photosynthesise and has an undulipodium for locomotion (×900)

More recent classification systems

Originally, the animal kingdom included single-celled organisms that had some animal-like features. The plant kingdom included single-celled organisms that had plant-like features. Better microscopes made it clear that in fact many single-celled organisms share some of the features of both plants and animals. Something had to be done.

Also, where do fungi fit in? Like plants, they do not move about, and their **hyphae** grow into the surrounding **substrate** in the same way as roots. However, they do not **photosynthesise**. They digest organic matter and absorb the nutrients – like animals.

The resulting upheaval in the world of **taxonomy** and classification ended with the adoption of the five-kingdom classification (see spread 2.3.6).

The most recent research uses a wider range of techniques, and has produced even more detailed evidence. Physiology is the study of how living things work – how muscles contract, how oxygen enters the blood, etc. Several more detailed branches of scientific study have grown out of physiology, including the study of biochemistry.

Using biochemistry in classification

Evidence from biochemistry can help to determine how closely related one species is to another. Certain large biochemical molecules are found in all living things. But they may not be identical in all living things. The differences reflect the evolutionary relationships.

Cytochrome *c*

A protein called cytochrome *c* is used in the process of **respiration**. All living organisms, except chemosynthetic prokaryotes, must respire, therefore all organisms must have cytochrome *c*. But cytochrome *c* is not identical in all species. **Proteins** are large molecules made from a chain of smaller units called **amino acids**. The amino acids in cytochrome *c* can be identified. If we compare the sequence of amino acids in samples of cytochrome *c* from two different species, we can draw certain conclusions:

- if the sequences are the same, the two species must be closely related
- if the sequences are different, the two species are not so closely related
- the more differences found between the sequences, the less closely related the two species Figure 2.

DNA

Certain biochemicals are found in all living organisms. All organisms use **DNA** or **RNA**. DNA always provides the genetic code – the instructions for producing proteins. That code is the same for all organisms. This means that a particular piece (or sequence) of DNA or RNA codes for the same protein in a bacterium as in any other organism.

Comparison of DNA sequences provides a way to classify species. The more similar the sequence, the more closely related the two species. This is probably the most accurate way to demonstrate how closely related one species is to another. Such evidence has largely backed up the evolutionary relationships that have already been worked out. We can also use it to clarify or correct relationships we are unsure about.

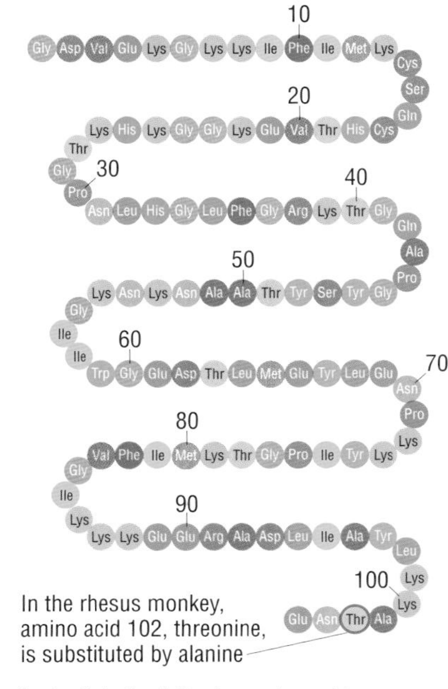

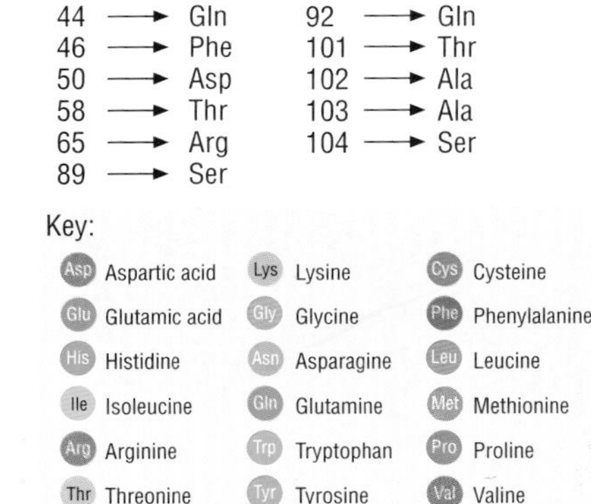

Figure 2 Primary sequence of human and chimpanzee cytochrome *c*

Three domains?

In 1990, Carl Woese suggested a new classification system. He based his ideas on detailed study of RNA. He divided the kingdom Prokaryotae into two groups: the Bacteria (Eubacteria) and the Archaeae (Archaebacteria). This division is based on the fact that the Bacteria are fundamentally different from the Archaea and the Eukaryotae. Some structural differences include the fact that bacteria have:

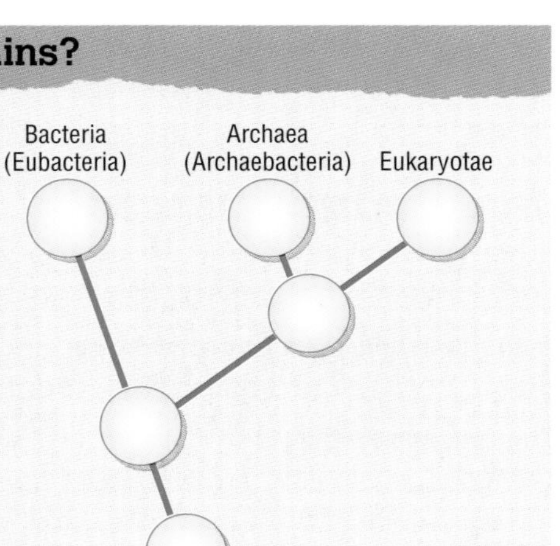

Figure 3 Three-domain tree

- a different cell membrane structure
- flagella with a different internal structure
- different enzymes (RNA polymerase) for building RNA
- no proteins bound to their genetic material
- different mechanisms for DNA replication and for building RNA.

The Archaea share certain features with the Eukaryotae:

- similar enzymes (RNA polymerase) for building RNA
- similar mechanisms for DNA replication and building RNA
- production of some proteins that bind to their DNA.

RNA and DNA are part of the basic mechanism that translates **genes** into visible characteristics. Carl Woese argued that these differences between the Bacteria (Eubacteria) and the Archaea (Archaebacteria) are fundamental. He suggested that these two groups are more different from each other than the Bacteria are from the Eukaryota. An accurate classification system must reflect this difference. Woese's three-domain system of classification is now widely accepted by most biologists.

Questions

1 How does the study of the structure of biological molecules help with classification?

2 What is the basis of the three-domain classification?

2.3 (10) Variation

By the end of this spread, you should be able to . . .

* **Define the term variation.**
* **Discuss the fact that variation occurs within, as well as between, species.**
* **Describe the differences between continuous and discontinuous variation, using examples of a range of characteristics found in plants, animals and microorganisms.**
* **Explain both the genetic and the environmental causes of variation.**

Key definitions

Variation is the presence of variety – of differences between individuals.

Genetic variation is caused by differences between the genes and the combinations of genes or alleles.

Continuous variation is variation in which there is a full range of intermediate **phenotypes** between two extremes.

Discontinuous variation is variation in which there are discrete groups of phenotypes with no or very few individuals in between.

Differences between individuals

The presence of differences between individuals is called **variation**. No two individuals are exactly alike, however similar they may look. Identical twins start as one cell that divides and then separates into two cells. Each of these two cells then develops into a separate person. While the two original cells had the same genetic information, the subsequent replication of **DNA** and cell divisions may have introduced changes to the DNA. Also, slight environmental differences in the womb or after birth can mean that the two individuals show physical differences.

Variation within species

It can be easy to forget that members of one species show variation. Like any other species, humans show variation. If you think of almost any characteristic, you will be able to find differences between members of the population. For example, eye colour, hair colour, skin colour and nose shape are all characteristics that show variation between different people.

Variation between species

The variation that occurs between **species** is usually obvious. In fact, it is that very variation that is used to separate members of one species from another.

Continuous and discontinuous variation

There are two forms of variation – **continuous** and **discontinuous**.

Continuous variation

Continuous variation is where there are two extremes and a full range of intermediate values between those extremes. Most individuals are close to the mean value. The number of individuals at the extremes is low.

Examples of continuous variation include:

- height in humans
- length of leaves on an oak tree
- length of stalk (reproductive **hypha**) of a toadstool.

Figure 2 Toadstools showing differences in lengths of stipes (reproductive stalks)

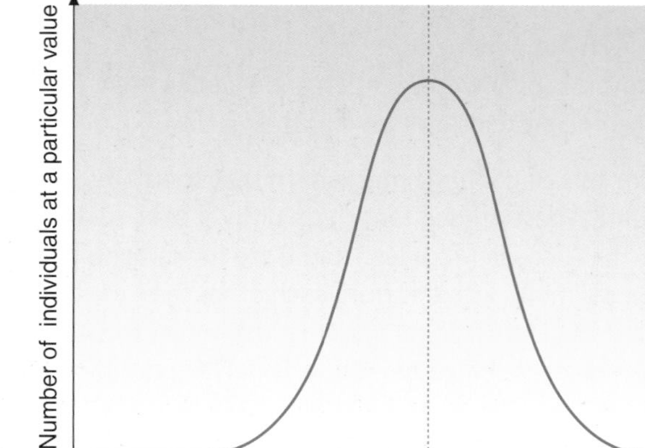

Figure 1 Continuous variation

Discontinuous variation

Discontinuous variation is where there are two or more distinct categories with no intermediate values. The members of a species may be evenly distributed between the different forms, or there may be more of one type than the other.

Figure 3
Discontinuous variation

Examples of discontinuous variation within species include:

- sex – mammals are either male or female; plants can be male, female or hermaphrodite
- some bacteria have flagella, but others do not
- human blood groups – you are blood group A, B, AB or O.

What causes variation?

There are two general causes of variation.

Inherited or genetic variation

The **genes** we inherit from our parents provide information that is used to define our characteristics. The combination of **alleles** (versions of genes) that we inherit is not the same as that in any other living thing (unless we have an identical twin). We may share many alleles with other members of our species, and we share genes with members of other species. However, there is never a complete match. Human cells contain approximately 25 000 genes, and many of these may have more than one allele. The chances of any two individuals having the same combination of alleles is remote. So the combination of characteristics each of us possesses is unique.

Environmental causes of variation

Many characteristics can be affected by the environment. For example, an overfed pet will become obese. A person's skin will tan and become darker with careful exposure to sunlight. A hawthorn tree usually grows upright to a height of about 6 m, but at very windy sites most of the branches will grow sideways in the direction away from the prevailing wind. The environment has affected the *direction* of growth. If a hawthorn grows in a rock crevice, where there is very little soil or water, it grows to height of only 150 cm – the environment has affected the *amount* of growth.

Environmental variation and genetic variation are linked. In the past century, humans have become taller as result of a better diet. But however good a diet you have, you are unlikely to grow very tall if all the rest of your family are short. This is because the height you can reach is limited by your genes.

Not all our genes are active at any one time. For example, when you reach puberty, many changes occur in your body because different genes are becoming active. Changes in the environment affect which genes are active. This is what brings about the changes you see.

Questions

1. List six characteristics that show continuous variation.
2. List six characteristics that show discontinuous variation.
3. Explain how the environment can cause variation.

2.3 (11) Adaptation

By the end of this spread, you should be able to . . .

* **Outline the behavioural, physiological and anatomical (structural) adaptations of organisms to their environments.**

Key definitions

An **adaptation** is a feature that enhances survival and long-term reproductive success.

Xerophytic plants are adapted to living in very dry conditions.

What is an adaptation?

All members of a species are slightly different from one another. They show **variation**. Any variation that helps the organism survive is an **adaptation**. The organism is adapted to its environment.

The process of evolution works by selecting particular adaptations to survive from one generation to the next. An individual that has a characteristic which helps it survive in its environment is more likely than an individual without that characteristic to live long enough to reproduce. Over a long period of time – very many generations – more and more individuals in the population will have that characteristic, as those that did not will have died out. The characteristic is an adaptation. We say that the adaptation has been selected.

Adaptations help the organism cope with environmental stresses and obtain the things they need to survive. A well adapted organism will be able to:

- find enough food or photosynthesise well
- find enough water
- gather enough nutrients
- defend itself from predators and diseases
- survive the physical conditions of its environment, such as changes in temperature, light and water levels
- respond to changes in its environment
- still have enough energy left over to reproduce successfully.

How do adaptations enhance survival?

Adaptations can work in different ways. They may be behavioural, physiological (or biochemical), or anatomical.

Behavioural adaptations

A behavioural adaptation is an aspect of the behaviour of an organism that helps it to survive the conditions it lives in. For example, when you touch an earthworm it quickly contracts and withdraws into its burrow. The earthworm has no eyes, so it cannot tell that you are not a bird about to eat it. Its rapid withdrawal is a behavioural adaptation to avoid being eaten.

Physiological/biochemical adaptations

A physiological or biochemical adaptation is one that ensures the correct functioning of cell processes. For example, the yeast *Saccharomyces* can respire sugars anaerobically or aerobically to obtain energy, depending on how much oxygen there is in its environment. Producing the correct **enzymes** to **respire** the sugars present in their environment is a physiological/biochemical adaptation.

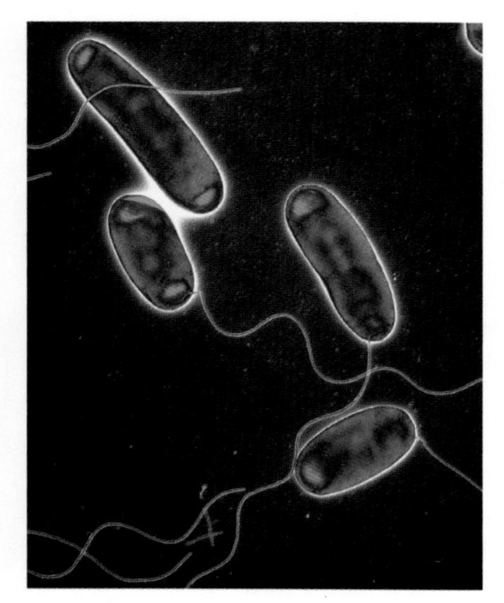

Figure 1 *Legionella* bacteria showing flagella (×16 000)

Anatomical adaptations

Anatomical means structural. Any structure that enhances the survival of the organism is an adaptation. For example, many bacteria (e.g. *Legionella*, Figure 1) have flagella that enable them to move independently. The flagellum is a structural adaptation.

An example of adaptation: xerophytic plants

Behavioural adaptations

You may not think that plants can show behaviour, but many certainly respond to their environment. **Xerophytes** respond to shortage of water in a number of ways.

- Some plants close their **stomata** when little water is available. This conserves water so that they do not wilt.
- Some plants open their stomata only at night. This conserves water because less will be lost from the leaves at night when it is cooler and the air is more humid.
- Some plants fold or roll their leaves when water is in short supply. This reduces water loss by trapping moist air in the folded leaf, so reducing the water vapour **potential gradient** for diffusion out of the leaf.
- Some plants even open their stomata when they are short of water. This causes the leaves to wilt and exposes less surface area to the sun.

Physiological/biochemical adaptations

These are the mechanisms by which a plant can open or close its stomata, fold its leaves or store water.

The saguaro cactus (*Cereus giganteus*) has a stem with an accordion-fold structure. During dry periods the folds tighten into ridges and become more pronounced. When water becomes available, the cactus absorbs it from the ground. The water fills the cells of the stem, where it may be stored for years. The cells expand and cause the stem to expand and become more rounded. The accordion folding becomes less obvious.

Figure 2 Saguaro cactus

Anatomical adaptations

These are structures that enable a xerophyte to survive in very dry conditions. There are many examples of structural adaptations.

- The roots may be shallow, but spread out over a wide area. This allows a plant to absorb a lot of water when it is available. A mature saguaro cactus can absorb almost 750 dm^3 of water during a brief storm.
- The roots may be very long. This enables the plant to reach water that is deep underground. The camel thorn tree (*Acacia erioloba*) has roots up to 40 m long.
- The stem or leaves may be fleshy – an adaptation to store water.
- The leaves may be reduced in size – this reduces the surface area for evaporation.
- The leaves may be very waxy, so moisture can leave the leaf only through the stomata.
- The leaves may be curled, folded, hairy, or have their stomata sunk in **pits**. These adaptations trap a layer of moist air next to the stomata, reducing the water potential gradient for water vapour to diffuse into the atmosphere. All these adaptations reduce water loss by **transpiration**.

Questions

1 List the adaptations of one organism chosen from the list below to its environment.
   (a) dog
   (b) holly bush
   (c) cod
   (d) *E. coli*
   (e) kelp
2 Describe how each adaptation enables it to survive.

2.3 (12) Natural selection

By the end of this spread, you should be able to . . .

* **Explain the consequences of the four observations listed by Darwin in proposing his theory of natural selection.**
* **Define the term speciation.**

Charles Darwin

Charles Darwin was a naturalist who spent much of his life observing and studying living organisms. The theory of **evolution** was not his idea. The idea that one organism might have developed from another over time was not new. Darwin managed to propose a mechanism for this process. This made it easier to believe in the theory of evolution, and caused a certain amount of upheaval in Victorian Britain. His proposed mechanism was **natural selection**.

Darwin's ideas came about after a 5-year trip around the world in a ship called the HMS *Beagle*. During this trip he visited the Galapagos Islands, where he discovered a large number of unusual **species**. Many of these species were similar to those found on the South American mainland. What interested Darwin was that there was clear **variation** between members of the same species found on different islands. He also noted that what appeared to be a wide variety of bird species were actually all closely related finches.

How Darwin's observations led to this new theory

Darwin made four particular observations:

- offspring generally appear similar to their parents
- no two individuals are identical
- organisms have the ability to produce large numbers of offspring
- populations in nature tend to remain fairly stable in size.

Darwin also read *An Essay on the Principle of Population* by Thomas Malthus. Malthus argued that the human population would outstrip its food supply. But he thought that constant competition for food and resources would keep the population in check.

Darwin realised that variation was the key to understanding how species change. He saw that when too many young are produced, there is competition for food and resources. As all the offspring are different, some may be better **adapted** than others. The better adapted individuals obtain enough food and survive long enough to reproduce. These individuals can pass on their characteristics to the next generation. The less well adapted individuals are likely to die before they reproduce. Therefore the population does not grow indefinitely.

Over a long period of time, a number of small variations may arise. If these variations are beneficial, they will help an organism survive. Over time, the species will accumulate many small variations and will change. Eventually one group of organisms belonging to one species could give rise to another species. This will happen if it becomes so different that it is unable to interbreed with the rest of the species.

Darwin's conclusions can be summarised as:

- there is a struggle to survive
- better adapted individuals survive and pass on their characteristics
- over time, a number of changes may give rise to a new species.

Darwin's finches, a group of birds that he thought were very different but turned out to be closely related. The differences were due to adaptation to eating different foods.

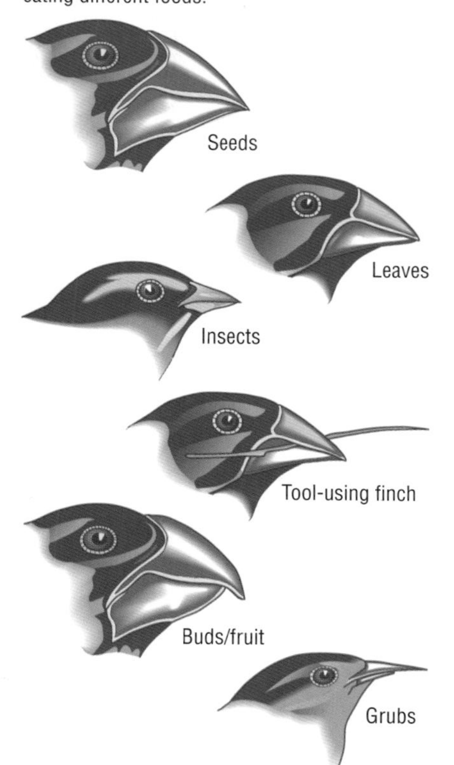

Figure 1 Darwin's finches

Natural selection

Natural selection is the process in which some environmental factor determines which individuals will survive. We say individuals are 'selected' from the population, or that they undergo **selection pressure**. If the individual has beneficial characteristics, it will be at an advantage. It will be 'selected' to survive and pass on its beneficial characteristics. However, if it does not have enough beneficial characteristics the individual will not be 'selected' to survive. It will struggle and may die, and will not pass on its characteristics.

Examples of factors in the environment that act as selective forces include:

- availability of suitable food – if an individual is adapted to eat the available food it is at an advantage (it has a selective advantage)
- predators – an individual adapted to avoid being seen and eaten, or to escape, has a selective advantage
- diseases – if an organism can survive a disease it has a selective advantage
- physical and chemical factors – if an organism can survive, for example, growing in a very shady place or in a desert, or living in a place with extremely cold winters and very hot summers, it has a selective advantage.

Key definitions

Natural selection is the 'selection' by the environment of particular individuals that show certain variations. These individuals will survive to reproduce and pass on their variations to the next generation.

Speciation is the formation of a new species.

New species

The formation of a new species from a pre-existing one is called **speciation**.

How long does speciation take?

Forming two closely related species from one does not occur suddenly. It is a long, slow accumulation of changes. These eventually mean that individuals can no longer interbreed freely to produce viable offspring. It is likely to take many generations. That is not to say it cannot happen overnight. Bacteria and single-celled organisms can pass through several generations in a few hours. This may be sufficient to allow speciation to occur.

How does speciation occur?

In order to form a new species from one original group of organisms, there must be some reproductive barrier. This means that some organisms are unable to breed with others in the group. Variations or changes that provide a benefit spread down the generations in a population through reproduction. If changes occur in part of the group, but cannot spread to the whole group, then only part of the group will benefit. A collection of small changes that cannot pass to the whole group means that some members become different from the others. They may become so different that they can no longer interbreed.

Reproductive barriers

A reproductive barrier is any factor that prevents effective reproduction between members of the species.

- Geographical separation will prevent effective interbreeding between the individuals of two populations. Different groups of the same species living on different islands will be unlikely to interbreed freely. So speciation is likely to occur. This is what allowed the evolution of new species in the Galapagos Islands. This is known as **allopatric** speciation.
- Sometimes a reproductive barrier may arise within the population. This may be due to a biochemical change that prevents fertilisation. It may be due to a behavioural change – perhaps a courtship dance is not recognised. Or it may be due to physical change, where the sexual organs of two groups of individuals are no longer compatible and they cannot mate. Any change that prevents one member of the population breeding with another can act as a reproductive barrier. This is known as **sympatric** speciation.

Questions

1 Explain how a particular colour of fur may be advantageous to a predator or prey species.

2 What factors may cause a struggle to survive amongst members of a population?

(13) The evidence for evolution

By the end of this spread, you should be able to . . .

* **Discuss the evidence supporting the theory of evolution with reference to fossil, DNA and molecular evidence.**

What evidence did Darwin have?

Even in Darwin's time, fossils clearly showed a number of interesting facts.

- In the past, the world has been inhabited by **species** that were different from those present today.
- Old species have died out.
- New species have arisen.
- The new species that appeared are often similar to the older ones found in the same place.

This led to questions:

- Why would one species die out?
- Why would a similar one replace it?
- Did one give rise to the other?

Brachiopods

Darwin studied a wide range of fossilised organisms. He found a group of animals called Brachiopods. These fossils appeared in rocks formed over a long period of time (500 million years). He realised that the brachiopods he found changed slowly over time. The rocks from different eras contained their own characteristic species of brachiopod. So brachiopods could be used to determine the relative age of the rocks.

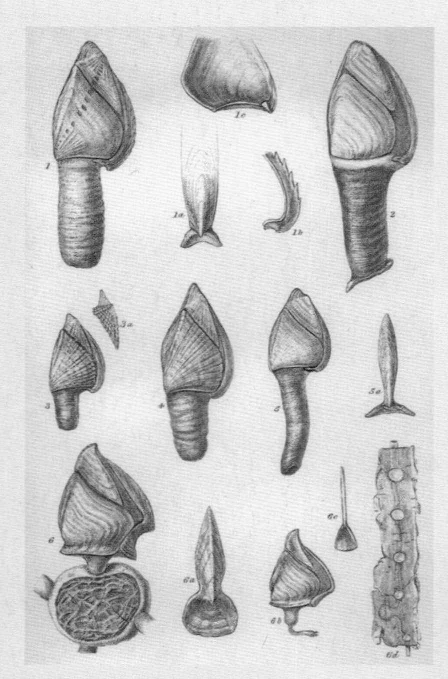

Figure 1 Brachiopods

Armadillos and Glyptodonts

Darwin was fascinated by the similarities he found between species living today and fossil species. He began to understand that fossil species gave rise to more modern species. He felt this must be because the more modern species had **variations** that meant they were better adapted to the environment. He was also struck by the differences between the fossil species and the modern species. Many of the fossil species were much larger than modern species, but otherwise appeared very similar. For example, some modern species of armadillo grow to only 15 cm long, while the glyptodont was many times this size.

Figure 2 A modern armadillo, which is only 15 cm long

Figure 3 A fossil glyptodont which is over a metre long

More recent fossil finds

Since Darwin's time, more fossils have been found. One of the main lines of evidence for evolution is that the general trend in the fossil record appears to be from smaller, more simple organisms in the distant past to larger, more complex organisms today.

There have been several important discoveries.

- The evolution of the modern horse (*Equus*) over the past 55 million years is well documented in the fossil record.
- The evolution of humans (*Homo sapiens*) can be seen in a series of fossils dating back 3–4 million years, but there are many gaps in this fossil record.
- Some particularly interesting finds provide a link between major groups. For example, the fossil of *Archaeopteryx* appears to be one of the earliest birds, and also shows many features typical of reptiles.

Figure 4 Fossil of Archaeopteryx, which could grow to about 50 cm in length

Gaps in the record

The problem with fossil evidence is that it is incomplete. Usually only the hard parts of an organism survive to become a fossil. Many living things don't have any hard parts, so they leave no fossils. Also, fossils form only under certain conditions. After they have been formed, they can still be damaged or destroyed by movements of the rocks.

More recent evidence

Biological molecules

Recent study of biological molecules can provide very strong evidence for evolution.

- The fact that certain molecules are found throughout the living world is evidence in itself. If one species gives rise to another, both are likely to have the same biological molecules. This suggests that all species arose from one original ancestor.
- Two closely related species will have evolved recently as separate species. The biological molecules in each are likely to be identical or very similar.
- In species that took separate evolutionary paths a long time ago, the biological molecules are likely to differ more.
- Evidence from molecules such as cytochrome *c* (see spread 2.3.9) and other proteins show this pattern of changes.

Protein variation

Vital **proteins** such as **DNA** polymerase and **RNA** polymerase are found in all living things. DNA polymerase is involved in DNA replication and RNA polymerase is involved in translating the genetic information in the DNA (**genes**) into proteins. It has been shown that the central part of these proteins is very similar across all living groups. Higher organisms have added extra subunits. These do not affect the basic function of the enzyme, but they appear to improve regulation of its action.

When the sequence of **amino acids** in cytochrome *c* is compared between many species, the number of changes is greater for species that are very different. As very different species took different evolutionary paths a long time ago, there has been more time for changes to accumulate.

DNA

The structure of DNA can be used in a similar way to that of cytochrome *c*. Genes can be compared by sequencing the bases in the DNA. This shows that closely related species evolved relatively recently as separate species. It also shows that more distantly related species have more differences in their DNA. Therefore they must have evolved as separate species further back in time. Comparisons of human DNA with that of other organisms shows the following evolutionary relationships:

- 1.2% of our coding sequence is different from that of chimpanzees
- 1.6% is different from that of gorillas
- 6.6% is different from that of baboons.

Questions

1. Explain how a knowledge of brachiopod evolution could help to determine the age of a rock.
2. Why was the discovery of the fossil of *Archaeopteryx* so important?

2.3 (14) # Evolution today

By the end of this spread, you should be able to . . .

* **Outline how variation, adaptation and selection are major components of evolution.**
* **Discuss why the evolution of pesticide resistance in insects and of drug resistance in microorganisms has implications for humans.**

How evolution works

1 Variation must occur before evolution can take place.
2 Once variety exists, then the environment can 'select'. It will select those variations that give an advantage.
3 Individuals with an advantage will survive and reproduce.
4 Therefore they pass on their advantageous characteristics (inheritance).
5 The next generation will be better adapted to their environment. Over time, the group of organisms becomes well adapted to its environment (adaptation).

It is *genetic variation* that is important for evolution. Variation due to environmental factors will not be passed to offspring.

Evolution and diversity

In some situations, a group of individuals within a species will evolve into a new, different species. This may occur if:

- a population of the species migrates to a new environment
- an environmental change affects only some populations of the species
- there is a reproductive barrier preventing some populations of the species from interbreeding – these populations then evolve along different paths according to their local environment, and these changes cannot be passed to the rest of the species.

If part of a species evolves into a new species, this increases **diversity**.

Evolution on a small scale occurs within a species, too. Some members of the species may have adaptations that make them more suited than other individuals of the same species to the local environment. Both groups of individuals can still interbreed – they are still the same species. Having a lot of local populations that are all genetically different increases diversity.

An example of this is wild salmon. Salmon return to the river of their birth after a long period living in the oceans. Scientists have discovered that salmon from a particular river are especially adapted to life in that river. Salmon from another river will not be able to compete so well with them for food or mates. When commercially bred salmon escape from their cages, they may breed with the wild salmon. Their offspring are not so well adapted to life in the river, so fewer of them survive. In some areas, local populations of wild salmon are becoming extinct. This will cause a loss of diversity.

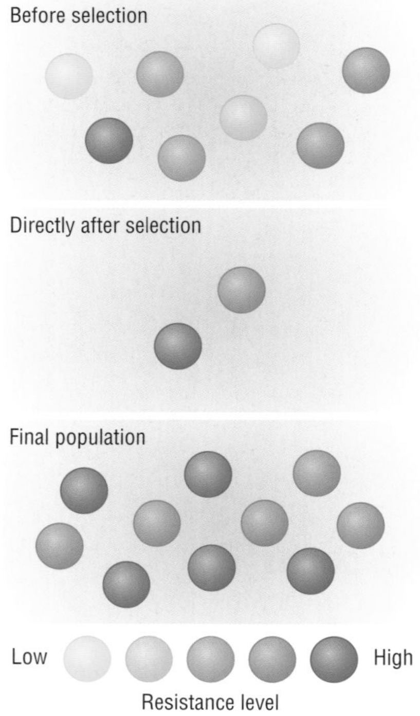

Figure 1 Development of resistance in populations of organisms

Evolution today

Is evolution still going on? Most certainly, yes. At any time, a species or a group of organisms can be placed under a new **selection pressure** (an external pressure that drives evolution in a particular direction), so different characteristics (variations) will be selected. Evolution will occur. This is most obvious in organisms that have a short life cycle.

The insects

Some insects are pests. They eat our food crops, or cause damage to them. They can also carry disease. Humans have devised ever more ingenious ways to kill insects. But some insects always survive.

Pesticides are chemicals designed to kill pests. Insecticides specifically kill insects. An insecticide applies a very strong selection pressure. If the individual insect is susceptible, it will die. If it has some form of resistance, the individual will survive. This will allow it to reproduce and pass on the resistance characteristic. So the resistance quickly spreads through the whole population.

Resistance to insecticides has developed in different ways:
- the insects may be able to break down the insecticide (using **enzymes**) – they metabolise them
- the target receptor protein on the cell membrane may be modified.

Mosquitoes, which carry the malaria parasite, have developed resistance to a number of insecticides. Pyrethroid insecticides are used to treat the mosquito nets used for protection over beds in countries where malaria occurs. Populations of mosquitoes in some areas have developed resistance to pyrethroids by metabolising them. DDT is used in household-spraying programmes. Insect populations have become resistant to both DDT and pyrethroids by evolving modified receptors.

Evolved resistance

In Lagos, Nigeria, mosquitoes were fully susceptible to both permethrin and DDT in 1999. But since 2001, populations in the area have developed resistance to both. Resistance to insecticides has seriously undermined our ability to control the spread of malaria and outbreaks of other insect pests.

Examiner tip

Never use the term 'immune to antibiotics' as this suggests that bacteria have immune systems. Bacteria are *resistant* to antibiotics. Also, remember that individuals don't evolve. Individuals are selected and the population evolves.

Another problem is the concentration of pesticides in the food chain. If insects are resistant, they survive applications of these chemicals. The insects may then be eaten by their predators. The predators receive a larger dose of the insecticide, and it is quite possible for the insecticide to move all the way up the food chain. In this way, humans may receive quite large doses of insecticide.

Microorganisms

The use of antibiotics is a very powerful selection pressure on bacteria. When you take antibiotics, most of the bacteria are killed. But there may be one, or a few, that are resistant to the antibiotic. They are rarely completely unaffected by the antibiotic – but they are more resistant than most. Once most of the bacteria have been killed, you tend to feel better. So many people stop taking the antibiotics before they have finished the prescribed course. This allows the resistant bacteria to survive and reproduce to create a resistant strain of bacteria. Overuse and incorrect use of antibiotics has led to strains of bacteria resistant to virtually all the antibiotics in use. Some doctors now prescribe multiple antibiotics. This greatly reduces the chances that some bacteria will survive.

Some bacteria have gained a particularly wide range of resistance. The so-called 'superbug', MRSA, is one. MRSA stands for methicillin-resistant *Staphylococcus aureus*. But it might as well stand for multiple-resistant *Staphylococcus aureus*. This bacterium has developed resistance to an ever-increasing range of stronger and stronger drugs. This is an example of an 'evolutionary arms race'. Medical researchers are struggling to develop new and effective drugs, but the bacterial populations rapidly become resistant to them.

Questions

1 Explain why evolution occurs in a shorter time in populations of microorganisms than in populations of mammals.
2 Explain why evolution tends to happen in short bursts.

2.3 (15) Conservation of species

By the end of this spread, you should be able to . . .

* **Outline the reasons for the conservation of animal and plant species, with reference to economic, ecological, ethical and aesthetic grounds.**

How human activities affect nature

Several thousand years ago, humans lived as hunter–gatherers in small numbers and had little effect on natural processes. In recent centuries, human society has developed:

- we have learned to use the environment to our advantage
- our numbers have risen dramatically – and continue to rise
- we are using more and more of the Earth's resources
- our activities often harm other **species**, either directly or indirectly
- loss of **biodiversity** occurs
- **extinction** may occur.

The latter three effects began to happen as long ago as the Palaeolithic, but they have greatly accelerated in recent times.

Extinction

Extinction occurs when the last living member of a species dies and the species ceases to exist.

Key definition

Extinction is when a species ceases to exist.

Humans started to spread widely over the Earth about 100 000 years ago. Since then, the rate of extinction of other species has risen dramatically. Some scientists believe that increasing human activity caused the extinction of animals such as the giant sloth (*Megatherium*) and the mammoths (*Mammuthus*) 10 000–14 000 years ago. These animals were hunted for food.

What is more certain is that:

- there have been 784 recorded extinctions since the year 1500
- up to 20% of the species alive today could be extinct by 2030
- a third of the world's primate species now face extinction – even our closest relatives, the great apes, could be extinct in 20 years
- some scientists believe that up to half the species alive today could be extinct by the year 2100
- the current rate of extinction is 100–1000 times the normal 'background' rate
- the current rate of extinction is at least as fast as in any previous extinction event.

The only conclusion is that we are at the start of a great mass extinction event. There have been other mass extinction events in the past – for example, when the dinosaurs and many other species became extinct. But this extinction event is being caused by human activity, rather than by dramatic climate change or natural disaster.

Figure 1 Passenger pigeon

Loss of biodiversity

The extinction of even one species reduces **biodiversity**. If we lose half the species in the world, this will cause a huge loss of biodiversity.

In an effort to produce more food, humans have cleared huge areas of natural vegetation. Natural vegetation is usually very diverse, with a well balanced range of species. Removal of any natural vegetation removes the **habitat** of many different organisms. As more habitats are removed, the chances of extinction for certain species become greater.

We often replace the natural habitat with a crop of very low diversity – often a **monoculture** (a crop of plants of a single species bred to be very similar). This makes the product easier to harvest. The oil palm, which is grown for palm oil, is a good example. Between 1985 and 2000, the development of oil palm plantations was responsible for 87% of deforestation in Malaysia. Oil palms now cover up to 40% of cultivated land in some tropical countries. A further 6 million hectares are scheduled to be cleared for oil palm plantations in Malaysia. In Indonesia, 16.5 million hectares are scheduled to become oil palm plantations.

Human activities

Activities that reduce biodiversity and cause extinction include:

- hunting for food (over-harvesting)
- killing for protection – attempting to kill insects that are vectors of disease (e.g. *Anopheles* mosquitoes) or to remove the threat of a predator
- killing to remove competitors for our food – use of pesticides to kill insects, fungi and other pests
- pollution
- habitat destruction, such as deforestation or clearing for development or agriculture
- inadvertent introduction of new predators and competitors to natural flora and fauna.

Figure 2 A monoculture of oil palms

Why we need to conserve species

There are many reasons why we should try to conserve the diversity of species alive today.

Economic and ecological reasons

One important reason to conserve species is that evolution has provided answers to many technological problems we face. For example, what is the best aerodynamic shape in water? The answer is provided by those animals that live and move in water. What is the best shape of an aerofoil or wing? What is the best design for distributing cooling fluid around a structure? All these problems have been solved by millions of years of evolution. If we allow species to go extinct, then we could be losing many solutions to new problems.

In 1997, an international team of economists and environmental scientists attempted to quantify the economic value of natural **ecosystems**. They came up with a figure of $\$33 \times 10^{12}$. They looked at all the ways in which natural ecosystems perform processes that are of value to humans. These include:

- regulation of the atmosphere and climate
- purification and retention of fresh water
- formation and fertilisation of soil
- recycling of nutrients
- detoxification and recycling of wastes
- crop pollination
- growth of timber, food and fuel.

These are all important ecological processes, particularly regulation of the atmosphere – the process of **photosynthesis** carried out by plants removes carbon dioxide from the air and replaces the oxygen. Without soil, we would not be able to grow food. Without the recycling of nutrients and wastes, the soil would become infertile. Without atmospheric oxygen, we would not be able to breathe.

Ethical and aesthetic reasons

All living organisms have a right to survive and to live in the way for which they have become **adapted**. The loss of habitat and biodiversity prevents many organisms from living where they should. This includes those human groups who still live in natural habitats.

We experience a feeling of joy and wellbeing when observing the infinite variations of nature. No human art or design can compete. Studies have shown that patients recover more rapidly from stress and injury when they are exposed to pleasing natural environmental conditions. It is clear that natural systems are very important for our wellbeing and for our physical, intellectual and emotional health.

Questions

1 List the human activities that affect nature.

2 What is meant by extinction?

2.3  The effect of global climate change

By the end of this spread, you should be able to . . .

* **Discuss the consequences of global climate change on the biodiversity of plants and animals, with reference to changes in patterns of agriculture and spread of disease.**
* **Explain the benefits for agriculture of maintaining the biodiversity of animals and plant species.**

The importance of genetic diversity

Genetic diversity makes it possible for a **species** to evolve. Without genetic **diversity**, plant and animal species will not be able to adapt to changes in the environment. Threats to a species with low genetic diversity would include:

- changes in the climate
- increase in levels of pollution
- emergence of new diseases
- arrival of new pests.

Humans affect the genetic diversity of natural **habitats**. As we clear natural vegetation, we reduce the size of natural habitats and reduce the population size of any species living in those habitats. This reduces the overall **gene pool** for the species. This decreases the genetic **variation** and hence the ability of the species to **evolve**.

Modern agriculture uses **monoculture** and selective breeding. This reduces the variation and genetic diversity of domesticated plants and animals. It also leads to the **extinction** of some varieties within a species – this is called **genetic erosion**.

Climate change

Species that have lost their genetic variation will be unable to evolve. As the climate changes, they will be unable to adapt to the changes in temperature and rainfall in the area where they live.

The only alternative will be for them to move. Plant and animal populations can move with changing conditions. This will mean a slow migration of populations, communities and whole **ecosystems** towards the poles – plants currently growing in southern Europe many soon grow in northern Europe.

However, there will be obstructions to this migration. Possible obstructions include:

- major human developments
- agricultural land
- large bodies of water
- humans.

Consider the plight of the golden toad of the Costa Rican cloud forest. This amphibian is already facing extinction due to climate change. As the climate warms, the toad moves uphill to stay in the most suitable habitat. What happens when it reaches the top of the hill? The toad will be faced with having to migrate through unsuitable habitats to reach another area with a suitable habitat. This will not happen.

Consider, also, the role of protected areas such as national parks, where hunting is not allowed. As the climate changes, the selected site no longer holds the conditions and plants that the protected animals need. The animals will migrate to live outside the protected area, where they will be at risk again from human activities.

Agriculture

Climate change could alter several factors in the environment that will affect agriculture and **biodiversity**. Some may be beneficial to agriculture, others may not. These include:

- higher carbon dioxide levels altering photosynthesis
- higher temperatures increasing growth rates
- longer growing seasons
- greater evaporation of water and therefore greater precipitation
- a change in the distribution of precipitation
- loss of land due to rise in sea level and increased salinity of the soil.

Domesticated plants and animals are particularly at risk. This is because we have selectively bred our crop plants and animals to provide the best yield in specific conditions. We have developed crops and animals that have little variation. Those species will be unable to evolve and adapt to a changing environment. As the climate changes, farmers will find that their traditional crops do not provide the same yield. There may be insufficient water, rainfall at the wrong time of year, or temperatures may be too high. Farmers will need to change the crops they grow and the varieties of animals that they keep. Crops from southern Europe may be grown in Britain, while parts of southern Europe may become desert.

Diseases

Crops being grown in new areas will encounter new diseases and pests. These crops will not have resistance to these specific diseases and pests. Higher temperatures may provide a longer growing season for crops, but the pests will also have more time in which to increase in numbers. More species of pests and diseases may be able to overwinter successfully and cause greater infestations earlier in the year. This will mean lower yields and less food for humans.

Human diseases will also migrate. The huge variety of tropical diseases that thrive in the warmth and moisture of the tropics may become a problem in Europe. *Anopheles* mosquitoes that carry malaria, and tsetse flies that carry the trypanosomes that cause sleeping sickness, may be able to live in new areas. This may cause the spread of these diseases to new areas.

The need for conservation of biodiversity

Allowing biodiversity to decline means that genetic diversity also declines. This means that we could lose the natural solution to some of our problems.

Wild animals and plants may hold the answers to problems caused by climate change. Populations of wild plants growing in an area have had thousands of years to evolve. They have adapted to overcome the problems presented by the environment. They may also have adapted to the pests and diseases found in that area. By careful selection and breeding from wild species, we may be able to breed new crop varieties that can cope with the new conditions created by climate change. Genetic engineering to produce **transgenic** species could also be used.

The number of potential new medicines to be found in wild plants is unknown. The range of possible vaccines that could be developed from wild microorganisms is also unknown. It is important to maintain the genetic diversity of wild plants and animals because of the potential that exists in the wide range of species currently alive.

Questions

1 Explain why loss of genetic diversity means the species can no longer evolve.
2 Explain why domesticated animals are particularly at risk.

2.3 (17) Conservation *in situ*

By the end of this spread, you should be able to . . .

* **Describe the conservation of endangered plant and animal species *in situ*, with reference to the advantages and disadvantages of this approach.**

Key definition

Conservation *in situ* means conserving a species in its normal environment.

Conservation *in situ*

Conservation *in situ* means attempting to minimise human impact on the natural environment and protecting the natural environment.

Legislation

It is possible to pass legislation to stop such activities as hunting, logging and clearing land for development and agriculture. The legislation is specific to a particular country. It can be difficult to persuade some countries that legislation is necessary. It can also be difficult to enforce such legislation – especially if the government is not in favour of it.

Conservation parks

It is often possible to stop unacceptable activities by establishing conservation areas such as national parks and nature reserves. But we should remember that these are not the only sites of *in situ* conservation. Land management agreements on private land and farm sites can also be used for conservation.

The principles for choosing a reserve or park must include:

- comprehensiveness – how many species are represented in the area and what are the prevailing environmental conditions?
- adequacy – is the area large enough to provide for the long-term survival of all the species, populations and communities represented?
- representativeness – is there a full range of diversity within each species and set of environmental conditions?

Designating an area as a reserve has a number of advantages:

- plants and animals are conserved in their natural environment
- it permanently protects **biodiversity** and representative examples of **ecosystems**
- it permanently protects significant elements of natural and cultural heritage
- it allows management of these areas to ensure that ecological integrity is maintained
- it provides opportunities for ecologically sustainable land uses, including traditional outdoor heritage activities and the associated economic benefits
- it facilitates scientific research
- it may be possible to restore the ecological integrity of the area.

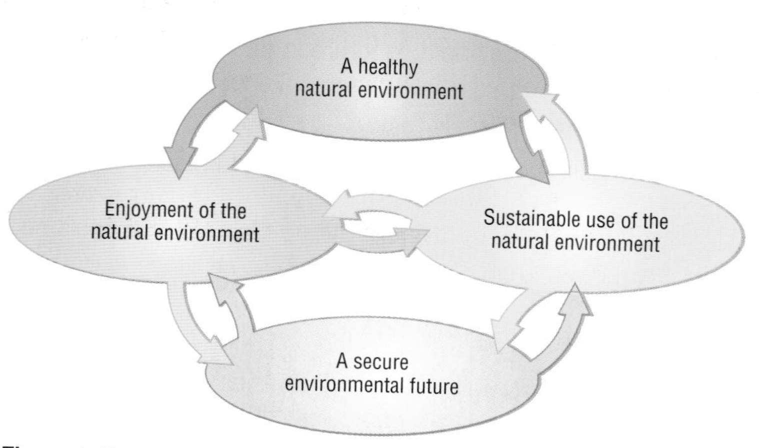

Figure 1 Flow chart to explain the benefits of *in situ* conservation

A conservation reserve should not mean excluding human activity. A reserve should meet the needs of the local indigenous people. They might use the land for traditional hunting, or for spiritual and religious activities. In the past, reserves have been set up without consideration of the local people, and this has led to conflict. The reasons why conflict arises could be due to:

- protected animals coming out of the reserve to raid crops – primates often raid farms for maize, mangoes and sugar cane
- people continuing to hunt protected animals for food
- illegal harvesting of timber and other plant products
- tourists feeding protected animals or leaving litter.

Conservation reserves in the UK

Various bodies in the UK work to conserve and enhance the natural environment – including the landscape, biodiversity, geology and soils, natural resources, cultural heritage and other features of the built and natural environment. At a national level, responsibility is held by the government bodies Natural England, Scottish Natural Heritage, the Countryside Council for Wales and the Environment and Heritage Service in Northern Ireland.

Much of England is protected by designated status.

- National Parks – there are 14 National Parks in the UK, covering many of the most beautiful and valued landscapes.
- National Nature Reserves (NNR) – at the end of 2004 there were 215 NNRs in England, covering nearly every type of vegetation found here.
- Sites of Special Scientific Interest (SSSIs) – there are over 6000 SSSIs in the UK. These are the country's very best wildlife and geological sites. They include some of our most spectacular and beautiful **habitats** – large wetlands teeming with waders and waterfowl, winding chalk rivers, gorse and heather-clad heathlands, flower-rich meadows, windswept shingle beaches and remote upland moorlands and peat bogs.
- Local Nature Reserves – often run by County Wildlife Trusts.

Figure 2 Repopulation of meadow grassland

Repopulation

Where biodiversity has been lost, it is possible to rebuild it. There are many examples of sites where recreated wildlife habitats have been made to work. In the UK, the numbers of bitterns and otters are increasing in new reed beds. Conifer crops are being cleared for wildlife habitat recovery and large areas of grazing land are being helped to revert to traditional meadow grassland.

In the Phinda Reserve of South Africa, work began in 1990 to clear away livestock and reintroduce natural fauna. More than 1000 wildebeest, zebras, giraffes and other ungulates were released between 1990 and 1992. Nearly 30 white rhinos and 56 elephants followed. Later in 1992, 13 lions and 17 cheetahs were released. This was a start towards recreating the rich mammal community that existed in the region before European colonisation.

Questions

1 What are the advantages of *in situ* conservation?
2 What are the difficulties associated with repopulation of an area with wild species?

2.3 18 Conservation *ex situ*

By the end of this spread, you should be able to . . .

* **Describe the conservation of endangered plant and animal species *ex situ*, with reference to the advantages and disadvantages of this approach.**
* **Discuss the role of botanical gardens in the *ex situ* conservation of rare plant species or plant species extinct in the wild, with reference to seed banks.**

Key definition

Conservation *ex situ* means conserving an endangered species by activities that take place outside its normal environment.

Conservation *ex situ*

Animal species

Traditional zoological collections held any animals that were collected by their owner or, perhaps, were unusual to the public. More recently, the role of zoos has changed, and many prefer to be known as 'wildlife parks'. These wildlife parks now play an important role in conservation. They concentrate on breeding endangered **species**. Such breeding programs can increase the numbers of individuals in an endangered species. They can also enable repopulation by introducing captive-bred animals to the wild.

These breeding programmes breed animals in captive surroundings. The animals are protected from predators, and their health can be maintained by veterinary science.

Figure 1 A white rhino

But captive breeding is an expensive process and can be very difficult for a number of reasons.

- The animals are often not in their natural environment and many animals fail to breed successfully.
- Space is limited and this limits the number of individuals, which restricts the genetic **diversity**.
- A decrease in the genetic diversity of a population results in a lack of **variation**.
- This means the species is less able to adapt to changing conditions, which can affect animals' ability to breed successfully.
- Even if reproduction is successful the animals have to survive reintroduction to the wild, where they need to find food and survive predation.
- There can also be difficulties with acceptance by the existing wild members of their species.

Modern techniques make it possible to preserve large amounts of genetic material by freezing sperm or eggs.

- Sperm freezing, artificial insemination, *in vitro* fertilisation and embryo-transfer techniques are used with domestic animals. These techniques can also be used with wild animals.
- Reproductive physiology is quite species-specific and further research into each endangered species is needed to ensure the techniques are used effectively.
- Carrying out research on a domestic species that is very similar to the target species can help to speed up progress.
- This also saves the rare, endangered individuals from experimental work until the last steps. The Arctic fox, the European reindeer and the European mink are all species that should soon benefit from these techniques.

Plant species

As with traditional zoos, many botanical gardens started life as a collection made by an affluent collector. Today, most botanical gardens are involved with conservation of endangered species.

The conservation of plants is, perhaps, a little easier than that of animals.

- As part of their life cycle, most plants naturally have a dormant stage – the seed.
- As seeds are produced in large numbers, they can be collected from the wild without causing too much disturbance to the **ecosystem** or damaging the wild population.
- These seeds can be stored and germinated in protected surroundings.
- Seeds can be stored in huge numbers without occupying too much space.
- Plants can often be bred asexually.
- The botanical garden can increase the numbers of individuals very quickly.
- This provides an ample supply of individuals for research.
- The captive-bred individuals can be replanted in the wild.

However, there are disadvantages.

- Any collection of wild seeds will cause some disturbance.
- The collected samples may not hold a representative selection of genetic diversity.
- Seeds collected from the same species from another area will be genetically different and may not succeed in a different area.
- Seeds stored for any length of time may not be viable.
- Plants bred asexually will be genetically identical – reducing genetic diversity further.
- Conclusions from research based on a small sample may not be valid for the whole species.

Seed banks

A seed bank is a collection of seed samples. The Kew Millennium Seed Bank Project at Wakehurst in Sussex is the largest *ex situ* conservation project yet conceived. Its aim is to store a representative sample of seeds from every known species of plant. By 2010, they hope to have collected seeds from 10% of the known dryland species. These will include examples of the rarest, most useful and most threatened species.

Figure 2 Wakehurst Place in Sussex, home of the Millennium Seed Bank

Seed banks contain seeds that can remain viable for decades and possibly hundreds of years. The seeds are not simply being stored. Some of them are being used to provide a wide range of benefits to humanity. These range from food and building materials for rural communities, to disease-resistant crops for agriculture. The seeds can also be used for habitat reclamation and repopulation. The collections held in the Millennium Seed Bank, and the knowledge derived from them, provide almost infinite options for their conservation and use.

Storage of seeds

In order to prolong their viability, seeds are stored in very dry or freezing conditions. Seeds are resistant to desiccation, and the level of moisture in each has a direct effect on storage. For every 1% decrease in seed moisture level, the life span doubles. For every 5 °C reduction in temperature, the life span also doubles. But seeds stored for decades may deteriorate. There is little use in storing seeds that die and will not be able to germinate. So it is essential to test the seeds at regular intervals to check their viability.

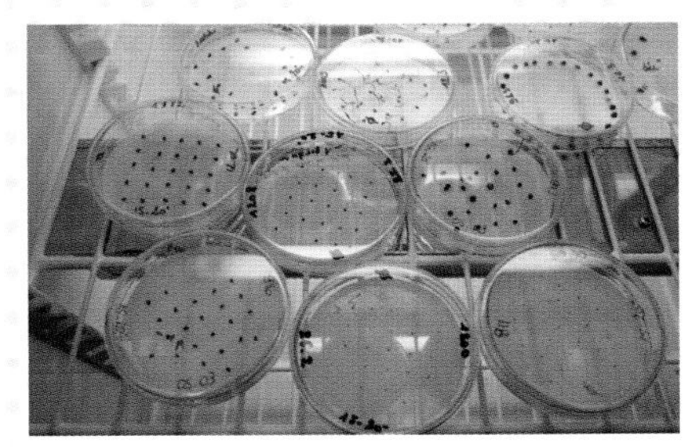

Figure 3 Germination tests

Scientists at Kew carry out some 10 000 germination tests each year.

- They must remove samples and germinate them periodically.
- They 'plant' the seeds in Petri dishes of nutrient agar and keep them in controlled conditions.
- Scientists measure the germination rate (percentage of seeds that germinate) and the success of germination.
- This enables scientists to monitor the condition of the stored seeds.
- Research continues into the physiology of seed dormancy and germination. With luck this will lead to discovery of the most effective methods of storage.

Questions

1 Explain how freezing sperm or embryos can be used in the conservation of animal species.

2 List the reasons why seeds need to be stored in facilities such as seed banks.

(19) International cooperation

By the end of this spread, you should be able to . . .

* **Discuss the importance of international cooperation in species conservation with reference to the Convention on International Trade in Endangered Species (CITES) and the Rio Convention on Biodiversity.**
* **Discuss the significance of environmental impact assessments (including biodiversity estimates) for local authority planning decisions.**

An international problem

The loss of **habitat** and the number of endangered species is a worldwide problem. It therefore needs a worldwide solution. International cooperation in **conservation** is essential. This can be achieved in a number of ways.

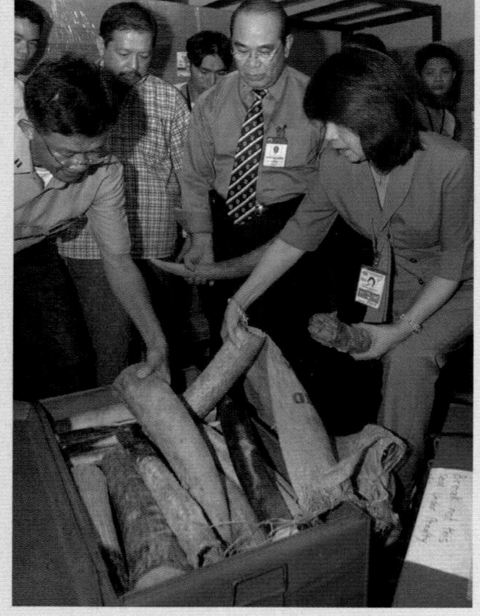

Figure 1 Elephant tusks seized by Customs

CITES

The Convention on International Trade in Endangered Species of Wild Fauna and Flora (CITES) is an international agreement made by the majority of governments in the world. It was first agreed in 1973. The overall aim is to ensure that international trade in specimens of wildlife does not threaten their survival. Over 25 000 species of plants and animals have been identified as being at risk from international trade.

CITES aims to:
- regulate and monitor international trade in selected species of plants and animals
- ensure that international trade does not endanger the survival of populations in the wild
- ensure that trade in wild plants is prohibited for commercial purposes
- ensure that trade in artificially propagated plants is allowed, subject to permit
- ensure that some slightly less endangered wild species may be traded subject to a permit, as agreed between the exporting and importing countries.

International trade policies can be very hard to enforce. Where there is demand for a product, there will be attempts to supply it. Smuggling of live plants and animals and their products is a constant problem.

Convention on Biological Diversity

Figure 2 CBD symbol

The Convention on Biological Diversity was signed by 150 government leaders at the 1992 Rio Earth Summit. The Convention is dedicated to promoting **sustainable development**. It recognises that biological **diversity** is about more than plants, animals, microorganisms and their ecosystems. It is also about people and our need for secure sources of food, medicines, fresh air and water, shelter and a clean and healthy environment in which to live.

The aims of the convention are:
- conservation of biological diversity
- sustainable use of its components
- appropriate shared access to genetic resources
- appropriate sharing and transfer of scientific knowledge and technologies
- fair and equitable sharing of the benefits arising out of the use of genetic resources.

The Convention encourages cooperation between countries and states. It encourages each partner to develop a national strategy for conservation and the sustainable use of biological diversity. More specifically, it states that partner states must adopt ***ex situ* conservation** facilities, mainly to complement ***in situ*** measures.

Zoos, botanical gardens and seed banks

Part of the cooperation between partner states is the sharing of genetic information and technology. Each member state must adopt *ex situ* conservation measures. These are wildlife parks, botanical collections and seed banks. Such facilities in different member states provide support for each other and share their technologies and genetic material.

Breeding programmes in wildlife parks are strengthened by importing animals from parks in other partner states. Time, expense and distress to rare animals can be reduced by importing genetic material. This means transporting the sperm, eggs or embryos and using artificial insemination or *in vitro* fertilisation techniques.

Similarly, breeding programmes in plants can be enhanced by sharing stored specimens. The Kew Millennium Seed Bank has partner projects in Australia, Botswana, Burkina Faso, Chile, China, Egypt, Jordan, Kenya, Lebanon, Madagascar, Malawi, Mali, Mexico, Namibia, Saudi Arabia, Tanzania, South Africa and America. These partners also duplicate the collections in case of unforeseen disaster.

The level of sharing between the partners is shown in the statistics at Kew. Since the year 2000:
- 3400 seed collections have been sent out
- these have been used in research into **biodiversity**, agriculture, water, energy, health, etc.
- 500 collections have been used by Millennium Seed Bank Project partners in reintroduction and restoration projects

Environmental impact assessment

Another aspect of the Convention on Biological Diversity is that all partner states must undertake an **environmental impact assessment** (see spread 2.3.3) before any major development. On an international level, the reasons for carrying out an environmental impact assessment are to:
- avoid or minimise any significant adverse affects on the biological diversity of an area
- ensure that any potential environmental consequences of a development are taken into account
- promote the exchange of information, consultation and notification of any development that might affect another partner state
- promote the notification of any grave danger or damage to biological diversity that may affect another partner state
- promote arrangements for emergency responses to activities or events that present a grave and imminent danger to biological diversity.

EIA is mainly carried out at local level. A local planning authority will consider whether a particular development requires an EIA. The criteria used fall into three categories:
- size of the development
- environmental sensitivity of the location
- types of impact expected.

Where an EIA is required, there are three broad stages to the procedure:
- the developer compiles an environmental statement, which includes an assessment of local biodiversity and the effect the development is likely to have on biodiversity
- the environmental statement is publicised
- the authority takes it into account when making a planning decision.

EIA is a means of assessing the likely significant environmental impact of a development. The EIA ensures that the local planning authority makes its decision in the knowledge of any likely significant effects on the environment. This helps to ensure that the importance of the predicted effects is properly understood by the public and the planning authority before it makes its decision.

EIA is also important to the developer. It can help to identify the likely effects of a particular project at an early stage. This can produce improvements in the planning and design of the development. In addition, developers may find EIA a useful tool for considering alternative approaches to a development. This can result in a final proposal that is more environmentally acceptable.

Key definition

Environmental impact assessment (EIA) is a procedure to assess the likely significant effects that a proposed development may have on the environment.

Questions

1. What might be considered as part of an environmental impact assessment?
2. Suggest why international trade policies may be hard to enforce.

2.3 Biodiversity and evolution summary

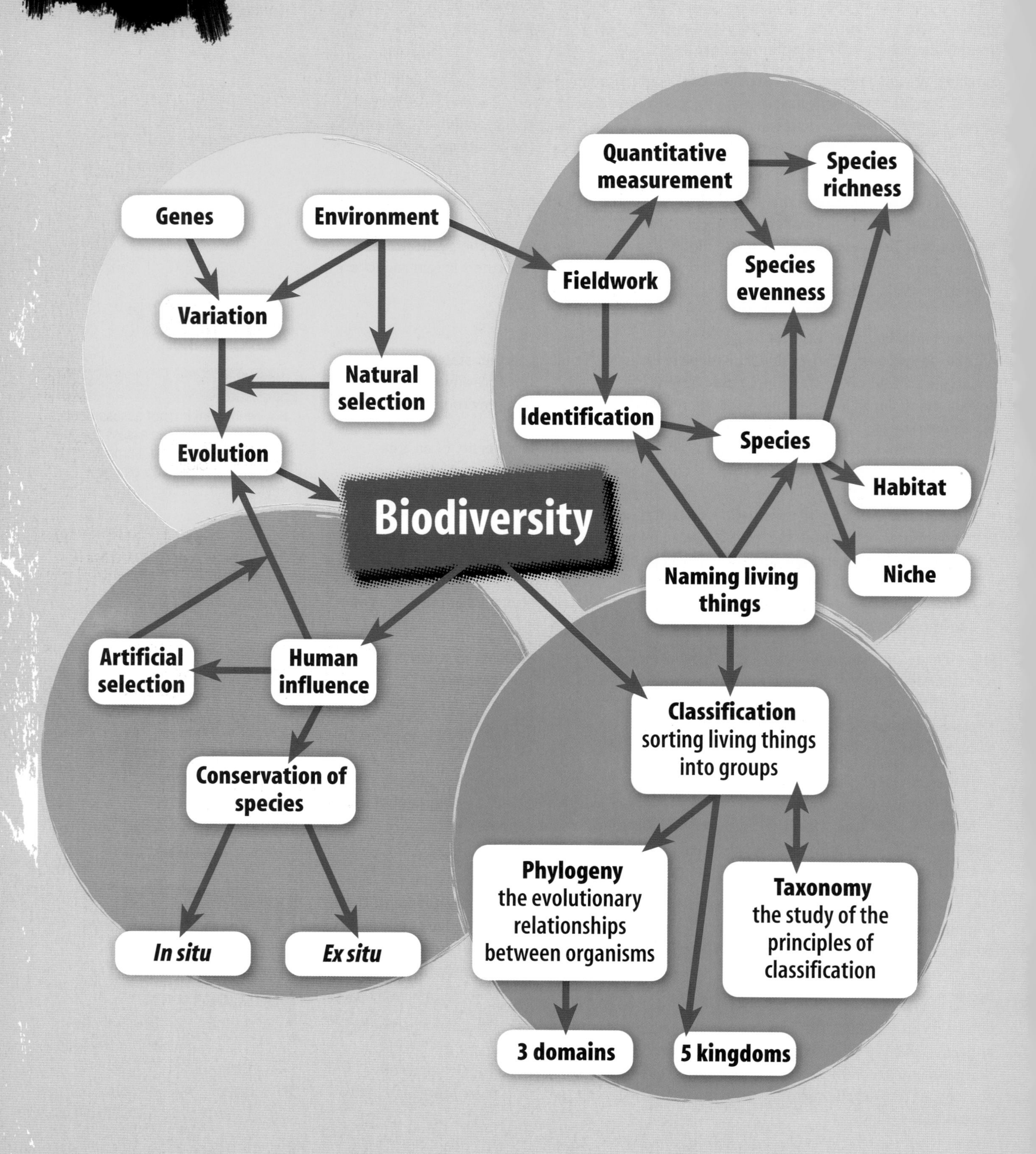

Practice questions

1. Complete Table 1 showing the main features of each kingdom.

Kingdom	Features
Prokaryotae	
	Eukaryotic, single-celled, wide diversity, some animal-like, some plant-like
Fungi	
	Multicellular, eukaryotic, autotrophic, have cellulose cell walls

Table 1 [6]

2. Explain the difference between continuous variation and discontinuous variation. [3]

3. (a) Describe what capture methods you might use to sample a meadow to find the diversity of animal life. [3]
   (b) Why is it important that your samples are collected at random? [2]

4. (a) Suggest why there is a range of estimates for global biodiversity. [2]
   (b) Explain what is meant by the terms:
   habitat
   biodiversity [3]

5. (a) State the key observations made by Charles Darwin. [4]
   (b) Explain the terms:
   selective pressure
   selective advantage [3]
   (c) Explain how fossils can be used as evidence for evolution. [3]
   (d) Figure 5 shows *Archaeopteryx*. Explain the significance of fossils such as *Archaeopteryx*. [2]

Figure 5

6. (a) Explain how you would use a quadrat and point frame to measure the ground cover of clover in a field. [3]

7. (a) What is meant by the term classification? [2]
   (b) What is meant by the term phylogeny? [2]
   (c) What is the relationship between natural classification and phylogeny? [2]

8. (a) Use Simpson's index to calculate the diversity of a habitat that contain the following organisms:
   20 woodlice + five mice + one shrew +
   32 earthworms + 15 grasshoppers + one owl. [4]
   (b) Explain why a habitat with a high diversity is thought to be more stable. [3]

9. Explain how DNA analysis and biochemistry can be used to clarify the evolutionary relationship between closely related species. [5]

1 Figure 1 shows two unicellular organisms labelled **D** and **E**. These organisms are members of different kingdoms.

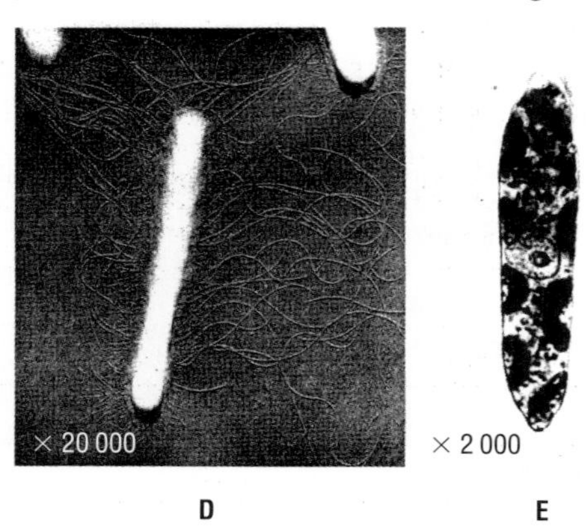

Figure 1

(a) (i) Study Figure 1 and identify the kingdom to which each organism belongs. Write your answers in Table 1 below.

(ii) Complete Table 1 by stating two features that are characteristic of the organism in the kingdom you have stated.

	unicell **D**	unicell **E**
kingdom		
features	1	1
	2	2

Table 1 [6]

In February 2001, the BBC reported that scientists had discovered a 'new species' of camel in a remote part of Asia. These camels differ from domesticated Bactrian camels in the following ways:

- 3% of their DNA base sequences are different
- their humps are further apart
- they have hairier knees
- there is no fresh water in the area and they survive by drinking salty water.

(b) (i) State which of these observed differences is most significant in deciding that this is a 'new species' of camel. [1]

(ii) Justify your answer. [2]

(c) Describe what further evidence is required to show that this 'new species' is a different species from the domesticated camels. [3]

[Total: 12]

(Jun 04 2804)

2 The snail *Cepaea nemoralis* lives on the ground among leaf litter and herbaceous vegetation.

- It exists in three different colours: brown, pink and yellow.
- In some of these snails there is a shell banding pattern on this background colour. Snails can be divided into banded and unbanded forms.
- The background colour and banding are controlled by genetic factors.

A group of students in central England carried out the following investigation.

- Samples of snails were collected from populations in two different habitats.
- The first habitat was mixed deciduous woodland where the leaf litter was a dark uniform colour.
- The second habitat was grassland, which is more variable in colour but predominantly pale yellow and green.

The main predator of the snail is the song thrush, which has excellent colour vision. It therefore acts as a major selection pressure on these populations.

Table 2 shows the percentage of yellow-shelled snails and unbanded snails found in the samples.

habitat	**sample**	**% of sample yellow**	**% of sample unbanded**
woodland	1	12	88
	2	21	77
	3	12	70
grassland	1	79	21
	2	58	14
	3	83	22

Table 2

(a) Explain the terms:
habitat [1]
selective pressure. [2]

(b) When the students compared their results with previous investigations in the same habitats, they found that the percentages were very similar.

Using the data in Table 2, describe how selection pressures, such as predation by the song thrush, can maintain different proportions of snail colours in the populations in the woodlands and grassland habitats. [6]

[Total: 9]

(Jun 06 2804)

Answers to examination questions will be found on the Exam Café CD.

3 Plant populations can be investigated by the use of quadrat sampling. Quadrat size varies, and different sizes are used in different situations.

(a) On Figure 3.1:

(i) draw a graph to show how the number of species recorded varies as the size of the quadrat increases; [1]

(ii) indicate the optimum size of quadrat to use. [1]

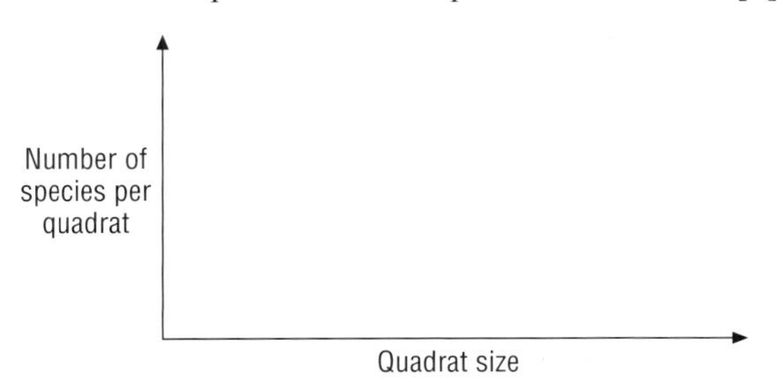

Figure 3.1

(b) Explain how you would use a quadrat to determine:
species frequency
percentage cover. [5]

(c) Outline the problems associated with assessing percentage cover. [3]

Figure 3.2 shows a point frame.

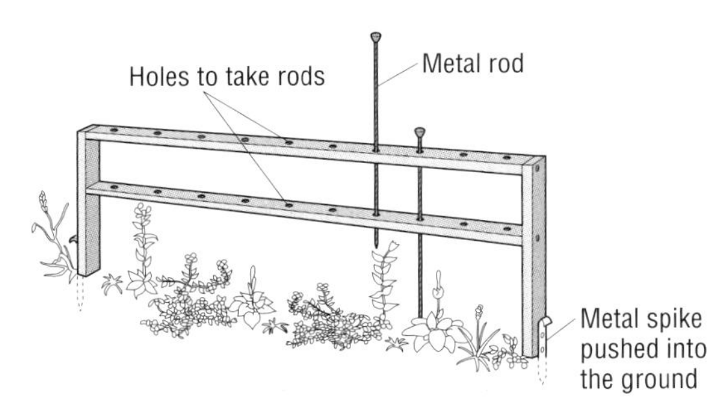

Figure 3.2

A group of students used a point frame to determine the percentage cover of plant species in the middle of some sand dunes. They placed the point frame in one position and lowered the metal rods. They recorded how many times the rods hit each species. Their results are shown in Table 3.

species	number of 'hits'
lady's bedstraw	14
clover	12
lesser hawkbit	18
ragwort	10
ribwort plantain	16
bird's foot trefoil	8
others	32

Table 3

Calculating the number of hits as a percentage of total hits is an estimate of percentage cover for a species.

(d) Using the results in Table 3, calculate the percentage cover for lesser hawkbit. Show your working and express your answer to the nearest 0.1%. [2]

[Total: 12]

(Jun04 2806-03)

Spread answers

1.1.1 Living organisms consist of cells

1. In observing thin sections of cork, Hooke saw that the sections were made of a number of repeating, small, fairly rectangular spaces. These had the appearance of the shape of rooms in which monks lived. These rooms were called cells.
2. Light cannot penetrate thick slices of tissue, so no detail would be seen.
3. The light microscope is easy to use and can be used to observe living cells such as single-celled organisms. It can also be used to observe the arrangements of tissues in organisms.

1.1.2 Cell size and magnification

1. 100/10 000 = 0.01 mm (observed size divided by magnification).
   This can be converted to μm by multiplying by 1000.
   0.01 mm = 10 μm.
   Or it can be calculated as 100 mm = 100 x 1000 μm
   Actual size = 100 000/10 000 μm = 10 μm
2.

μm	mm	m
5	0.005	0.000005
300 000	300	**0.3**
23 000	**23**	0.023
75	0.075	0.000075

1.1.3 Electron microscopes and cell details

1. Both are microscopes and so enable us to see a magnified image of a small object. Both use a radiation source passed through the specimen; the radiation beam is focused using lenses. The radiation passed through the specimen is used to generate an image.
   Light microscopes are easier to use, portable and can be used to observe living specimens. To use electron microscopes requires a great deal of skill and training. Preparation of specimens is difficult and the specimen has to be placed in a vacuum. It is not possible to observe living specimens. It is not possible to see colours of an electron micrograph – although colour can be added to give a false colour electron micrograph. Electron microscopes have much greater resolution and so can give greater magnification whilst giving a clear image.
2. Both types of microscope are used widely because they can give different information about living things. Although electron microscopes have greater resolution and magnification, they cannot be used to observe living cells and tissues. The light microscope gives us good detail of living processes in cells. The magnification of the electron microscope has allowed us to investigate the internal components of cells in great detail.
3. The inability to observe living material and the preparation processes required to observe specimens are the main limitations of the electron microscope.
4. Low resolution so cannot magnify much above ×1500 and still give a clear image.

1.1.4 Cells and living processes

1. The sharing of tasks that need to be performed between specialists in each of the tasks. In cells, different organelles perform different functions. Each function is performed well by the organelle responsible, and the organelle performing its function contributes to the overall life of the cell.
2.

Movement	The cytoskeleton includes contractile proteins; cilia and flagella are also capable of movement.
Respiration	Mitochondria are responsible for generating ATP in cellular respiration.
Sensitivity	Some of the proteins made in the cell are packaged in the Golgi apparatus and placed on the cell surface membrane. Such proteins may act as sensors, for example to detect nutrient molecules in the surroundings.
Nutrition	Cells needs nutrients to provide for their energy and growth needs, nutrients can be moved around cells in vesicles. Nutrients in the cell can be broken down (digested) in lysosomes.
Excretion	Waste products need to be moved out of the cell, many of these can be moved out of the cell in vesicles.
Reproduction	Making new cells requires instructions – these are stored in the nucleus.
Growth	Growth requires energy and therefore needs mitochondria. It also requires nutrients, brought into the cell, which can be moved around in vesicles.

3. (a) To gather maximum sunlight energy or to avoid being damaged by excess UV light.
   (b) To detect and take up foreign molecules and invading microorganisms in the blood.

1.1.5 Organelles – structure and function

1 Nucleus, chloroplast, mitochondrion, endoplasmic reticulum, centriole, lysosome, ribosome
2 The cells in the leaf called palisade mesophyll cells.
3 Muscle contraction requires a lot of energy in the form of ATP, whereas fat storage cells have very low energy requirements. (Fat cells that store brown fat have many mitochondria, as these cells use their fat to generate heat energy.)
4 Cells that are responsible for the production of enzymes, hormones or growth regulators. Hormones and growth regulators require processing and packaging, for export, in the Golgi apparatus.

1.1.6 Organelles at work

1 A cell surface membrane made of phospholipids, instructions in the form of DNA, ribosomes, cytoplasm, both can perform all the characteristics of living things.
2 Prokaryotic cells are much smaller, they do not have any internal membranes and so do not have a nucleus, mitochondria, chloroplasts, endoplasmic reticulum or lysosomes. Prokaryotic ribosomes are smaller.
3 It is thought that chloroplasts and mitochondria have evolved from prokaryotic cells, and that the DNA and ribosomes they retain are derived from this evolution. This is known as the 'endosymbiont theory'.

1.1.7 Biological membranes – fluid boundaries

1 The phosphate head group cannot pass through the hydrophobic region in the centre of the bilayer.
2 To support rapid respiration in order to supply ATP for contraction.
3 Different membranes are permeable to a variety of substances and impermeable to a variety of other substances. Semi-permeable suggests 'half-permeable', which is unlikely to be the case in any membrane.

1.1.8 The fluid mosaic model

1 (a) Externally placed proteins and glycoproteins.
(b) Transmembrane protein pores (channels).
(c) Enzymes, which are proteins.
2 It must be constructed from amino acids with hydrophobic R-groups.
3 Eventually the membrane would rupture and break down completely.

1.1.9 Communication and cell signalling

1 They are able to attach (by their surface antigens) to a normal cell surface protein receptor found in these cells.
2 Liver cells and muscle tissue cells (and most cells, as it can increase glucose uptake and rate of respiration in many cells).
3 Different toxins will have slightly different shapes, and so will attach more or less strongly to their target membrane proteins.

1.1.10 Crossing membranes 1 – passive processes

1 Carbon dioxide will diffuse into photosynthesising plant cells when there is light, oxygen will diffuse out of the plant cell at this time. Some of the oxygen produced during photosynthesis will be used for respiration, but the rate of photosynthesis will exceed that of respiration for much of the day. In darkness, carbon dioxide will diffuse out of the cell because respiration continues, so oxygen will diffuse into the cell at this time.
2 An amoeba is very small, so has a large surface-area-to-volume ratio. All the amoeba's cell contents are close to the oxygen supply in the surrounding water and so diffusion is sufficient for the organism's needs.
3 Both simple and facilitated diffusion involve the movement of materials from a place of their higher concentration to a place of their lower concentration. Facilitated diffusion requires a protein channel to 'allow' the diffusion to take place. Simple diffusion is where molecules are small enough, or are not charged, and can diffuse through the lipid bilayer. Large, fat-soluble molecules can also pass by simple diffusion through the lipid bilayer.

1.1.11 Crossing membranes 2 – active processes

1 Both involve protein carriers. In active transport, energy (ATP) from the cell is required to change the shape of the carrier, so transport occurs only one way across the membrane. Active transport can thus work against concentration gradients. Facilitated diffusion relies only on the kinetic energy of the diffusing molecules.
2 Active transport proteins use energy to ensure materials can pass only one way through the membrane – against the concentration gradient. Facilitated diffusion carriers allow the movement of molecules both ways through the membrane, so the concentration of molecules will distribute evenly across the membrane over time.
3 Endophagocytosis/endocytosis.

1.1.12 Water is a special case

1 Temperature – increased temperature increases the rate of diffusion and osmosis because it increases the kinetic energy of the molecules involved.
Mixing – increases the rate of diffusion and osmosis.

Concentration gradient – increased concentration gradient increases the rate of diffusion; in water increased water potential gradient increases the rate of osmosis.
2 Solute dissolves and water molecules cluster around the solute molecules. This reduces the capacity for water molecules to move freely and the water potential decreases.
3 The water potential of the cell cytoplasm is lower than that of the surrounding water. This means that water will move into the cell, down the water potential gradient, by osmosis. *Amoebae* move excess water into a membrane-bound vacuole called a 'contractile vacuole'. This vacuole periodically empties water out of the cell, a process called exocytosis or exopinocytosis. This is an active process that requires ATP.
4 The water potential will increase.

1.1.13 New cells – parent and daughter cells

1 (a) 64 (b) 64
2 0 as there is no nucleus.
3 12 – as an onion cell has 12 chromosomes.

1.1.14 Two nuclei from one

1 The chromatids separate in anaphase. So during anaphase and telophase, there are effectively 24 chromosomes in the cell.
2 Replicated sister chromatids remain held together at the centromere, they attach to the spindle and line up down the middle of the cell. When the centromere splits, the sister chromatids are pulled in opposite directions towards opposite poles of the cells.
3 Sketch should show 2× value in mitosis, 1× value in cytokinesis and 2× value again in mid interphase.
4 The formation of a cell wall in plant cells stops cells being able to divide effectively. Meristematic cells have very thin cell walls.

1.1.15 Cell cycles and life cycles

1 Genetically identical cells/organisms.
2 Genetically identical crops will grow at the same rate and so be ready to harvest at the same time; they will also produce a known crop product because they lack any genetic variation. If a new pest or disease, to which the crop plant is susceptible, appears, then *all* the plants will succumb and the whole crop may be lost.
3 Asexual reproduction means plants can reproduce quickly even when there are no other plants around to reproduce sexually with. Asexual reproduction saves the energy associated with flower, fruit and seed production.

1.1.16 Cell specialisation

1 Because the nucleus is lost, the shape changes. This gives the erythrocytes a larger surface-area-to-volume ratio, so oxygen can be taken in and released more efficiently.
2 Two of: aerobic respiration, protein synthesis, lipid metabolism, DNA replication.
3 Neutrophils contain many lysosomes that have digestive enzymes to break down the foreign materials that they ingest. Sperm cells have many mitochondria to produce ATP which provides energy for movement. They have an undulipodium for movement. They also have enzymes to digest the membrane around the egg and so aid fertilisation. Root hair cells have a large surface area to maximise uptake of minerals and water.

1.1.17 Organising the organism

1 Photosynthesis requires water in order to generate carbohydrates. Cells also need to be turgid to carry out their metabolic functions.
2 Palisade layer is packed with many chloroplasts. Palisade cells are close to the surface in a 'flat' layer. The cells are long and thin, so more can be packed in.
Palisade mesophyll cells are cylindrical, so there are air spaces between cells to maximise carbon dioxide availability in a thin layer close to palisade cells.
There are large air spaces between spongy mesophyll cells, so carbon dioxide can diffuse from stomata to palisade cells.
Palisade cells have thin cell walls – short diffusion distance for carbon dioxide.
Large vacuole in palisade cells means that chloroplasts are close to the cell wall – short diffusion distance for carbon dioxide.
Cell walls of palisade cells very transparent for light to reach chloroplasts.
Lower epidermis with stomata to allow carbon dioxide entry and oxygen exit to/from leaf.
A leaf vein system containing xylem vessels and phloem tissue to bring water to the leaf cells and take away photosynthetic products.
The vein system supports the leaf, holding it up to maximise light capture.
3 In growing shoots and leaves, carbohydrates may need to be moved to the leaf to support growth. In mature leaves, carbohydrate needs to be moved away from the leaf to other parts of the plant. Carbohydrates may have to be moved from leaves to roots during summer, but from roots to growing shoot regions after winter.

1.2.1 Special surfaces for exchange

1 Large, active organisms need a lot of oxygen for respiration. They cannot supply enough through their skin surface, as their surface-area-to-volume ratio is too low. A special surface for exchange gives a greater area, so more oxygen can be absorbed and more carbon dioxide can be removed.

2

Feature	How it helps efficiency
large surface area	more space for molecules to diffuse
thin barrier	short diffusion distance
permeable membrane/ barrier	to allow desired molecules through
good supply of molecules to exchange surface	keeps concentration high so diffusion gradient high
removal of molecules from other side	keeps concentration of molecules low so diffusion gradient high

3 The tip of a plant root has specialised epithelial cells. These are called root hair cells. The root hair cells have a long extension, which increases their surface area. The wall and membrane are permeable to water and minerals.

1.2.2 The lung as an organ of exchange

1 Large numbers of alveoli increase surface area; the wall of each alveolus is very thin; there is a very good blood supply.

2 Our tissues need a lot of oxygen; the blood must be as close to the air in the alveoli as possible so that oxygen can move into the lungs by diffusion quickly. A thin barrier means a short diffusion distance.

3 Breathing movements (ventilation) bring fresh air into the lungs. This carries oxygen into the alveoli and removes carbon dioxide. Blood flowing in the capillaries carries the oxygen away from the alveoli and brings more carbon dioxide to the alveoli. Therefore there is a concentration gradient of oxygen being high in the air of the alveoli and lower in the blood. There is a concentration gradient of carbon dioxide being high in the blood and lower in the air of the alveoli. The diffusion gradient is kept high by the short distance between the air in the alveoli and the blood in the capillaries.

1.2.3 Tissues in the lungs

1 Cartilage in the trachea and bronchi hold the airways open against air pressure as we breathe in.

2 The mucus in the lungs is sticky and will catch any bacteria or other pathogens in the lungs. The cilia are able to move the mucus up the airways towards the throat, so that it can be swallowed.

3 Antagonistic means that they work against each other. The smooth muscle contracts to narrow the lumen of the bronchioles. As this happens, the elastic fibres are deformed. When the muscle relaxes, the elastic fibres recoil to their original shape and extend the muscle fibres again.

1.2.4 Measuring lung capacity

1 Diffusion

2 (a) tidal volume = 0.3 dm^3
breaths per minute = 13
(b) tidal volume = 1 dm^3
breaths per minute = 22

3 2 $dm^3\ min^{-1}$

4 Larger tidal volume and more breaths per minute

5 To allow the tidal volume and breathing rate to increase

1.2.5 Transport in animals

1 A good transport system has a pump, vessels and a fluid to transport the materials that are needed around the body. It will also have exchange surfaces to load substances into the transport system and to remove them where required. It will be adaptable to cope with changes in demand.

2 Substances enter the blood at the exchange surface. A large organism may need to transport these substances a long distance to ensure they reach all the living tissue. Diffusion alone will not be rapid enough to supply all the tissues – especially if the organism is active.

3 Fish have a single circulatory system. Blood flows through the gills to be oxygenated and to lose carbon dioxide. From there it flows to the body tissues, delivering oxygen for respiration and collecting carbon dioxide. It then flows to the heart and back to the gills.
Mammals have a double circulatory system. Deoxygenated blood returns from body tissues to the right side of the heart. It is pumped to the lungs to become oxygenated. It returns to the left side of the heart and is pumped to the body tissues. A double circulation is more efficient.

4 In a double circulatory system, blood returns to the heart after being oxygenated. It is possible to boost the pressure of the blood after it has been oxygenated. This means blood can flow faster and deliver more oxygen to the tissues. In a single circulatory system, blood does not return straight to the heart after oxygenation. It loses

pressure in the organ of gaseous exchange. Therefore it leaves the organ of gaseous exchange at low pressure and travels more slowly to the tissues. Therefore it cannot deliver as much oxygen to the tissues.

1.2.6 The structure of the mammalian heart

1 The atrioventricular valves stop blood flowing back up from the ventricles to the atria when the ventricles contract.
2 The tendinous cords attach the valves to the sides of the ventricles. They prevent the valves from opening backwards, which would allow blood to flow up into the atria.
3 The left ventricle pumps blood further. It pumps blood all the way around the body – therefore the wall must be thicker to create more pressure and overcome the resistance of the systemic vessels. The right ventricle pumps blood to the lungs and less pressure is needed to reach the lungs – so the walls of the right ventricle can be thinner.
4 If the right ventricle created too much pressure, it could damage the thin membranes in the lungs. This would lead to inefficient gaseous exchange.

1.2.7 The cardiac cycle

1 The chambers of the heart fill as the muscular walls relax and recoil to their original size. When the heart is full, the atria contract, pushing extra blood into the ventricles. Then the ventricles contract, which pushes the atrioventricular valves closed and increases pressure in the ventricles. The pressure pushes open the semilunar valves and blood leaves the ventricles to flow into the major arteries (aorta and pulmonary artery).
2 The increasing pressure in the ventricles pushes the atrioventricular valves closed. Once the pressure in the ventricles is higher than in the atria, the valves close.

1.2.8 Control of the cardiac cycle

1 The sinoatrial node creates electrical stimuli at regular intervals. These are what start or initiate the heartbeat.
2 Fibrillation is when the atria do not beat in a synchronised way. This means the chambers are unable to fill properly and so the heart cannot pump the correct amount of blood at each beat.
3 The arteries that carry blood away from the heart open at the top of the ventricles. The ventricles start to contract at their base (apex) so that the blood is squeezed upwards to the openings of the major arteries.
4 Trace drawing looks like the diagram for ECG trace of a normal person. P = excitation (contraction) of atria, QRS is excitation (contraction) in ventricles and T is diastole (ventricle recovery).
One small wave (P) followed by a larger QRS complex wave and another smaller T wave.
5 The QRS complex indicates the stimulation of the ventricles. The ventricles have more muscle and this is reflected in the greater electrical signal.

1.2.9 Blood vessels

1 The circulatory system of a locust is open. There is a contractile vessel (heart) that carries blood towards the head. It opens into a blood space that contains all the organs of the locust. Blood circulates slowly from the head to the rear of the animal and eventually back into the heart. Fish have a closed circulatory system. Blood flows in vessels through the gills to be oxygenated and to lose carbon dioxide. From there it flows to the body tissues, delivering oxygen for respiration and collecting carbon dioxide. It then flows to the heart and back to the gills.
2 An open system cannot contain any pressure; the blood flow is slow and is not directed to particular organs. In a closed system, the blood can be pumped under pressure so it flows more quickly. Also, the vessels carrying the blood can carry it direct to specific organs and back to the heart. A closed system can be more responsive to changes in demand by increasing pressure to carry more oxygen; also it can dilate vessels to the organs that demand most oxygen and constrict the flow to other organs.
3

Feature	Arteries	Capillaries	Vein
wall thickness	thick	very thin (one cell thick)	thin
muscularity	thick muscle layer	no muscle	thin muscle layer
elasticity	thick layer of elastic tissue	no elastic tissue	little elastic tissue
inner surface	smooth endothelium, often folded	one layer of cells	smooth endothelium, not folded
shape	round	round	flattened
size of lumen	small	tiny (7 μm diameter)	large
valves	none except at exit from ventricles	none	pocket valves all along their length

4 Arteries carry blood at high pressure. They have a thick wall to withstand the pressure. The elasticity allows them to expand and recoil to smooth out changes in blood

pressure. The muscular tissue enables constriction to maintain blood pressure or in smaller arteries (arterioles) to restrict flow. The endothelium is folded to allow expansion of arteries when a surge of blood enters. Capillary walls need to be thin to ensure gases and nutrients can diffuse across easily.
Veins do not need thick walls to withstand pressure, but need valves to ensure the blood continues to flow in the correct direction. The lumen is wide to ease the blood flow (reduce friction).

1.2.10 Blood, tissue fluid and lymph

1

Feature	Blood	Tissue fluid	Lymph
location	in blood vessels – arteries, capillaries and veins	surrounding tissue cells	in lymph vessels
direction of flow	from heart to tissues in arteries; from tissues to heart in veins	no flow	from tissues to blood system
cause of flow	pressure created by heart in arteries; pressure created by action of skeletal muscles in veins	–	pressure from tissue fluid and valves to ensure one-way flow

2 Blood contains plasma proteins, which are too large to leak out of capillaries.
3 The heart – contraction of left ventricle.
4 The hydrostatic pressure of the blood is higher than that of the tissue fluid. The blood fluid is forced through small pores in the capillary wall from blood to tissue fluid.

1.2.11 Carriage of oxygen

1 Red blood cells pick up oxygen in the lungs and release it in respiring tissues.
2 The iron in the haem group binds to the oxygen.
3 Affinity means attraction for.
4 The oxyhaemoglobin dissociation curve is an S-shape or sigmoid shape. This is because more oxygen will attach as the pO_2 in the air to which haemoglobin is exposed increases. At low pO_2 it is hard for oxygen molecules to attach. As the pO_2 rises, and some oxygen molecules attach to the haemoglobin, this makes it easier for more oxygen molecules to attach. This is due to a conformational change in the shape of the haemoglobin. However, as the haemoglobin becomes saturated it becomes more difficult for the oxygen to attach to the haemoglobin.
5 A fetus is not exposed to the outside air. It gains oxygen from its mother's blood. It must be able to pick up oxygen at a pO_2 that is low enough to make the maternal haemoglobin release oxygen.

1.2.12 Carriage of carbon dioxide

1 Dissolved as CO_2 in the plasma, combined with haemoglobin as carbaminohaemoglobin and as hydrogencarbonate ions in the plasma.
2 CO_2 enters the red blood cell and is combined with water to form carbonic acid. The enzyme carbonic anhydrase ensures that this occurs. The carbonic acid dissociates to hydrogen ions and hydrogencarbonate ions.
3 The hydrogencarbonate ions are negatively charged. They diffuse out of the red blood cell, leaving it with a positive charge. Chloride ions are moved into the red blood cell to neutralise this charge.
4 The hydrogen ions released from the dissociation of carbonic acid compete with oxygen for a space on the haemoglobin molecule. If there are more hydrogen ions then more oxygen will be released from the haemoglobin.
5 Actively respiring tissue releases more CO_2. This forms more carbonic acid so more hydrogen ions are formed. More hydrogen ions will cause more oxygen to be released from the haemoglobin.

1.2.13 Transport in plants

1 Water is transported upwards in the xylem from the roots to the leaves.
2 Phloem, in between the arms of the xylem in the central stele of roots and on the outer edge of the vascular bundles of stems.

1.2.14 Xylem and phloem

1 A xylem vessel has thick, lignified walls. They have pits to allow movement of water from one tube to another. There are no cell contents and no cross-walls to obstruct flow.
2 Pits allow sideways movement of water from one tube to another.
3 Sieve tube elements have a thin layer of cytoplasm to reduce resistance to the flow of sap.
4 Sieve elements do not have nuclei. They have incomplete end walls that are perforated to form sieve plates.

5

xylem	phloem
tubular vessel	tubular vessel
dead	living
transports water and minerals up the plant	transports sugars up and down the plant
one-way flow	can flow either way
no contents	few contents, thin cytoplasm
no end walls	end walls adapted as sieve plates
no companion cells	possess companion cells to carry out metabolic functions and actively load the phloem

1.2.15 Plant cells and water

1 Turgid means full of water and firm.
Plasmolysis means that the cell contents have shrunk enough to pull the cell membrane away from the cell wall.

2 The cell with a water potential of –1000 kPa will lose water to the cell with a water potential of –1200 kPa.

3 The space between the cell wall and cell membrane of a plasmolysed cell is filled by the same solution as is outside the cell wall.

1.2.16 Water uptake and movement up the stem

1 The Casparian strip blocks the apoplast pathway. Water cannot pass between the cells or through the cell walls. Water must pass across the endodermis, so it must pass into the cytoplasm or into the symplast pathway.

2 Water lost at the top of the tree must be replaced. Water enters the leaf cells from the xylem. As molecules move out of the xylem, a tension force pulls a continuous chain of water molecules up the xylem. These water molecules are held together by cohesion forces, and the whole chain of molecules is put under tension by the loss of molecules at the leaf.

1.2.17 Transpiration

1 Water moves from the xylem to the mesophyll cells by osmosis. It evaporates from the surface of the mesophyll cells and diffuses out via the stomata.

2 A potometer measures the uptake of water by a leafy shoot. The water uptake can be measured by the movement of a meniscus along a capillary tube. The volume equals the cross-sectional area of the tube multiplied by the distance moved. If the time taken to lose a known volume is measured, then the rate of uptake is calculated from the formula volume/time.
The potometer must be set up under water to ensure no extra air bubbles get into the system. All joints must be air-tight and water-tight. The leaf area must be dried and measured. The potometer and shoot must be left to acclimatise before any readings start.

3 Environmental factor	4 How it affects water loss
Increased temperature	Increases rate of evaporation of water, so water vapour potential in the leaf air space is higher. Also decreases relative humidity of the air, so the air can contain more water vapour.
Increased wind speed	Moves water vapour away from leaf surface so the water vapour potential immediately outside the leaf is reduced.
Decreased relative humidity	With less water vapour in the air, there is a lower water vapour potential, this allows water vapour to diffuse out of the leaf more easily.
Increased light intensity	Increased light intensity opens the stomatal pores wider. This allows water vapour to diffuse down its water vapour potential gradient more easily.
Water availability	If water cannot be taken up from the soil, the stomata will close, reducing water loss.

1.2.18 Reducing water loss – xerophytes

1 Stomata remain open during the day to allow gaseous exchange to occur. Carbon dioxide must get into the leaf for photosynthesis.

2 The water vapour potential inside the leaf is higher than that outside. There is a gradient in the water vapour potential. Water vapour will move down this gradient from the leaf to the outside via the open stomata.

3 Feature of xerophytes	How it reduces water loss
Waxy leaf	Reduces evaporation from surface.
Hairs on leaf	Hold water vapour close to leaf surface, so water vapour potential remains high outside the stomata and reduces the water potential gradient.
Stomata in pits	Pits hold water vapour close to stomata, so water vapour potential remains high outside the stomata and reduces the water potential gradient.
Smaller leaves	Less surface area over which water vapour loss can occur.
Spongy mesophyll densely packed	Reduces surface area over which water is evaporated.
Ability to close stomata in daytime	Reduces water loss, as no space through which to diffuse out of leaf.
Rolling leaves	Traps air and water vapour next to stomata, so water vapour potential remains high outside the stomata and reduces the water potential gradient.
Low water potential in leaf cells	Reduces evaporation of water from cell surfaces.

4 Marram grass has a waxy cuticle, its leaves are rolled, it has hairs on its epidermis and its stomata are in pits.

1.2.19 The movement of sugars – translocation

1 Translocation is the movement of assimilates (sugars) up and down the plant. It is essential to transport sugars to regions where they can be used in respiration or stored as starch.
2 A source is a region where sugars are (actively) loaded into the phloem. A sink is a region where the sugars are removed (by diffusion) from the phloem, to be used or stored.
3 Active loading means using ATP (energy) to transport sugars into the phloem.
4 Hydrogen ions are used to help in active loading. They are pumped out of the companion cells and allowed to diffuse back in. As they diffuse back in, they come through special cotransporter proteins that allow them through only if they are attached to a sugar molecule.
5 Translocation is an active process that loads sugars in at one end of the phloem and decreases the water potential. This means that water enters the phloem by osmosis and increases the hydrostatic pressure. The water moves from a region of high hydrostatic pressure to a region of low hydrostatic pressure, carrying the dissolved sugars with it. Transpiration is driven by evaporation of water from the leaf surface. This reduces the hydrostatic pressure of the water at the top of the plant, creating tension that allows water to be pulled up the plant. This is aided by cohesion between the water molecules.
6 An aphid uses its proboscis to pierce the phloem and drink the sap. Once an aphid has placed its proboscis in the phloem, the aphid can be removed to show that the sugary sap drips from the cut end of the proboscis. Further investigation can reveal that the other end of the proboscis is in the phloem.
7 As the hydrogen ions are pumped out of the companion cell, this leaves fewer hydrogen ions in the cells. This means a higher pH.
8 The sieve plates could be used to block flow of sugars if the phloem is cut and sugars are being lost.

2.1.1 Biochemistry and metabolism

1 Many nutrients are large polymers. The organism must break these down to the basic monomers and then use the monomers to manufacture its own specific polymers.
2 The combined rate of all of the reactions going on in the body, including all anabolic and catabolic reactions. The metabolic rate can be estimated by measuring the amount of oxygen uptake per unit time in an individual, as this gives an indication of respiration that provides the ATP required to drive metabolism.
3 A diet that contains all the nutrients in proportions that match the needs of the individual. It also contains sufficient energy for the individual's metabolism and level of activity, and water and fibre.

2.1.2 Biochemicals and bonds

1 0=C=0
2 Enzymes
3 Heat energy

2.1.3 Carbohydrates 1: simple sugars

1 $C_3H_6O_3$
2 Soluble, sweet, crystalline
3 glucose + glucose → maltose + water
or
$C_6H_{12}O_6 + C_6H_{12}O_6 \rightarrow C_{12}H_{22}O_{11} + H_2O$

4 On α-glucose the –OH group on C_1 is below the plane and on β-glucose the –OH group on C_1 is above the plane. This changes the overall shape of the molecule.

2.1.4 Carbohydrates 2: energy storage

1 Animals have a higher/more rapid demand for energy to support movement of muscles. Increased branching in glycogen means more 'ends' giving access to retrieve glucose from the storage molecule more quickly.
2 Glucose can enter the respiratory pathway that generates ATP. Glycogen and starch must first be broken down to glucose – so taken 'out of storage' before the glucose can be respired. Glycogen and starch are insoluble in water so do not cause osmosis in the cells where they are stored.

2.1.5 Carbohydrates 3: structural units

1 It forms long, rigid fibres with high tensile strength.
2 Humans do not possess the enzymes required to break down cellulose. This means cellulose cannot be digested. It gives the food going through the digestive system 'bulk' and acts as material for the muscles of the digestive system to squeeze against. (Herbivore mammals also do not have cellulose enzymes, but they have bacteria and protoctists in their guts that do have these enzymes.)
3 In turgid cells, the cell contents push against the walls. When many turgid cells are pushing against each other, this gives great strength and stability.

2.1.6 Amino acids – the monomers of proteins

1 Because they are not essential parts of the diet. They are still essential for protein synthesis, but they can be made in the body from other amino acids.
2 (a) $H_2N-CHR-CO-NH-CHR-CO-NH-CHR-CO-NH-CHR-COOH$
(b) Three
(c) Because it consists of several monomers joined together – in this case, many amino acids joined by peptide bonds.

2.1.7 Proteins from amino acids

1 Because they are made from amino acids joined end-to-end. In all cases, the amino group of one amino acid joins to the acid group of the next. So in all cases, the amino group of the first amino acid appears at one end and the acid group of the last amino acid appears at the other end.
2 The equivalent of one water molecule.
3 In the lysosomes.

2.1.8 Levels of protein structure

1 Covalent (disulfide) bonds, because these are not broken by heat, only by reducing agents.
2 The shape of facilitated diffusion and active transport proteins – channels or carriers. The part of the protein in the central part of the bilayer has hydrophobic R-groups facing outwards as these are next to the lipid bilayer, and hydrophilic R-groups facing inwards as these are close to the water-soluble molecules passing through the channels. Antigens and receptors on the cell surface membrane will each have a specific 3D shape.
3 Diagram should note that hydrophobic amino acids would be found in parts of the protein in contact with the central part of the phospholipid bilayer, and hydrophilic groups facing towards the inside of the channel.

2.1.9 Proteins in action

1 Both are proteins with a unique primary, secondary and tertiary structure. Both have quaternary structure. Haemoglobin is a globular protein with a metabolic function and contains a prosthetic group. Collagen is a fibrous protein with a structural/support function and does not have a prosthetic group. Collagen contains fewer types of amino acid.
2 Both molecules are fibrous, giving rigid structures used in support of the organism; the molecule fibres are strengthened by bundling together; multiple hydrogen bonds are responsible for holding the bundles of fibres together.
3 Cellulose is a polysaccharide; collagen is a protein.

2.1.10 Lipids are not polymers

1 Lipids are formed from three fatty acids and one glycerol, not from repeating monomer units.
2 Unsaturated fats contain fatty acids where some of the carbons are double-bonded. Saturated fats are made from fatty acids that have no double bonds.
3 Triglycerides contain twice the energy per unit mass of carbohydrates, and can be stored in large quantities away from the metabolically active cytoplasm of the organism. This makes them ideal for long-term energy storage.
4 Four – one glycerol and three fatty acids.

2.1.11 Essential oils?

1 Lack of lipid can lead to delicate organs being unprotected and to a lack of energy stores. It can also lead to deficiency of fat-soluble vitamins. Excess lipid can accumulate around vital organs, affecting their function, and can also accumulate in other places such as blood vessel linings, impairing their function and increasing the risk of heart disease and stroke.
2 The lipid store means a large energy reserve for the camel. Placing this store in a hump rather than evenly around the skin means the camel is not over-insulated and so less likely to overheat. Respiration of lipid also produces larger volumes of water than respiration of carbohydrate.
3 In fat storage cells (vitamins A and D also stored in the liver).
4 A water molecule is released in a condensation reaction, but ester bonds join two different subunits together rather than two monomers as with glycosidic and peptide bonds.

2.1.12 water – a vital biological molecule

1 The oxygen end is slightly negatively charged, the hydrogen ends are slightly positively charged, so there is an uneven charge distribution across the molecule.
2 Non-polar molecules such as fatty acids, because their even charge distribution means that polar water molecules cannot easily cluster around them.
3 Temperatures on the Earth vary, as do temperatures within organisms due to living processes. Metabolic reactions require liquid water.
4 Endopinocytosis results in the movement of liquids into a cell (exopinocytosis moves liquids out of a cell).
5 Osmosis

2.1.13 Practical biochemistry – 1

1 The gaining of electrons (may also be gaining of hydrogen or loss of oxygen).
2 The biuret test is positive only in the presence of peptide bonds. A solution of separate amino acids contains no peptide bonds.
3 A precipitate means that solid particles are coming out of solution. An emulsion is an even mixture of two substances, where one does not dissolve in the other but droplets are small enough to disperse evenly.
4 It would give a positive result as the β-glucose molecules can reduce the Benedict's solution.

2.1.14 Practical biochemistry – 2

1 Sketch should show absorption increasing as transmission decreases.
2 Transparent to let light through, careful handling to avoid fingerprints and grease marks from skin that affect light passing through.
3 Carry out Benedict's test using fixed volumes and concentrations of the two juices with fixed amounts of Benedict's solution. Filter the precipitates out of each solution. Place the filtrates into separate cuvettes and measure transmission of light through them. Compare the samples with the calibration graph to read off the concentration of reducing sugar present. Repeat the experiment a number of times to find the mean for each result.

2.1.15 Nucleotides – coding molecules

1 Two.
2 The sugar molecule is the same in a nucleic acid polymer for all the constituent nucleotides. One type of polymer contains deoxyribose, the other ribose – this makes it easy to distinguish the two types.
3 Phosphate at one end and a sugar at the other.

2.1.16 DNA – information storage

1 Parallel, but with chains running in opposite directions.
2 TAATCCGATA
3 20% A, 30% C, 30% G.
4 When the copying error occurs during mitosis for growth/replacement of cells, cancers may occur. When a mutation happens when gametes are formed, any genetic disease involving point mutations can occur (e.g. haemophilia, sickle cell anaemia, phenylketonuria).

2.1.17 Reading the instructions

1 Complementary RNA nucleotides are lined up against each base on the template strand, producing a complementary strand. As base-pairing rules apply, this lining up will be the same as it appears on the coding strand – apart from U in RNA replacing T in DNA.
2 UAAGCGCAAUUA and ATTCGCGTTAAT
3 RNA is single-stranded and so less stable, as nucleotide bases are exposed and not paired. It also contains uracil instead of thymine, which may contribute to the lower stability of the molecule. mRNA results in the production of proteins. If the cell is to control protein production, the disintegration of mRNA stops too much of a certain protein being made, and so allows for regulation of the protein levels in a cell.

4 Table should note the following.
Similarities: nucleotide polymer, same phosphate, both have bases A C and G. Both are instruction molecules carrying a code.
Differences: sugar DNA has deoxyribose and RNA has ribose. Base differences – DNA has T and RNA has U. RNA is single-stranded and DNA is double-stranded. There is one form of DNA, three forms of RNA (m, t and ribosomal) DNA is very stable, RNA less so.

2.1.18 Enzymes are globular proteins

1 The catalytic part of the enzyme is the active site; this has a very specific shape so only one or a very few molecules are able to fit into the active site.
2 Only proteins can produce such a wide variety of shapes. This is because proteins are made from 20 different amino acid sub-units.
3 (a) Carbohydrases.
(b) Lipases.

2.1.19 Inside and out – where enzymes work

1 The enzymes produced are not lost to the environment and so can be retained and recycled. The environment of the internal system can be regulated to give the optimum conditions for the enzymes' activity.
2 The role of many white blood cells is to take in and destroy foreign organisms and debris. The destruction of this material is achieved by potent digestive enzymes in the lysosomes of these cells.
3 Enzymes regulate metabolic process by catalysing reactions at a rate appropriate to the organism.

2.1.20 Enzyme action

1 In the induced fit hypothesis the enzyme molecule changes shape to hold the substrate molecule more tightly. In the lock and key hypothesis, there is no shape change of the enzyme molecule.
2 The enzyme molecule holds the substrate molecule in such a way that the reaction proceeds more easily.
3 Most of the reactions essential to life do not take place at a rate sufficient to sustain life without enzymes.

2.1.21 Enzymes and temperature

1 Increased kinetic energy leads to increased numbers of collisions between enzyme and substrate molecules and so increases rate of reaction.
2 Disulfide bonds.
3 In exercising, heat is generated by the organism in the muscles which takes some time to remove, also some infections lead to fevers. If the normal body temperature was too near to the optimum temperature for the enzymes, then in situations of increased temperature many enzymes may be denatured. A higher body temperature would require more energy and might denature other proteins.

2.1.22 Enzymes at work – pH

1 Changing pH means changes to the hydrogen ion concentration. Since the tertiary structure of a protein is held in part by bonds derived from charge differences, then changes to the hydrogen ion concentration will affect tertiary structure and so active site shape.
2 There are many hydrogen bonds; each one is weak, but lots of them together are very stabilising.
3 The tertiary structure of each enzyme is held in place by a number of charge-based bonds. In each case, the optimum tertiary structure is achieved with the hydrogen ion concentration (and so positive charges in solution) given by the optimum pH for the enzyme.
4 Even a small change in pH results in a significant change in hydrogen ion concentration. This has an effect on the enzyme tertiary structure and so the active site shape.

2.1.23 Enzymes at work – concentration effects

1 Sketch shows rate increasing in proportion to increase in concentration.
2 A straight line increase – as the dashed line on Figure 2.
3 Enzymes can catalyse reactions very quickly and are reusable, so only very small concentrations are needed. Low concentration also means they are easier to control.

2.1.24 Enzymes at work – inhibitors of action

1 Carry out the reaction with a range of substrate concentrations. If the rate of reaction increases up to the same as that given without inhibitor present, then the inhibitor is a competitive inhibitor.
2 The competitive inhibitor may be a large molecule with a small part the same shape as the substrate.
3 Cells must be able to control the concentration of various molecules in the cell. Where these are products of enzyme-controlled reactions, it is important to regulate enzyme activity so that the optimum level of product is achieved and maintained.

2.1.25 Enzymes at work – coenzymes and prosthetic groups

1 The amount required is low because nicotinamide as a coenzyme is 'reusable' so can take part in enzyme-controlled reactions over and over again.
2 The haem group.
3 The Golgi appatatus.

2.1.26 Interfering with enzymes – poisons and drugs

1 Enzymes are proteins, and the stomach contains potent protease enzymes. The enzymes would be broken down, and so, useless.
2 The kidney is able to retain glucose in the blood when urine is formed. In a diabetic there is so much glucose in the blood that the kidney is unable to retain it all so some is lost.
3 The fungus releases antibiotics to destroy other organisms so that the other organisms cannot take up its food supply.
4 The protease enzymes from viruses differ in overall shape from the proteases found in humans. Protease inhibitors can be specific to the parts of the molecule that differ.

2.1.27 Investigating enzyme action – 1

1 Living tissues will have variable concentrations of enzyme in them depending on a number of factors including age, genetic make-up, metabolic processes occurring. For example, the level of catalase found in potato tissue will be higher in regions that are growing because growth involves metabolic processes that result in the production of hydrogen peroxide as a by-product. Different potatoes may be genetically different in the levels of enzymes in their cells and older potatoes may have reduced enzyme concentrations as enzyme molecules deteriorate over time.
2 A thermometer gives the actual temperature, the thermostat on a water bath can be set at specific temperature, but the actual temperature will fluctuate above and below this set point.
3 The independent variable (IV) is the factor that the experimenter changes, e.g. temperature. The dependent variable (DV) changes as a result of the IV changing, e.g. rate of reaction. The experimenter measures the DV.

2.1.28 Investigating enzyme action – 2

1 A range of pH buffers could be used, keeping temperature, enzyme and substrate concentration constant. A colorimeter could be used to read how rapidly the milk clears.
2 A set of wells should be cut and water added in place of enzyme at each pH.

2.1.29 Enzymes and metabolism – an overview

1 ATP production will occur at a rate suitable to the organism's needs.
2 Inhibiting an enzyme at the start of the sequence means that intermediate products in the sequence do not build up as well as the end product. This gives control over the whole sequence.
3 Destruction of nerve cell components means that the impulses from the brain to the muscles cannot get to the muscles so they cannot be stimulated to contract.
4 Some phenylalanine is needed in the synthesis of some proteins. It is an essential amino acid.
5 The damage caused by the build-up of phenylalanine is to growing nervous tissue. In adults, the nervous tissue is mature and so unaffected by higher levels in the blood.
6 Individuals with phenylketonuria (PKU) can't change phenylalanine to tyrosine, so they cannot make melanin.
7 It is easy to screen all newborns and then to prevent PKU developing by putting them on a special diet. If we waited to see who was affected, it would be too late, as there is no cure. The care of affected individuals is costly and lasts for the rest of their lives, which would have little quality.

2.2.1 Nutrition

1 Carbohydrates, fats, proteins, minerals, vitamins, water, fibre.
2 If sufficient energy is consumed, the person will have enough energy for their activities and should remain a constant weight. If too little energy is consumed, the person will lose weight. If too much energy is consumed, the person will gain weight.

2.2.2 Diet and CHD

1 Most foods contain Na and Cl so it is not necessary to add salt. Salt in food is absorbed into the blood. It will decrease the water potential of the blood fluid. This leads to retention of more water in the blood, causing a greater fluid volume. This will increase blood pressure.
2 HDLs carry cholesterol back to the liver. In the liver, the cholesterol is used in metabolism or broken down. Therefore a high concentration of HDLs can help to reduce cholesterol concentrations in the blood and is associated with reduced deposition in the artery walls. LDLs carry cholesterol from the liver to the tissues. They are associated with increased deposition in the artery walls.

2.2.3 Improving food production

1 Natural selection is selection of particular varieties that are well adapted to the environment. They are selected by environmental forces or pressures. Artificial selection is selection by humans to achieve the exaggeration of certain features.

2 Pesticides will kill pests that consume the crop or damage the crop in some way. Removing the pest allows the crop to grow to its full yield.

2.2.4 Microorganisms and food

1 Food spoilage is caused when microorganisms grow on food. Microorganisms take nutrients by releasing enzymes onto food and then absorbing the digested food. The action of the enzymes spoils the food, causing it to 'go off'.

2 Heating food can kill microbes that are already on the food. If it is heated to kill microbes, then packaged to prevent further contamination, food will not become spoilt. Cooling food slows down the rate of reproduction of the microorganisms. It also slows down their metabolic processes. It slows the rate of enzyme action. This reduces the effect that microorganisms have on the food, which it will last longer before becoming spoilt.

2.2.5 Organisms that cause disease

1 A parasite is an organism that lives in or on its host for all or part of its life; it causes harm and gains nutrition from the host. A pathogen is any organism that causes disease.

2 Virus: cold, influenza; bacteria: tuberculosis, pneumonia; fungi: plant rust, athlete's foot; protoctist: sleeping sickness, malaria.

2.2.6 Transmission of diseases

1 Any disease of the respiratory tract, such as cold, influenza, pneumonia, TB.

2 AIDS is caused by the HIV virus. This virus attacks the cells of the immune system – in particular the T helper cells and macrophages. These cells normally help to mediate the immune response. If they are killed by viral action, the person will have a weak immune system and will be more susceptible to infections.

2.2.7 The worldwide importance of certain diseases

1 Nutrition may not be as good so their immune systems are not as strong. There may be insufficient funding to provide vaccinations to protect all the people. Water may not be treated to remove infectious pathogenic organisms. Food may not be cooked well enough to kill pathogens. Education about the causes of disease and how to avoid them may not be very good. Health authority response to an epidemic may not be well organised. People live in remote villages so cannot reach care centres. Housing conditions may be poor and may promote the transmission of pathogens.

2 Prevalence means the number of people with the disease at any one time. Incidence means the number of new cases of the disease recorded in a particular time period.

2.2.8 Non-specific response to diseases

1 A neutrophil will engulf a foreign body by surrounding the body. The cell membrane folds inwards and forms a vesicle around the foreign body. Lysosomes move to the vesicle and empty enzymes into the vesicle. The enzymes digest the foreign body and the nutrients are absorbed into the cytoplasm.

2 Non-specific means that it will attack any pathogen or foreign body.

2.2.9 Antibodies

1 An antigen is a molecule that can stimulate an immune response. An antibody is a protein that can attach to and neutralise antigens.

2 An antibody has a variable region that is specific in shape to the antigen it attacks. It is a complementary shape to the antigen. It has a constant region that can attach to a phagocyte. It consists of four polypeptide chains that are held together by disulfide bridges. It also has hinge regions that allow some flexibility in the shape to allow attachment to more than one antigen or more than one pathogen.

2.2.10 Communication between cells

1 Cell surface receptors are specific in shape to the molecule that they bind to. They must be specific so that they bind only to that specific molecular messenger. This ensures that the cell responds only to the correct molecular messenger.

2 Our own cells carry antigens, just like any other cell. The antigens associated with our own cells are recognised as 'self'. This is due to a process during development that destroys any cells from the immune system that have receptors complementary in shape to the antigens on our own cells. Therefore, once the immune system is mature, it contains no cells with receptors that attach to our own antigens and stimulate a response.

2.2.11 The specific immune response

1 T helper cells release cytokines that help coordinate the immune response; they activate B cells and stimulate phagocytes. The T killer cells attack and destroy infected host cells by injecting toxins into cells that show signs of infection.

2 Memory cells remain in the body for years after an infection. They can mount a more rapid and more vigorous secondary immune response when the body is invaded for a second or third time by the same pathogen.

2.2.12 Vaccination

1 It is the deadliest strain of flu that has been seen so far and is also able to infect a range of species. If it mutates to be able to pass from one human to another, it could cause a pandemic that would kill many people.
2 If we could vaccinate every person in the world, or at least most people, then the virus would have no-one to infect. It could not be transmitted and would be unable to spread further.

2.2.13 Finding new drugs

1 Poppies grow among other crops. The seeds may be incorporated into food products, such as bread, by accident. They may even be added to give extra taste or extra nutrients. The effects of poppy seeds may have been discovered by accident – someone who ate bread containing seeds found that a wound did not hurt as much as before.
2 He was conducting experiments on bacteria. He noticed that his agar plates did not contain bacteria in a region that had been accidentally inoculated with a fungus. The fungus appeared to prevent the growth of the bacteria.

2.2.14 The effects of smoking

1 Carbon monoxide, nicotine and tar.
2 Tar coats the lining of the lungs. It stops the action of the cilia and stimulates the goblet cells to produce more mucus. The mucus is not moved by the cilia, so it collects in the lungs. It eventually builds up to block small airways. It must be coughed out.
3 The tar collecting in the lungs is sticky. It catches any bacteria or other pathogens that may be in the air. As the mucus is not removed, the pathogens can remain in the lungs and multiply. They will eventually cause an infection.

2.2.15 Smoking – nicotine and carbon monoxide

1 Nicotine mimics the transmitter substances in nerves and makes the smoker feel more alert. It causes the release of adrenaline, which speeds up the heart rate and raises blood pressure by constricting the arterioles. It reduces blood flow to the extremities. It also makes the platelets more sticky.
2 The process of atherosclerosis is advanced by smoking and this roughens the walls of the arteries. The platelets are more sticky due to the nicotine. Both these effects mean that blood clotting is more likely.
3 If an artery wall is damaged, the body may try to repair it by laying down fatty substances and smooth muscle in the damaged area. Macrophages ingest LDLs in the blood and become foam cells. These sink into lesions in the artery walls. This deposition bulges into the lumen of the artery. This will narrow the lumen, reducing blood flow. If this occurs in the coronary arteries, the flow of blood and oxygen to the heart muscle is reduced. This can lead to insufficient supply of oxygen to the heart muscle during exercise. This puts the heart under stress,

2.2.16 Cardiovascular diseases

1 A stroke is felt as a sudden numbness or weakness of the face, arm or leg on one side of the body. It may also be felt as sudden confusion, trouble with speaking, trouble seeing clearly, loss of balance or coordination, or just a severe headache.
2 A multifactoral disease is one that has no single cause; it has a number of factors that contribute to its onset.
3 There are many risk factors for CHD.
Age – older people have more fatty deposits in their arteries.
Sex – men are more likely to develop CHD, although after the menopause the risk for women is equal to that for men.
Cigarette smoking increases deposition of fatty substances in the walls of the arteries, increases blood pressure, and increases the likelihood of blood clots.
Obesity places more stress on the heart.
Hypertension increases deposition of substances in the artery walls.
High blood cholesterol concentration increases deposition.
Physical inactivity allows the circulatory system to lose efficiency.
High level of animal fats in the diet increases blood cholesterol.
High salt in the diet increases blood pressure.
Continual stress raises blood pressure.

2.2.17 The evidence linking smoking to disease

1 The experimental evidence consists of tests on dogs that were forced to inhale cigarette smoke. They developed lung cancer. The tar from smoke smeared on the skin of mice caused cancer in their skin.
2 There are so many risk factors for CHD that it is difficult to show that any particular one has a significant input to causing the disease.
3 A health authority might use the information provided by epidemiologists to identify those at risk, screen those at risk, target research to find the cause of a disease, target funding to educate those at risk, give advice to those at risk.

2.3.1 Biodiversity

1 A habitat is a place where an organism lives.
2 Possible response: lawn, tree, shrub, pond, etc.
3 The fundamental biochemistry of life is the same for all species. The biochemistry of cell membranes, many cell organelles, respiration, protein synthesis, etc. is the same, and requires the same or similar proteins and enzymes. The codes for these proteins and enzymes are all held in the DNA that is a part of every living cell.

2.3.2 Sampling plants

1 A habitat may be too large and the numbers of organisms too great to make an actual count. A sample is much quicker and gives a representation of the whole habitat.
2 If a sample is not random, it will not be representative. It is easy to be distracted by something that looks more interesting and include it in the sample – but this may lead to misrepresenting the frequency of larger and more colourful species.
3 Random samples using a quadrat. Take a measure of percentage cover in each quadrat. The percentage cover of each species found should be recorded. The number of quadrats used should be dependent on the size of the habitat and the diversity found there.

2.3.3 Sampling animals

1 Collect the leaf litter and sift through to find any larger animals. Place the leaf litter in a Tullgren funnel and collect any animals in a beaker of alcohol below the funnel.
2 Disturbance to larger animals such as birds, deer, badger; even mice and other small rodents may be disturbed. These animals may avoid the area. Trapping of small mammals can cause starvation if the traps are not emptied sufficiently regularly. Some small animals can become 'trap-friendly' and gain their food from traps as they realise they will not be harmed. If trapping is too vigorous, it may affect the population size. Trampling of vegetation may cause temporary damage or even permanent damage to sensitive vegetation. Taking samples for identification may reduce the population of rare plants.
3 Loss of habitat. Effect of water run-off from parking area (run-off may be polluted). Change in the drainage may affect soil moisture in surrounding area. Increased use of motor vehicles could pollute air. Increased disturbance from traffic may scare away larger animals. Light pollution from lights at night could scare away animals.

2.3.4 Measuring biodiversity

1 Species richness is the number of species in an area. Species evenness is a measure of how many individuals of each species are present.
2 Ensure you sample at different depths in the water. Ensure you sample the mud. Ensure you sample at different distances from the bank. Ensure you have adequate footwear and don't fall in!
3 In a habitat with a high diversity, there is a complex feeding structure and any one species relies on many others. If one food species disappears, predators can feed on others. In a simple, low-diversity habitat, one species may rely entirely upon another – if the food species disappears, the predators will have no food and will also disappear.

2.3.5 Classification and taxonomy

1 A common ancestor is an ancestor that is shared by two or more taxonomic groups.
2 Taxonomy is the study of the principles behind classification – the study of the differences between species. Classification is the process of sorting things into groups.
3 It helps us to understand our own evolution. Studying those organisms that are more closely related to ourselves can help us understand our own biology and behaviour.

2.3.6 The five kingdoms

1 Fungi are saprophytic – this means they gain nutrients by causing decay. They have no chlorophyll and cannot photosynthesise. Either they are single-celled or they have multinucleate cytoplasm inside a mycelium. Plants are multicellular and have chlorophyll. They photosynthesise, to manufacture sugars, using sunlight energy.

2

Prokaryotes	Protoctista
single-celled	single-celled
prokaryotic	eukaryotic
no true nucleus	nucleus
DNA naked	DNA associated with proteins
DNA in rings	DNA straight (as a chromosome)
smaller cells	larger cells
few organelles, none being membrane-bound	cell divided into a range of organelles, most being membrane-bound

2.3.7 Classifying living things

1 A spider has eight legs and two body parts; an insect has six legs and three body parts.
2 A species is the basic unit of classification. It is a group of organisms that are essentially the same. Apart from minor variations, all members of the species look the same and share the same adaptations. They can also breed freely to produce viable offspring.

2.3.8 Naming living things

1 People in different countries, or even different parts of one country, could use the same common name for different organisms. Also, the same organism could be given different names in different countries.
2 Identifying living things allows us to classify them. Once we know its identity, we can find out more about an organism. Particularly, we can find out how common or rare it may be. This will have significance as part of an environmental impact assessment.

2.3.9 Modern classification

1 Biological molecules have a structure that is coded for by DNA. A similar molecular structure in two species indicates that the DNA must be similar. Therefore the species must be closely related.
2 The kingdom Prokaryota has two very different groups in it. The Archaea have many similarities to the eukaryotes – similar membrane structure, flagellum structure and RNA polymerase, and they have proteins on their DNA. The Bacteria (*Eubacteria*) are very different, with different membrane structure, different flagella structure, RNA polymerase, etc. These differences are fundamental and suggest that a split is needed. The Archaea are more closely related to the eukaryotes than are the Bacteria.

2.3.10 Variation

1 Hair length, hair colour, skin colour, height, weight, length of branches on a tree.
2 Sex (male/female), ability to roll tongue, albinism, number of fingers, ability to distinguish red and green.
3 Environmental factors can affect the phenotype and cause variation: for example, intense sunlight can burn the skin and will cause tanning. Eating too much will cause an increase in weight, starvation will cause a loss of weight and will affect health.

2.3.11 Adaptation

A dog:

1 Adaptation	2 How enables it to survive
fur	keeps it warm
long legs	can run quickly
eyes at front of face	help judge distance and depth
claws	rip flesh of prey
sharp teeth	catch food and kill it
carnassial teeth	cut flesh
short digestive system	because meat needs little digestion
very acid stomach	kills pathogens in food

2.3.12 Natural selection

1 Fur colour can provide camouflage. Prey can avoid being spotted by a predator; predator can avoid being seen by its prey. Fur colour can also be used as a signalling device to help identification. It may also enhance heat absorption or reflection.
2 Overcrowding and a shortage of some resource. Whenever there is competition for a limited resource, there will be a struggle to obtain enough to survive.

2.3.13 The evidence for evolution

1 Brachiopods changed slowly through time. The earliest forms are found in the oldest rocks and the later forms are found in more recent rocks. If a rock of unknown age is found to contain brachiopods, it can be dated by comparison with other rocks containing the same brachiopods.
2 Evolution tends to occur quickly when the environment changes. Therefore organisms that represent a link between one species and another are rare. This means that fossils of such organisms are very rare. *Archaeopteryx* represents an organism that links the reptiles with the birds.

2.3.14 Evolution today

1 Microorganisms have a short life cycle. With every generation, mutations can arise that cause variation. A selective force can cause a selection pressure which changes the species in a few generations. In mammals the life cycle is so much longer that mutations are less frequent, but selection pressures still need a few generations to cause change.

2 When the environment changes, this places new selective pressures on the species. If there are variations that are advantageous, these are selected and the species changes. Once the species is well adapted to the new environment, change slows down or stops.

2.3.15 Conservation of species

1 Building, polluting, farming, travelling, deforestation, hunting, and many more.
2 Extinction is when the last member of a species dies.

2.3.16 The effect of global climate change

1 If the genetic diversity is reduced, there is less variation within the species. This means there is less likely to be an individual that possesses advantageous characteristics if the environment changes. Evolution can occur only if a variation provides a selective advantage over other variations.
2 Domesticated animals have been adapted by artificial selection to have certain characteristics that are required by humans. These are unlikely to be advantageous if the environment changes. Also, the genetic diversity of the domesticated animals has been reduced so that all are efficient at producing what humans want.

2.3.17 Conservation *in situ*

1 Conserving species in their natural habitat means that all the conditions they require are already present. They are well adapted to living in that habitat, and should be able to survive and breed successfully. No special provisions need to be made.
2 If a species has been made extinct in a particular region, there must be a reason. This could be lack of food or suitable nest sites, or perhaps competition. Reintroduction of the species must be accompanied by removal of the problem that made it extinct. Repopulation often uses captive-bred individuals. These may be unable to find food in the wild, or may be unable to avoid predation. They may not be accepted by other members of the species that are already present in the region.

2.3.18 Conservation *ex situ*

1 Freezing sperm means that it can be transported easily and one male can be used to fertilise a large number of females. This helps to increase the population size quickly. It also means that a small population can maintain genetic diversity by importing sperm from another population. The sperm can also be saved for use some years later. Freezing embryos allows a population to be maintained even when there are no suitable habitats for them to survive in. The embryos can be kept until suitable habitats have been created. If there are only a small number of adults, they can only look after a small number of offspring. Extra embryos can be frozen to be used once there are some more adults to look after the young.
2 Seeds are stored to: maintain the genetic diversity of each species; maintain the diversity of species; supply seeds for growing food in times of famine; maintain a genetic bank for hybridisation; enable repopulation and habitat reclamation; enable continued research.

2.3.19 International cooperation

1 Environmental impact assessment must consider the effect a development will have on the environment near the development. It must include water resources, drainage, pollution, damage to habitat, disturbance, loss of species diversity, loss of habitat.
2 Not all governments agree with trade policies. Some governments may be corrupt. Many conservation areas are out in the wild, and it is very difficult to stop poachers or hunters. Not all governments would have the will or the finances to police the areas adequately. If goods are exported they will be in unmarked boxes and therefore difficult to spot.

Practice answers

1.1 Cells

1 Growth and repair (and cancer/tumour formation); replacement of cells
2 B-6, C-2, D-5, E-1, F-4
3 (a) A Phospholipid molecule
B Protein molecule
C Transmembrane protein (pore)
D Glycoprotein
E Glycolipid
F Cholesterol
(b) 7–10 nm
(c) (i) Leaf mesophyll cell in distilled water – water moves in – cell does not burst
Red blood cell in concentrated salt solution – water moves out – cell does not burst
(ii) Red blood cell in distilled water – water moves in – cell bursts
(d) Passive: small molecules such as gases and lipid-based molecules can diffuse across membranes. Other, hydrophilic, substances pass through the membrane down a concentration gradient through

pores formed by channel proteins or are allowed through by a carrier protein (both facilitated diffusion).
Active: protein pumps use energy from ATP to transport substances across the membrane against a concentration gradient.
Endo- and exocytosis move bulk quantities by fusing vesicles to the membrane and releasing the contents outside the cell, or surrounding the substance outside the membrane and pinching off a vesicle to transport it inside.

4 Formation of, new/more, cells;
formation of genetically identical cells; clone
growth;
refer to major growing time;
asexual reproduction;
e.g. (of method of asexual reproduction);
repair/replacement of damaged cells/formation of scar tissue;
specific location, e.g. meristems;
maintain chromosome number

5 **(a)** Nuclear envelope
(b) XY is 65 mm = 65 000 μm.
Magnification is $\frac{65\,000}{7}$ = 9286×.

6 Change in the water potential surrounding the cell linked to changes in salt concentration as tide comes in and out. Changes affect the water potential gradient between the cell and its surroundings, sometimes water will be moving in to the cell, sometimes out of the cell, at other times there will be no net movement when water potential of surroundings is equal to that of the cell contents so the problems are to get rid of excess water or to prevent loss of too much water from the cell.

1.2 Exchange and transport

1 **(a)** A = epidermis
B = cortex
C = phloem
D = xylem
(b) Root.

2 **(a)** Cardiac muscle.
(b) Muscle of left ventricle is thicker than right to create more pressure as it must move the blood further against resistance.
(c) **(i)** The aorta carries blood away from the left ventricle.
The pulmonary artery carries blood away from the right ventricle.
(ii) The pulmonary artery carries blood back from the lungs to the left side of the heart.
The venae cavae carry blood back from the body to the right side of the heart.

3 **(a)** *Two from:*
large surface area
thin barrier
well supplied with blood vessels
well ventilated.
(b) Enters mouth or nose → down trachea → along bronchi → along bronchioles → into alveoli.
(c) Alveoli.

4 **(a)** Exercise increases heart rate.
(b) Tidal volume of lungs increases during exercise.
(c) *Any two from:*
digestive system/small intestine/ileum
sinusoids in liver
root hairs
hyphae of fungi.

5 Oxygenated, haemoglobin, heart, oxygen, dissociation, respiration, red blood, hydrogen carbonate.

6 **(a)** Allows alveolus wall to stretch to allow more air to enter alveolus.
Recoils to help expel air from alveolus.
(b) Ciliated (columnar) epithelium.
Goblet.
(c) Synchronised movement (Mexican wave).
To move/waft mucus up to back of throat.

7 **(a)** The palisade tissue in a leaf.
The cortex of a root.
(b) Phloem.

8 **(a)** The demand for oxygen is high.
All tissues demand oxygen.
Surface area of body is not large enough to supply enough oxygen.
(b) Carbon dioxide enters the red blood cell.
Carbon dioxide combines with water to form carbonic acid.
Under influence of enzyme carbonic anhydrase carbonic acid dissociates to form hydrogen ions and hydrogencarbonate ions.
Hydrogen ions compete with oxygen for space on the haemoglobin molecule.
More oxygen molecules released from oxyhaemoglobin.

9 **(a)** Active pumps remove hydrogen ions from companion cells.
Hydrogen ions diffuse back into companion cells through special cotransporter proteins carrying sucrose into companion cell.
Sucrose diffuses through into sieve tube element via plasmodesmata.

(b) Air movement carries water vapour away from the stomata.
Water vapour potential (or relative humidity) outside the leaf is lower than inside.
Water vapour diffuses out of leaf via stomata down water vapour potential gradient.

10 (a) From the atrioventricular node the wave of excitation travels down the septum in the Purkyne tissue to the base (apex) of the ventricles.
The wave of excitation then spreads up the walls of the ventricles from the apex upwards causing contraction which pushes the blood upwards towards the arteries.
(b) Coronary arteries lie over surface of heart.
Supply oxygen and nutrients to heart muscle.
Oxygen and nutrients needed for respiration.
To supply energy for contraction of heart muscle.

2.1 Biological molecules

1 Deoxyribose.

2 5

3 (a) Thymine; (b) Uracil.

4 solvent, liquid, dense, insulates, hydrogen, surface tension

5 Starch molecules and protein molecules have different shapes. A part of the starch molecule can fit into the enzyme's active site because its shape is complementary to that of the active site.

6 The solution must be boiled for several minutes in hydrochloric acid or heated to 40 °C with sucrase enzyme.

7

type of molecule tested	reagents used	positive result	negative result
protein	**biuret reagent**	**lilac colour seen**	blue solution
lipids	alcohol and water	white emulsion	clear liquid
starch	**iodine (in potassium iodide solution)**	**blue-black**	yellow solution

8 Reduces the rate of reaction *('inhibits' is not worth a mark).*
Fits into a site on the enzyme which is away from the active site.
Changes the shape of the active site.
Substrate can no longer bind with the active site.
Increasing substrate concentration has no effect.

9 (a) Eventually all the starch has been broken down to maltose, so the level of maltose reaches the same value in both cases because the amount of starch was the same at the start of each experiment.
(b) At 23 °C the enzyme and substrate molecules have greater kinetic energy than they do at 18 °C. This means the molecules move around more rapidly and so collide more frequently. Increased frequency of collisions between enzyme and substrate molecules leads to increased rate of reaction, so product is formed more quickly.
(c) Glycosidic bond.

10 Two new molecules are made. Each contain 1 old strand of DNA and 1 new strand.

11 (a) (i) Both contain six carbon atoms, 12 hydrogen atoms and six oxygen atoms.
(ii) Glucose forms a six-sided ring whereas fructose forms a five-sided ring.
(b) Diagram should show glycosidic bond formed, as in the structure of maltose, and a molecule of H_2O removed.

12 (a) Arachidonate is the substrate for the enzyme. This comes from phospholipids source in membranes nearby. The prostaglandins produced can be stored and transported from the ER for use as required. The location keeps the reaction separate from other reactions in the cell that may be important.
(b) Ibuprofen acts as a competitive inhibitor because it is blocking the active site so the substrate cannot enter. Aspirin acts as a non-competitive inhibitor. It changes the shape of the active site by attaching to the enzyme close to, but not in, the active site. In both cases, an enzyme–substrate complex cannot form.

2.2 Food and health

1 (a) Human immunodeficiency virus or HIV.
(b) By unprotected sexual intercourse.
By blood-to-blood contact.
By sharing unsterilised hypodermic syringes.
Through unscreened blood products.
By needle stick or similar accident.
By re-using unsterilised surgical equipment.
Across placenta.
In breast milk.

2 Pesticides are used to kill pests, e.g. fungicides kill fungal pests, insecticides kill insect pests. If left alone, these pests can damage the crop and reduce yield. Fertilisers are a source of nutrients, e.g. nitrates, phosphates and potassium. These are used by plants to increase growth and rate of growth. This can

increase yield. Organic fertilisers can improve the structure of the soil. This helps the growth of plant roots, holds water in the soil, and releases nutrients slowly as organic matter decays.

3 (a) Quorn™ is a mycoprotein, a protein made by fungi.
(b) Fungi such as *Fusarium venenatum*, which produces Quorn.
Kluyveromyces, *Scytalidium* and *Candida* can also make protein.
Lactobacillus bacteria are used to make yoghurt.
Lactobacillus bacteria are also used to make cheese.

4 (a) Microorganisms can start to digest food. This happens externally and the food may begin to smell sweet. This is the smell of sugars released from polysaccharides.
Other microorganisms may release their waste products into the food; these may act as a toxin.
The visible presence of fungi or bacterial colonies on our food is also considered spoilage.
(b) Packaging helps to prevent microorganisms coming into contact with food.
Canning involves heating the food to denature proteins and kill microorganisms, and then sealing the food in a tin.
Osmotic methods such as salting, drying or mixing with sugar will dehydrate any microorganisms that land on the food and so prevent any further growth.
Cooling and freezing reduce the rate of growth and reproduction of microorganisms so they do not spoil the food.
Irradiation kills microorganisms by damaging their DNA.
Pickling food uses an acidic pH to kill microorganisms or inhibit enzyme action.

5 Skin prevents entry of pathogens.
The stomach has an acid environment, which will kill microorganisms in the food.
The lungs are lined by mucus, which can trap organisms and is moved out of the lungs by the action of the cilia.

6 (a) Neutrophils and macrophages.
(b) The phagocyte invaginates to engulf the foreign matter.
The foreign matter is trapped in a vacuole.
Lysosomes fuse with the vacuole.
Lysosomes release digestive enzymes into the vacuole.
Foreign matter is digested.
Component parts of foreign matter are absorbed into the cytoplasm of the phagocyte.

7 (a) Influenza is spread easily by droplet infection.
The virus regularly mutates to form new strains.
Vaccines are only effective against one strain.
Memory cells and antibodies are only effective against the strain that has invaded a person previously.
Therefore when a new strain arises, and it spreads easily, people are not immune to it.
(b) Avian flu spreads easily.
Occasionally it spreads to people.
It does not usually spread from one person to another.
If it mutates in a way that allows it to spread from person to person, it could cause a mass epidemic.
The H5N1 strain is of great concern as it is particularly deadly and seems able to infect a wide range of species.

8 (a) Production is very rapid.
Production can be much more efficient than growing animals.
Production is flexible – it is easily increased or decreased.
Microorganisms can grow on waste products.
(b) Palatability of the end product – the protein may not be very tasty.
The food must be isolated and purified thoroughly to avoid potential for infection with microorganisms.
The production process can easily be infected with unwanted microorganisms.
People may not want to eat food made by microorganisms.

9 Antigens are molecules that can start an immune response; they are often on the surface of cells and are used to identify the cell.
Antibodies are large protein molecules that are produced by the immune system to combat antigens and pathogens.

10 (a) Clonal selection is the selection of a particular lymphocyte that is capable of providing immunity to a specific antigen. The clone may be a T lymphocyte, which could differentiate into a T killer cell or into a T helper cell. The clone could also be a B lymphocyte, which will differentiate into plasma cells and memory cells.
(b) The B cell selected is capable of making a specific type of antibody. It differentiates and multiplies into a number of plasma cells. Each plasma cell manufactures antibodies with a specific amino acid sequence. This gives the antibody a particular 3D shape, which is complementary to the antigen.
(c) The variable region of the antibody is the part that provides specificity. There are hinges in the polypeptide chains that enable the molecule to be flexible – this allows it to bind more easily. The four polypeptide chains are held together in a particular way by disulfide

bridges. At the opposite end of the molecule from the variable region is a constant region. This provides a site to which phagocytes can bind.

11 (a) TB has many strains. It is slow growing and intracellular.
The bacterium can mutate and may become resistant to antibiotics.
It takes a long course of antibiotics to cure.
Often a wide range of antibiotics is needed to ensure the bacterium is killed.
Some people stop taking the antibiotics before the course is complete.

(b) Many disease-causing organisms need to enter cells. They do this by binding to cell surface membrane receptor sites, e.g. the HIV virus binds to the CD4 receptor on helper T lymphocytes. If the receptor is blocked by a molecule, the virus cannot bind and cannot enter the cell.
Other medicines may block receptor sites that are used by hormones or other cell signals. Opiates block sodium channels to prevent nervous action.

12 There are always new strains of disease-causing organisms evolving. Each may cause a new range of symptoms. There are many potential drugs to be found in plants. If global diversity is allowed to decline, some of the potential drug-producing plants may be lost. As microorganisms evolve, they develop resistance to drugs such as antibiotics. Antibiotics come from microorganisms. If we allow species of microorganisms to die out, we may be losing potential sources of new antibiotics. These new antibiotics may be effective against pathogens that have developed resistance to all other antibiotics.

13 (a) Multifactoral means that many factors contribute to the risk of developing the disease. No single factor will cause CHD on its own.

(b) high levels of salt in diet;
high levels of saturated fats in diet;
not enough antioxidants (vitamins A, C and E) in diet;
not enough fibre in diet;
not enough polyunsaturated fats / monounsaturated fats in diet;
lack of vitamin D;
eating too much.

(c) The lining of the artery (endothelium) gets damaged. White blood cells attempt to repair the damage. They deposit fatty substances or low-density lipoproteins under the endothelium. (White blood cells ingest fats and become foam cells. They sink into the lesions in the endothelium.) The deposition thickens the wall of the artery and protrudes into the lumen.

14 (a) Tar accumulates on the lining of the lungs. It can increase the diffusion pathway of oxygen into the blood, so blood is less well oxygenated. The tar contains carcinogens which cause cancer in the cells lining the lungs. The tar irritates the mucus-secreting (goblet) cells, causing more mucus to be produced. It also reduces the activity of the cilia and may damage cilia so that the mucus is not removed from the lungs.

(b) Nicotine causes constriction of the peripheral arterioles. This raises blood pressure. It also raises the heart rate. Finally, nicotine makes the platelets sticky and increases the chance of blood clots forming.

(c) Experimental evidence includes signs of lung cancer in dogs forced to smoke. Tar from cigarette smoke smeared onto the skin of mice caused cancerous growths in the skin.
The other evidence is epidemiological – this means studies such as one comparing the percentage of smokers with lung cancer to the percentage of non-smokers with lung cancer.

15 (a) Passive immunity is achieved when antibodies are acquired from an external source such as an injection or through breast milk. Active immunity is achieved when the immune system is stimulated to produce its own antibodies.

(b) An attenuated pathogen is a pathogen that has been weakened so that it will not cause a disease.

(c) An attenuated pathogen or a preparation of antigens from the pathogen can be injected into the blood. It is detected by the immune system and activates the production of antibodies and memory cells. The memory cells remain in the blood for a long time, providing immunity.

2.3 Biodiversity and evolution

1 (a) Prokaryotae: no true nucleus, few organelles, DNA naked, cells smaller than eukaryotic cells
Protoctista: eukaryotic, single-celled, wide diversity, some animal-like, some plant-like
Fungi: are eukaryotic, have a mycelium, the mycelium consists of hyphae, have walls made of chitin, have cytoplasm that is multinucleate
Plantae: multicellular, eukaryotic, autotrophic, have cellulose cell walls
Animalia: multicellular, eukaryotic, heterotrophic, able to move around

2 Continuous variation has two extremes, with a full range of variations in between.
Discontinuous variation has two or more distinct categories, with nothing in between.

3 (a) Pitfall traps – a cup buried so that its lip is just below the level of the soil. This will collect insects and small soil organisms.
Sweep netting to capture anything in the grass.
Observation for signs of larger organisms.
(b) Non-random samples are subjective and may give inaccurate results.

4 (a) We have not found all the species alive today.
New species are being formed.
Species are becoming extinct.
It can be difficult to tell if some individuals are members of different species or just variations within one species.
(b) Habitat – the place where individuals of a species live.
Biodiversity – the range of species, habitats and ecosystems in existence today.

5 (a) Offspring generally look similar to their parents.
No two individuals are identical.
Organisms have the ability to overproduce.
Populations in nature tend to be fairly stable in size.
(b) A selective pressure is some factor in the environment that means some individuals are less likely to reproduce than others (it may mean some are more likely to die than others). For example, it may force the individuals of a species to compete (e. g. a shortage of suitable food) or it may be a form of predation (better camouflaged individuals survive).
A selective advantage is a variation that enables some individuals to compete more strongly. These individuals are better adapted. These individuals are more likely to reproduce.
(c) Fossils show that species from the past were not identical to the species of today. A range of slight variations can be observed in some fossils as the species changed over time, e.g. the brachiopods. Fossils from more recent times show more similarities with current species, while older fossils are more different. Some rare fossils may show a glimpse of 'missing links' – the species that arose from one group of organisms that eventually gave rise to a new group of organisms.
(d) *Archaeopteryx* was an animal that showed many of the features of reptiles of its time. However, it also showed the presence of feathers – a feature of birds. This is evidence that the birds arose from the reptiles. Such 'missing link' fossils are very strong evidence that species change over time and that one group of organisms can give rise to another.

6 Place the quadrat at random by using random number tables;
place the point frame over the quadrat and lower the points onto the vegetation;
note how many points touch clover;
use the frame 10 times (100 points) – each point that touches clover is 1%;
repeat several times depending on size of the field.

7 (a) Classification is the process of sorting living things into groups. Natural classification does this by grouping things according to how closely related they are.
(b) Phylogeny is the study of the evolutionary relationships between organisms.
(c) Natural classification groups things according to how closely related they are – this should match the evolutionary tree produced by considering how recently organisms shared a common ancestor.

8 (a) $D = 1 - [\Sigma(n/N)^2]$
$N = 74$
$D = 1 - [(20/74)^2 + (5/74)^2 + (1/74)^2 + (32/74)^2 + (15/74)^2 + 1/74)^2]$
$D = 1 - [(0.073 + 0.00456 + 0.00018 + 0.187 + 0.041 + 0.00018)]$
$D = 0.694$
(b) With a high diversity there is greater species richness.
More species means that each species relies on a number of others.
If one species is affected by some change, the others may be less affected.
A species dependent on the one that is affected will have others to fall back upon.

9 Evolution occurs as a result of variation and selective pressure.
One cause of variation is changes in the DNA or genetic material.
Changes to the DNA occur randomly.
Species with a recent common ancestor will have few differences in their DNA.
Species with a more distant common ancestor will have more differences.
Changes to the DNA cause changes to the biological molecules, such as proteins.
When the DNA or protein structure of two species is analysed, two closely related species will have DNA and protein structures that are more similar than those of two less closely related species.

Glossary

α-helix A protein secondary structure – a right-handed spiral held in place by hydrogen bonds between adjacent C=O and NH groups.

Abundance The frequency of occurrence of plants in a sampled area, such as a quadrat.

Activation energy The level of energy required to enable a reaction to take place. Enzymes reduce the amount of energy required to allow a reaction to proceed.

Active immunity Immunity that is acquired by activation of immune system.

Active site The area on an enzyme molecule to which the substrate binds.

Active transport Movement of substances across membranes against their concentration gradient, requiring the use of energy in the form of ATP. Active transport usually involves the use of transport proteins.

Adaptation A feature of a living organism that increases its chances of survival, for example thick fur on an animal that lives in a cold habitat.

Adenine A nitrogen-containing organic base found in nucleic acids. It pairs with thymine in DNA and with uracil in RNA.

Adhesion Force of attraction between molecules of two different substances.

Adipose Describes tissue consisting of cells that store fat/lipid.

Affinity An attractive force between substances or particles.

Allele A version of a gene.

Allopatric Speciation due to organisms of a species being separated by geographical barriers so that over time members of the two populations become so different that they cannot interbreed and are considered to be two different species.

Alveoli Small air sacs in the lungs.

Amino acid An organic compound that contains both an amino group (–NH_2) and a carboxyl group (–COOH). Amino acids are the monomers of protein molecules.

Amylase An enzyme that catalyses the hydrolysis of starch to maltose.

Amylose Part of a starch molecule, consisting of many thousands of glucose residues bonded together.

Anaphase In mitosis, the stage when the newly separated chromatids are pulled towards opposite poles of the nuclear spindle.

Anomalous Describes a result/data point that does not appear to fit the pattern of the other results. It may be assumed to be anomalous if the experimenter has made an error or if the apparatus used is not suitable for the measurements being taken.

Antibiotics Molecules produced by microorganisms that kill or limit the growth of other microorganisms

Antibodies Protein molecules released by the immune system in response to an antigen, which are capable of neutralising the effects of the antigen.

Antigen A foreign molecule (which may be protein or glycoprotein) that can provoke an immune response. Organisms have antigens on their plasma (cell surface) membranes.

Antigen-presenting cell A macrophage that has ingested a pathogen and displays the pathogen's antigens on its cell surface membrane.

Apoplast pathway The route taken by water between the cells or through the cell walls in a plant.

Arteriosclerosis Hardening of the artery walls and loss of elasticity caused by atherosclerosis and deposition of calcium.

Artificial immunity Immunity acquired as a result of deliberate exposure to antigens or by the injection of antibodies.

Artificial selection Also called selective breeding – the process of improving a variety of crop plant or domesticated animal by breeding from selected individuals with desired characteristics.

Asexual reproduction The production of genetically identical new organisms by a single 'parent' organism.

Assay The use of comparative studies or samples to determine the concentration or quantity of a substance in a sample.

Assimilation Incorporation. Usually applied to the process of incorporating simple molecules of food produced by digestion into the living cells of an animal for use in metabolism. In plants, refers to the incorporation of carbon from carbon dioxide into organic substances during photosynthesis. The newly formed compounds may be referred to as assimilates.

Atherosclerosis / atheroma The process of deposition of fatty substances in the lining of arteries to form atheroma, which may eventually lead to arteriosclerosis (hardening of the arteries).

ATP Adenosine triphosphate – a molecule used to store energy temporarily in organisms. The molecule is broken down to adenosine diphosphate + phosphate to release energy to drive metabolic processes.

Atrioventricular node (AVN) A patch of tissue in the

septum of the heart that conducts the electrical stimulus from the atria in the heart through to the Purkyne fibres.

Atrioventricular valves Valves between the atria and ventricles that prevent backflow of blood.

Atrium One of the upper chambers in the heart.

Autotroph An organism that makes its own food from simple inorganic molecules, such as carbon dioxide and water. Some (photoautotrophs), e.g. plants, use light as the source of energy. Some (chemoautotrophs), e.g. some bacteria, use chemical energy. Autotrophs are the producers in a food chain.

Base-pairing rules Complementary base-pairing, between nitrogenous bases in nucleic acids. Adenine pairs with thymine (or uracil). Guanine pairs with cytosine.

Benedict's test A test for reducing sugars. The substance is heated to 80 °C with Benedict's reagent. If a reducing sugar is present, the Benedict's reagent changes from blue to red/red precipitate.

Binary fission Method of cell division in bacteria. The DNA replicates and the cell divides into two, each having the same DNA as the parent cell. It does not involve mitosis.

Binomial system A system of naming living things using two Latin words – the genus name and the specific name.

Biodiversity The number and variety of living things to be found in the world, in an ecosystem or in a habitat.

Biuret test A biochemical test for the presence of proteins.

Body mass index Numerical value found by dividing an individual's mass in kg by the (height in m)^2 and used to assess if the individual is underweight, acceptable weight, overweight or obese.

Bohr shift / Bohr effect The effect of carbon dioxide concentration on the affinity of haemoglobin for oxygen.

Bronchi Airways in the lungs that lead from the trachea to the bronchioles.

Bronchioles Airways in the lungs that lead from the bronchi to the alveoli.

Buffer A chemical system that resists changes in pH by maintaining a constant level of hydrogen ions in solution. Certain chemicals dissolved in the solution are responsible for this.

Calibration To determine the quantity of a substance in a solution by taking readings from solutions containing known amounts of the solution (e.g. by colorimetry) and constructing a calibration curve on a graph. This can then be used to determine the amount of that substance in solutions of unknown concentration. Also, to determine the value of intervals of a scale on an instrument, such as a thermometer.

Cambium Plant tissue in the stem and root that contains dividing cells.

Carbaminohaemoglobin The molecule resulting from combination of carbon dioxide and haemoglobin.

Carbohydrate A class of biological molecules with the general formula $C_x(H_2O)_x$. It includes sugars, starches, glycogen and cellulose.

Carcinogen A substance that causes cancer.

Cardiac cycle The sequence of events making up one heartbeat.

Cardiac muscle The muscle found in the heart. It has its own intrinsic heartbeat (it is myogenic).

Carnivore An animal that eats meat.

Carrier protein A protein found in membranes, which is capable of carrying a specific molecule or ion through the membrane by active transport.

Cartilage A flexible, slightly elastic connective tissue.

Cartilage ring A flexible ring of cartilage that holds the airways open.

Casparian strip A strip of waterproof material (suberin) in the cell walls of root endodermis cells. It blocks the apoplast pathway.

Catalyst A substance that increases the rate of a reaction but does not take part in the reaction, and so is re-usable.

Cell signalling Processes that lead to communication and coordination between cells. Hormones binding to their receptors on the cell surface membrane are an example.

Cell surface membrane *See* Plasma membrane.

Cellulose A carbohydrate polymer (of β-glucose) that forms plant cell walls.

Centriole An organelle from which the spindle fibres develop during cell division in animal cells.

Centromere The region of a chromosome where two sister chromatids are joined together, and where the spindle fibre attaches during cell division.

Channel protein A protein pore that spans a membrane, through which very small ions and water-soluble molecules may pass.

Chemotaxis The movement of cells or organisms towards or away from a particular chemical.

Chloride shift The movement of chloride ions into red blood cells to balance the loss of hydrogencarbonate ions.

Chlorophyll Pigments found in chloroplasts of plant (and some protoctist) cells. Each molecule consists of a hydrocarbon tail and a porphyrin ring head with a magnesium atom. Chlorophyll absorbs red and blue light, trapping the energy, and reflects green light

Chloroplast An organelle found in plants, which contains chlorophyll and is responsible for photosynthetic activity in the plant.

Cholesterol A lipid molecule (not a triglyceride) found in all cell membranes and involved in the synthesis of steroid hormones.

Chromatid A replicated chromosome appears as two identical strands in early stages of cell division. Each strand is a chromatid.

Chromatin Material staining dark red in the nucleus during interphase of mitosis and meiosis. It consists of nucleic acids and proteins. Chromatin condenses into chromosomes during prophase of nucler division.

Chromosome A linear DNA molecule wrapped around histone proteins found in the nucleus. Chromosomes become visible in prophase of cell division.

Cilia Short extensions of eukaryotic cells, typically 2–10 μm long and 0.25 μm in diameter. They may be used for locomotion or to move fluids or mucus over a surface, for example in the mammalian respiratory tract.

Ciliated epithelium Epithelial cells that have cilia on their cell surface.

Class Taxonomic group used in classification of living organisms. Members of the same class share some characteristics. Within each class are orders, consisting of families, genera and species. Similar classes are grouped into a phylum.

Classification The organisation of living organisms (or other items) into groups according to their shared similarities.

Clonal expansion The division of selected cells by mitosis to increase their numbers.

Clonal selection The selection of cells (of the immune system) with a specific receptor site. These cells will undergo clonal expansion as part of the immune response.

Clones Genetically identical cells or individuals.

Coenzyme An organic non-protein molecule that binds temporarily with substrate to an enzyme active site. It is essential for enzyme activity.

Cofactor A molecule or ion that helps an enzyme to work. It may be an inorganic ion or a coenzyme.

Cohesion The attraction between water molecules due to hydrogen bonding.

Collagen A structural fibrous protein found in connective tissue, bones, skin and cartilage. It accounts for 30% of body protein.

Companion cell A cell in the phloem involved in actively loading sucrose into the sieve tube elements. The companion cell is closely associated with a phloem sieve element, to which it is linked by many plasmodesmata.

Competitive inhibitor A substance that reduces the rate of an enzyme-controlled reaction by binding to the enzyme's active site.

Complementary (base/structure) Refers to structures that fit together because their shapes and/or charges match up. For example, adenine and thymine are complementary bases in DNA.

Concentration gradient The difference in concentration of a substance between two regions.

Condensation A type of chemical reaction in which two molecules are joined together by means of a covalent bond to form a larger molecule, and at the same time a water molecule is released.

Connective tissue A type of tissue that consists of separate cells held together by a ground substance (matrix).

Conservation *ex situ* Conservation in areas other than the natural habitat.

Conservation *in situ* Conservation in the natural habitat.

Constrict To make narrow. For example, vasoconstriction is the narrowing of blood vessels.

Continuous variation Variation between living organisms where there is a range of intermediates, such as height, hair colour and intelligence in humans. These characteristics are determined by many genes that interact. The expression of these genes is also influenced by the environment.

Control Part of an experimental investigation – set up to show that the variable being investigated is responsible for the change observed.

Coronary arteries Arteries that carry blood to the heart muscle.

Cortex Tissue in plant roots and stems between epidermis and vascular tissue.

Cotransporter protein A protein in a cell membrane that allows movement of one molecule when linked to the movement of another molecule in the same direction by active transport.

Covalent bond A chemical bond formed by the sharing of one or more electrons between two atoms.

Crenation State of animal cells when they have been immersed in a solution of lower water potential and have lost water by osmosis. They become shrivelled.

Cristae The folds found in the inner membrane of a mitochondrion. Stalked particles containing ATP synthase are found on cristae.

Cytokines Hormone-like proteins produced by vertebrate (including mammalian) cells, which are used for communication between cells, allowing some cells to regulate the activities of others.

Cytokinesis The division of the cell, following nuclear division, to form two new cells.

Cytosine A nitrogen-containing organic base found in nucleic acids. It pairs with guanine in DNA.

Cytoskeleton The network of protein fibres and microtubules found within the cell that gives structure to the cell and is responsible for the movement of many materials within it.

Deamination Removal of amine (NH_2) group from an amino acid

Denaturation An irreversible change in the tertiary structure of a protein molecule. It leads to loss of function in most proteins.

Deoxygenated Blood with haemoglobin that carries no or little oxygen.

Deoxyribose The 5-carbon sugar in DNA nucleotides.

Diaphragm A sheet of muscular and fibrous tissue separating the chest cavity from the abdominal cavity.

Diastole The period when the heart muscle in the ventricles is relaxing and blood pressure is at its lowest.

Differentiation The development and changes seen in cells as they mature to form specialised cells.

Diffusion The net movement of molecules or ions in a gas or liquid from an area of high concentration to an area where they are less concentrated.

Diffusion gradient The gradient in molecular concentration (the difference in concentrations) that allows diffusion to occur.

Dilate To make wider. For example, vasodilation is when the lumens of blood vessels become wider.

Dipeptide A molecule consisting of two amino acids joined by a peptide bond.

Diploid Cells or organisms that have two copies of each chromosome in their nuclei.

Disaccharide A molecule consisting of two monosaccharide sugars joined by a glycosidic bond.

Discontinuous variation Variation between living organisms within a species, where there are discrete categories and no intermediates, e.g. blood grops A, B, AB or O in humans. This type of variation is determined by one gene.

Disease A departure from full health.

Dissociation The breakdown of a molecule into two molecules, atoms or ions. For example, the release of oxygen from oxyhaemoglobin.

Dissociation curve (oxyhaemoglobin) The curve on a graph showing the proportion of haemoglobin that is saturated with oxygen at different oxygen tensions.

Diversity Being diverse – usually used in the context of biodiversity – where there are many different types of organisms present, or genetic diversity within a population of organisms that have genetic variation.

Division of labour Any system where different parts perform specialised functions, each contributing to the functioning of the whole.

DNA Deoxyribonucleic acid – a polymer of nucleotide molecules that form the instructions for the synthesis of proteins found within organisms. These nucleotides contain the 5-carbon sugar deoxyribose.

Domain Classification group. Carl Woese's three-domain classification system divides the Kingdom Prokaryotae into two domains and places all Eukaryotes in the third domain.

Double circulatory system A transport system in which blood travels twice through the heart for each complete circulation of the body.

Double helix Describes the structure of DNA, a twisted helix of two strands with bases joining the strands.

Ecosystem All the living organisms and all the non-living components in a specific area, and their interactions.

Elastic fibres Long fibres of the protein elastin that have the ability to stretch and recoil.

Elastic tissue Tissue containing the protein elastin, which is able to stretch and recoil.

Electrocardiogram Trace (graph) showing the electrical activity of the heart muscle (atria and ventricles) during a cycle.

Emulsion A suspension of one material in another as droplets, because it does not dissolve. For example, fat droplets dispersed in water.

Endemic Describes a disease that is always present in an area. May also mean a species that is found only in a particular area and nowhere else.

Endocytosis The process of taking materials into a cell by surrounding them with part of the plasma membrane, which then pinches off to form a vesicle inside the cell. This is an active process requiring ATP.

Endodermis A ring of cells between the cortex of a root and the area housing the xylem and phloem.

Endoplasmic reticulum (ER) A series of membrane-bound, flattened sacs extending from the outer nuclear membrane through the cytoplasm. It may appear rough (rough ER) when ribosomes are attached to the outer surface, and it is involved with synthesis of proteins. It may appear smooth (smooth ER) when ribosomes are not attached, and it is involved with lipid metabolism or membrane formation.

Endothelium A tissue that lines the inside of a structure, such as a blood vessel.

Endotherm An animal that produces heat within its cells, from respiration, to maintain a constant body temperature.

End-product inhibition The regulation of metabolic pathways where the last product in a sequence of enzyme-controlled reactions becomes an inhibitor of one of the enzymes earlier in the sequence.

Environmental impact assessment An assessment of the damage that may be caused to the (local) environment by a proposed development.

Enzyme A protein molecule that acts as a biological catalyst.

Enzyme–product complex The intermediate structure in which product molecules are bound to an enzyme molecule.

Enzyme–substrate complex The intermediate structure formed when a substrate molecule binds to an enzyme molecule.

Epidemic Describes a disease that spreads to many people quickly and affects a large proportion of the population.

Epidemiology The study of patterns of disease and the factors that influence their spread.

Epidermis Outer layer(s) of cells of a multicellular organism. Plants have a single layer surrounding the tissues of roots, stems and leaves. Invertebrates have an epidermis made of a single layer of cells that secrete a cuticle. Vertebrate (including mammals) epidermis consists of several layers, the outer layer being made of dead cells.

Epithelium A tissue that covers the outside of a structure.

Erythrocytes Red blood cells.

Ester bond The bond formed when fatty acid molecules are joined to glycerol molecules in condensation reactions.

Ethanol emulsion test A biochemical test for the presence of lipids.

Eukaryote An organism having cells with a nucleus and membrane-bound organelles.

Eukaryotic cell Cells that have a nucleus inside a nuclear envelope, and other membrane-bound organelles.

Evolution Gradual process by which the present diversity of living organisms arose from simple primitive organisms that were present about 4000 million years ago. New species have arisen by natural selection.

Evolutionary distance A measure of how far apart two organisms are on the evolutionary scale. Closely related species will be a short distance apart while distant relatives will be further apart.

Exchange surface A specialised area adapted to make it easier for molecules to cross from one side of the surface to the other.

Exocytosis The process of removing materials from the cell by fusing vesicles containing the material with the plasma membrane (cell surface membrane).

Extinction The death of the last individual in a species.

Extracellular Outside the cell. Extracellular enzymes/digestion work outside the cell.

Facilitated diffusion The passive movement of molecules across membranes down their concentration gradient, which is aided by transport (carrier) protein molecules. No metabolic energy is required.

Family Taxonomic group used in the classification of living organisms. Related genera are placed in the same family.

Fat Mixture of lipids, mainly triglycerides with saturated fatty acids, that is solid at body temperature. In living organisms they act as an energy store, insulation, waterproofing the outer layer and may give buoyancy.

Fatty acid A molecule consisting of a fatty (hydrocarbon) chain and an acid (carboxylic acid, –COOH) group.

Fermenter A vessel used to grow microorganisms in large numbers.

Fertiliser A substance added to soil to enhance the growth of plants.

Fibrillation A state in which the chambers in the heart contract out of rhythm.

Fibrous protein A protein with a relatively long, thin structure, which is insoluble in water and metabolically inactive, often having a structural role within the organism.

Flaccid A term used to describe plant tissue where the cells have lost turgor and are not firm.

Fluid mosaic (model) The model of cell membrane structure proposed by Singer and Nicholson – a phospholipid bilayer with proteins 'floating' in it.

Food tests Simple tests that show the presence of various biological molecules in samples or structures.

Gamete Sex cells, usually haploid (one set of chromosomes). Male and female gametes can fuse, during sexual reproduction, to form zygotes (diploid).

Gaseous exchange The movement of gases by diffusion across a barrier such as the atreous wall.

Gated channels Protein channels found in cell membranes, which can be opened or closed in response to cell signals.

Gene A length of DNA that carries the code for the synthesis of one (or more) specific polypeptides.

Gene pool The sum total and variety of all the genes in a population or species at a given time.

Genetic erosion The loss of genetic variation due to (artificial) selection.

Genome All the genetic material inside an organism (or cell).

Genus Taxonomical group used in the classification of living organisms. Species that are similar are placed in the same genus.

Globular proteins Proteins with relatively spherical molecules, soluble in water, often having metabolic roles in organisms.

Glucose A 6-carbon monosaccharide sugar.

Glycerol A 3-carbon (alcohol) molecule. It forms the basis of lipids when fatty acids are bonded to it.

Glycogen A polysaccharide found in animal cells. Formed from the bonding together of many glucose molecules, used as a store of glucose/energy.

Glycolipid A lipid with carbohydrate molecules attached.

Glycoprotein A protein with carbohydrate molecules attached.

Glycosidic bond The covalent bond formed when carbohydrate molecules are joined together in condensation reactions.

Goblet cells Mucus-secreting cells in epithelial tissue.

Golgi body Membrane-bound organelle in eukaryote cells. Its functions are: to modify proteins, made at the rough endoplasmic reticulum, into glycoproteins; to package proteins for secretion outside the cell; to make lysosomes; in plant cells to secrete carbohydrates that make up the cell walls.

Guanine A nitrogen-containing organic base found in nucleic acids. It pairs with cytosine.

Guard cells In pairs, these form the stomatal pore in the epidermis plants. They control the opening and closing of the pore by changes in their turgidity.

Habitat The place where an organism or population lives. It includes the climatic, topographic and edaphic factors as well as the plants and animals that live there.

Haem The iron-containing prosthetic group found in haemoglobin.

Haemoglobin The protein that carries oxygen in the red blood cells.

Haemoglobinic acid The acid produced when haemoglobin takes up hydrogen ions.

Haemolysis (first observed in red blood cells.) The rupturing of animal cell surface membranes and subsequent release of their contents, when animal cells are placed in a solution of higher water potential and water enters by osmosis.

Haploid A cell or organism that has one set of chromosomes/one copy of each chromosome.

Health Complete mental, physical and social wellbeing.

Herbivore An animal that eats plant material.

Heterotroph Organism that has heterotrophic nutrition – it gains nutrients from complex organic molecules. These molecules are digested by enzymes to simple soluble molecules and then built up into the complex molecules that the organism requires. Heterotrophs are consumers or decomposers in a food chain.

Histone Type of protein associated with DNA in eukaryotes. DNA is wound around histone proteins to form chromatin.

Homologous Chromosomes that have the same genes at the same loci. Members of an homologous pair of chromosomes pair up during meiosis. Diploid organisms, produced by sexual reproduction, have homologous pairs of chromosomes – one member of each pair from the male parent and the other member from the female parent. (Can also be used to refer to structures that have different functions but have a common evolutionary origin, such as human arm and a bird wing.)

Hormone Chemicals made in endocrine glands that are carried in the blood to target cells/tissues/organs. They act as chemical messengers and are associated with developmental changes of the organism. Most are polypeptides but some are steroids.

Hydrocarbon chain A chain of carbon atoms bonded together with hydrogen atoms bonded onto the carbons.

Hydrogen bond A weak bond formed when partially positively charged groups come close to partially negatively charged groups. It is seen in water molecules, and in the secondary and tertiary structure of proteins.

Hydrolysis A reaction in which a molecule is broken down into two smaller molecules by the addition of a water molecule and the breaking of a covalent bond.

Hydrophilic Associating with water molecules easily (water-loving).

Hydrophobic Water-repelling (water-hating).

Hydrostatic pressure Pressure created by a fluid pushing against the sides of a container.

Hyphae The strands that make up the body of a fungus.

Ileum The second, longer part of the small intestine.

Immune response A response to an antigen, which involves the activation of lymphocytes.

Immunological memory Ability of the immune system to respond very quickly to antigens that it recognises as they have entered the body before.

Incidence The number of new cases of a disease in a certain time period.

Induced fit (hypothesis) The theory of enzyme action in which the enzyme molecule changes shape to fit the substrate molecule more closely as it binds to it.

Inhibition/inhibitor The slowing of an enzyme-controlled reaction. An inhibitor slows down or prevents the formation of enzyme–substrate complexes.

Initial reaction rate Rate of reaction at the beginning of the reaction.

Intercostal muscles Muscles between the ribs, responsible for moving the rib cage during breathing.

Interferon A group of factors with non-specific antiviral activity. They also affect the immune system.

Interleukin Cell-signalling chemicals. Some are involved in activating cells of the immune system.

Interphase The phase of the cell cycle where synthesis of new DNA and organelles takes place.

Intracellular (enzymes / digestion) Inside the cell (intracellular enyzmes/digestion are found inside the cell).

Ion An atom (or group of atoms) carrying a positive or a negative charge.

Ionic bond Attraction between oppositely charged ions.

Keratin Fibrous protein found in skin, hair and nails.

Keratinocytes Cells that make keratin.

Kinetic energy Energy of movement.

Kingdom Taxonomic group. Living organisms are grouped into one of five kingdoms: Prokaryotae, Protoctista, Fungi, Plantae and Animalia.

Lactate A compound containing lactic acid – the product of anaerobic respiration in mammals and some bacteria.

Lacteal A blind-ending branch of the lymph system found in each villus of the small intestine.

Leucocytes White blood cells.

Lignin A waterproofing substance that impregnates the walls of xylem tissue. Lignin gives wood its strength.

Limiting factor A variable that limits the rate of a process. If it is increased, then the rate of the process will increase.

Lipase An enzyme that catalyses the breakdown of lipid molecules.

Lipids A diverse group of chemicals that includes triglycerides, fatty acids and cholesterol.

Lock-and-key hypothesis The theory of enzyme action in which the enzyme active site is complementary to the substrate molecule, like a lock and a key.

Lumen A cavity surrounded by a cell wall in cells, such as xylem vessels, which have lost their cell contents. Also used for the central cavities of blood vessels.

Lymphatic system A system of lymph nodes and lacteals with lymph fluid.

Lymphocyte A type of white blood cell activated as part of the immune response.

Lysosomes Membrane-bound vesicles made by pinching off from the Golgi body. They usually contain digestive enzymes.

Macromolecule A very large molecule.

Macrophages Large, phagocytic, *Amoeba*-like white blood cells that engulf, ingest and digest bacteria, damaged cells and worn-out red blood cells.

Magnification The number of times greater an image is than the object.

Maltose A disaccharide molecule consisting of two α-glucose molecules bonded together.

Marker-assisted selection A mechanism used by animal and plant breeders to help select individuals with the desired genotype. The desired gene is linked (marked) to a section of DNA that is easy to identify in a young individual.

Meiosis Nuclear division that results in the formation of cells containing half the number of chromosomes of the adult cell.

Memory cells B and T cells that remain in the body after an immune response. Their presence enables a much faster and greater second immune response.

Meristem cells Undifferentiated plant cells capable of rapid cell division.

Messenger RNA (mRNA) A type of RNA polynucleotide involved in protein synthesis. Carries the information coding for a polypeptide from the nucleus to the ribosomes in the cytoplasm.

Metabolism All the chemical reactions that take place in the cells of an organism.

Metaphase The phase of mitosis where the chromosomes line up at the equator of the spindle.

Microtubule motors Proteins associated with microtubules. The proteins can move along microtubules. Kinesin moves towards the (+) end of the microtubules and dynein moves towards the (–) end.

Microtubules Components of the cell cytoskeleton. They have a diameter of about 24 nm and length varying from several micrometres in most cells to possibly several millimetres in some nerve cells. Microtubules are involved in mitosis, cytokinesis and movement of vesicles within cells.

Microvilli Folds in the membrane of a cell that increase its surface area.

Mitochondrion (pl: mitochondria) The organelle found in cells in which most of the ATP synthesis occurs. It is the site of aerobic respiration.

Mitosis Nuclear division that results in the formation of cells that are genetically identical to the parent cell.

Monoculture A crop of plants of a single species bred to be very similar.

Monocytes Large, phagocytic white blood cells.

Monokines Chemical produced by monocytes to signal to other cells. Also called lymphokines.

Monomer A small molecule that is one of the units bonded together to form a polymer.

Monophyletic A group of organisms is said to be monophyletic if they all share a common ancestor and therefore belong to the same classification group.

Monosaccharide A simple sugar molecule. The monomer of polysaccharides.

Morbidity The proportion of people in a population who are ill with a particular disease at any one time.

Mortality The number of people who die from a disease in a certain time period.

Mucus A slimy substance secreted by goblet cells in animal epithelial tissues. It is made up mostly of glycoproteins (proteins bonded to carbohydrates) and is used to protect and/or lubricate the surface on to which it is secreted.

Multinucleate Describes cytoplasm that is not divided into cells but contains many nuclei.

Mutation A change in the structure of DNA, or in the structure or number of chromosomes.

Mycelium The filaments (hyphae) that make up the body of a fungus.

Myogenic Describes muscle tissue (heart muscle) that generates its own contractions.

Myoglobin A respiratory pigment (protein) with a higher affinity for oxygen than haemoglobin.

Natural immunity Immunity acquired through exposure to disease during the normal course of life.

Natural selection The best-adapted organisms in a population can outcompete those that are less well-adapted. They are selected by the environment to survive and reproduce, so passing on the favourable alleles that have made them well-adapted. Over time this produces a change in the proportions of alleles in the gene pool and evolution occurs. Natural selection is the mechanism for evolution.

Neutrophils Phagocytic white blood cells. They engulf and digest bacteria. Neutrophils have a many-lobed nucleus, and a granular cytoplasm due to the large numbers of lysosomes present.

Niche The exact role of an organism in the ecosystem – its use of the living and non-living components of the ecosystem.

Non-competitive inhibitor An inhibitor of an enzyme-controlled reaction that binds to the enzyme molecule in a region away from the active site.

Nuclear envelope The double membrane structure surrounding the nucleus in eukaryotic cells.

Nucleic acid A polymer of nucleotide molecules.

Nucleotide The monomer of nucleic acids consisting of a phosphate, a sugar and an organic base.

Nucleus A large, membrane-bound organelle found in eukaryotic cells, which contains the genetic material in the form of chromosomes.

Nutrition The total substances taken into an animal or plant for use in metabolism (the sum total of its diet).

Oestrogen Steroid hormone made in ovaries.

Omnipotent See totipotent.

Omnivore An animal that eats plant and animal material.

Opportunistic infection Infection caused by an organism that infects a host with a weakened (compromised) immune system.

Optimum (temperature / pH) The condition that gives the fastest rate of reaction in enzyme-controlled reactions.

Order Taxonomic group used in classification of living organisms. Similar families are placed in the same order.

Organ A collection of tissues that work together to perform a specific overall function or set of functions within a multicellular organism.

Organelle Structure inside a cell. Each organelle has a specific function.

Organic base Nitrogenous base in nucleic acid: adenine, thymine, uracil, cytosine, guanine.

Osmosis The movement of water molecules from a region of higher water potential to a region of lower water potential across a partially permeable membrane.

Oxygen tension The amount of oxygen in the air expressed as the pressure created by the presence of oxygen, expressed in kilopascals (kPa).

Oxygenated Describes blood carrying oxygen in the form of oxyhaemoglobin.

Oxyhaemoglobin Haemoglobin with oxygen molecules attached.

Pandemic Describes a disease that is spreading worldwide or over continents.

Parasite An organism that lives in or on another living organism (its host), deriving nutrition from the host, benefiting at the expense of its host.

Parenchyma Relatively unspecialised plant cells. They have living contents and thin, permeable cellulose cell walls. They may be able to photosynthesise, store food or support young plants.

Partial pressure The proportion of total pressure provided by a particular gas as part of a mixture of gases.

Partially permeable membrane A membrane that will allow some molecules to pass through but will not allow some others to pass through.

Passive immunity Immunity acquired without activation of the lymphocytes. It is provided by antibodies that have not been manufactured by stimulating the immune system, such as through the placenta or breast milk, or by injection.

Pathogen An organism that causes disease.

Peptide A molecule consisting of a small number of amino acids bonded together by (covalent) peptide bonds.

Peptide bond The covalent bond formed when amino acids are joined together in condensation reactions.

Pericycle A layer of cells in the root that lies just inside the endodermis. It usually consists of meristematic cells whose division gives rise to lateral roots.

Peristalsis Muscular contractions of muscle layers of gut to squeeze food along.

pH Gives measure of acidity/alkalinity of a solution. It is the reciprocal of the logarithmic value of the hydrogen ion concentration. So pH 1–6 are acidic (lots of hydrogen ions), 7 is neutral, and 8–14 are alkaline.

Phagocyte A cell that can carry out phagocytosis and ingest bacteria or small particles. Macrophages and neutrophils are phagocytes.

Phagosome A vacuole inside a phagocyte which is created by an infolding of the plasma (cell surface) membrane to engulf a foreign particle. The foreign particle is held inside the phagosome.

Phloem A tissue in plants that is used to transport dissolved sugars and other substances.

Phospholipid A molecule consisting of a glycerol molecule, two fatty acid molecules and a phosphate group covalently bonded together. Phospholipids form the basis of cell membranes.

Photosynthesis The process by which plants, some bacteria and some protoctists make food using carbon dioxide, water and sunlight energy.

Phylogeny The evolutionary relationships between organisms.

Phylum A taxonomic group used in classification of living organisms. Similar classes are grouped into the same phylum.

Pinocytosis The process of endocytosis involving the bulk movement of liquids into a cell.

Pits (or bordered pits) Thin areas in the lignified walls of xylem tissue cells that allow communication between adjacent cells.

Plaque Fatty material built up under the endothelium of an artery.

Plasma cells Mature β-lymphocytes (white blood cells) that secrete a specific kind of antibody.

Plasma membrane / cell surface membrane The membrane that surrounds every cell, forming the selectively permeable boundary between the cell and its environment. It is made up of a double layer of phospholipids with embedded proteins.

Plasma proteins Proteins made in the liver that are found in blood plasma.

Plasmid Small, circular piece of DNA present in some bacterial cells. Plasmids may have genes for antibiotic resistance. Plasmids can also be used as vectors in genetic engineering.

Plasmodesma (pl: plasmodesmata) A fine strand of cytoplasm that links the protoplasm of adjacent plant cells through a thin area of cell wall called a pit.

Plasmolysis Detachment of the plasma membrane from the cell wall as the cytoplasm shrinks when water is lost from a plant cell.

Platelets Fragments of cells in the blood that play a part in blood clotting.

Pluripotent Stem cells capable of differentiating to become a limited number of cell types found in the organism (e.g. cells of an early embryo). See also Totipotent/Omnipotent.

Polymer A large molecule made up of many/repeating similar, smaller molecules (monomers) covalently bonded together.

Polynucleotide A polymer consisting of many nucleotide monomers covalently bonded together (DNA and RNA are polynucleotides).

Polypeptide A polymer consisting of many amino acid monomers covalently bonded together.

Polysaccharide A polymer consisting of many monosaccharide monomers covalently bonded together.

Potometer Apparatus used to measure water uptake in a leafy shoot and so to estimate rate of transpiration.

Precipitate A suspension of small solid particles in a liquid, produced by a chemical reaction.

Prevalence The number of people with a particular disease at a certain time.

Primary defences The defences that prevent the entry of a pathogen into the body.

Primary structure The sequence of amino acids found in a protein molecule.

Prokaryote An organism with cells that do not contain a true nucleus.

Prophase The phase of mitosis where the chromosomes become visible as a pair of sister chromatids joined at the centromere.

Prostaglandin A chemical made in the body that is involved in inflammatory reactions.

Prosthetic group A non-protein organic molecule that forms a permanent part of a functioning protein molecule.

Protease An enzyme capable of digesting proteins.

Protein A polymer consisting of many amino acid monomers covalently bonded together.

Pulmonary circulation The circulation of the blood through the lungs.

Pulmonary vein The vein carrying oxygenated blood from the lungs to the left atrium of the heart.

Purine Adenine and guanine – nitrogenous bases consisting of a double ring structure.

Purkyne tissue (Purkinje tissue) Specialised tissue (muscle fibres) in the septum of the heart that conducts the electrical stimulus from the sinoatrial node to the ventricles.

Pyrimidine Thymine, cytosine and uracil – nitrogenous bases consisting of a single ring structure.

Quadrat A square frame used for sampling in field work.

Qualitative A study is qualitative if it does not involve quantity (numbers). For example, simple observations to see if a particular species lives in a selected area is qualitative.

Quantitative A study is quantitative if it involves quantity (numbers). For example, if you count the number of individuals of a species in a selected area, the study is quantitative.

Quaternary structure Protein structure where a protein consists of more than one polypeptide chain. Haemoglobin and insulin both have a quaternary structure.

Receptor sites Protein or glycoprotein molecules on cell surfaces, used for attachment of specific substances such as hormones or viruses.

Reducing sugar A carbohydrate monomer or dimer that gives a positive result in Benedict's test because it is able chemically to reduce copper sulfate in solution.

Reduction Chemical reaction involving transfer of electrons from one reactant to another. The substance that gains electrons is reduced.

Resolution The ability to distinguish two separate points as distinct from each other.

Respiration The process in which energy is released from complex molecules, such as glucose, within cells and transferred to molecules of ATP.

Ribose The 5-carbon (pentose) sugar found in RNA nucleotides.

Ribosomal RNA (rRNA) RNA found in ribosomes.

Ribosome The organelle, made of two subunits, on which proteins are synthesised inside the cell.

Risk factor A factor that increases the risk or chance that you may develop a particular disease.

RNA Ribonucleic acid – a single-stranded polynucleotide molecule that exists in three forms. Each form plays a part in the synthesis of proteins within cells.

Root hair cells Cells in the epithelium of roots that have long extensions to increase surface area for the absorption of water and minerals.

Secondary defences Defences that attempt to kill or inactivate pathogens that have already invaded the body.

Secondary structure The coiling or folding parts of a protein molecule due to the formation of hydrogen bonds formed as the protein is synthesised. The main forms of secondary structure are the α-helix and β-pleated sheets.

Secretion The release of a substance made inside the cell using the process of exocytosis.

Selection pressure An external pressure that drives evolution in a particular direction.

Semi-conservative replication The replication of a DNA strand where the two strands unzip, and a new strand is assembled onto each 'conserved' strand according to the complementary base-pairing rules. The replicated double helix consists of one old strand and one newly synthesised strand.

Semilunar valves Valves between the ventricles and the main arteries leading out of the heart, which prevent backflow of blood.

Septum The wall separating the ventricles of the heart.

Sexual reproduction The production of a new individual formed by the fusing of gametes from two different parent organisms. The offspring have unique combinations of alleles inherited from both parents.

Sieve tube element A cell found in phloem tissue through which sap containing sucrose is transported. It has very little cytoplasm, no nucleus, and non-thickened cellulose cell walls, with the end walls perforated to form sieve plates through which the sap passes from element to element.

Simple diffusion The movement of molecules from a region of their higher concentration to a region of their lower concentration.

Single circulatory system A circulation in which blood flows through the heart once during each circulation of the body.

Sink A part of a plant that removes sugars from the phloem.

Sinoatrial node (SAN) The patch of tissue that initiates the heartbeat by sending waves of excitation over the atria.

Smooth muscle A type of muscle (involuntary muscle) found mostly in certain internal organs and involved in involuntary movements such as peristalsis.

Solute A solid that dissolves in a liquid.

Solute potential (φ_s) The component of water potential that is due to the presence of solutes – the potential energy of a solution provided by the solutes.

Solution A liquid with dissolved solids.

Solvent A liquid that dissolves solids.

Source A part of the plant that releases sugars into the phloem.

Speciation The formation of a new species.

Species A group of organisms whose members are similar to each other in shape (morphology), physiology, biochemistry and behaviour, and can interbreed to produce fertile offspring.

Spindle A structure consisting of protein fibres found in eukaryotic cells during cell division. Chromosomes become attached to the spindle at their centromeres, and spindle fibres guide the movement of chromosomes to opposite end of the cell at telophase.

Starch A polysaccharide found in plant cells. It is formed from the covalent bonding together of many glucose molecules.

Stem cells Undifferentiated cells that are capable of becoming differentiated to a number of possible cell types (e.g. omnipotent, totipotent, pluripotent).

Stoma (pl: stomata) Pore in leaf epidermis, surrounded by two guard cells. Changes in turgidity of the guard cells can open or close the stoma. Stomata allow gaseous exchange in plants and also allow transpiration.

Stroma The gel-like matrix found in chloroplasts. The membranes of the thylakoids/grana are embedded in the stroma.

Substrate The substance that is used up in an enzyme-controlled reaction, leading to the formation of product. It fits into the active site of the enzyme at the start of the reaction.

Surface tension The 'skin' on the surface of water formed as a result of hydrogen bonding in water molecules pulling the surface molecules downwards.

Surfactant A chemical that can reduce the surface tension of a film of water.

Sustainable development Development that does not cause excessive harm to the surrounding environment. The local biodiversity (species diversity, habitat diversity and ecosystems) and the local people are able to continue to live and operate alongside the development.

Sympatric Speciation that occurs within one area – some factor other than geographical separation has prevented free interbreeding between members of the species.

Symplast pathway The route taken by water through the cytoplasm of cells in a plant.

Systemic circulation The circulation that carries blood around the body, excluding the circulation to the lungs.

Systole The stage of the heart cycle in which heart muscle contracts to pump blood.

Taxon (pl: taxa) A taxonomic group, such as a class or a family, used to aid classification.

Taxonomy The study of the principles behind classification.

Telophase Final phase of mitosis. Two new nuclear envelopes form around the two new nuclei.

Tendinous cords String-like tendons used to attach the atrioventricular valves of the heart to the sides of the ventricle wall. Sometimes called heart strings.

Tertiary structure The overall three-dimensional shape of a protein molecule. It is the result of interactions between parts of the protein molecule such as hydrogen bonding, formation of disulfide bridges, ionic bonds and hydrophobic interactions.

Testosterone Steroid hormone made in the testes.

Thrombus A blood clot.

Thylakoid Flattened membrane sacs found in chloroplasts, which hold the pigments used in photosynthesis and are the site of the light-dependent reactions of photosynthesis. A stack of thylakoids forms a granum.

Thymine A nitrogen-containing organic base found in DNA. It pairs with adenine.

Tissue A group of similar cells that perform a particular function.

Tissue fluid The fluid, derived from blood plasma, that surrounds the cells in a tissue.

Totipotent Undifferentiated cell that is capable of differentiating into any kind of specialised cell. All cells in an embryo are totipotent, as are meristem cells in plants. Embryonic stem cells are totipotent.

Trachea The windpipe leading from the back of the mouth to the bronchi.

Transcription The assembly of an mRNA molecule that is a copy of the DNA coding strand (and complementary to the template strand).

Transect A line through a habitat used to help take samples and study the habitat.

Transfer RNA (tRNA) A type of RNA polynucleotide involved in protein synthesis. It transports amino acids to the ribosomes to be added to the growing polypeptide chain.

Transgenic Organism that has genetic material from another organism, usually by genetic engineering.

Translocation The movement of sucrose and other substances up and down a plant.

Transmission The way in which a microorganism or other pathogen travels from one host to another.

Transpiration The loss of water vapour from the aerial parts of a plant due to evaporation.

Triglyceride A molecule consisting of a glycerol molecule and three fatty acid molecules covalently bonded together.

Turgid Describes a cell that is full of water as a result of entry of water due to osmosis. When the pressure of the cell wall prevents more water entering, the cell is said to be turgid.

Ultrastructure The detailed structure of the internal components of cells as revealed by the electron microscope rather than by the light microscope. Sometimes called fine structure.

Uracil A nitrogen-containing organic base found in RNA.

Urea Chemical made in the liver from amine groups from deaminated amino acids and carbon dioxide. It is toxic and is removed from the body in urine.

Vaccine A preparation of antigens given to provide artificial immunity.

Vacuolar pathway The pathway taken by water in plants as it passes from cell to cell via the cell cytoplasm and vacuole.

Variable (dependent variable, independent variable) Condition in an experimental investigation. Independent variables are altered by the experimenter; dependent variables change as a result of the changes in the independent variable.

Variation The differences between individuals.

Vascular tissue / bundle The transport tissue in a plant – usually found as a bundle containing both xylem and phloem.

Vector An organism that carries a disease-causing organism (pathogen) from one host to another. The term is also used to describe an agent (such as a plasmid) that transfers genetic material from one cell to another.

Vena cava Either of two large veins that carry deoxygenated blood from the body back to the heart.

Ventilation Breathing – movement of diaphragm and rib cage that bring air into and out of the lungs.

Ventricles The lower chambers in the heart.

Vesicle A membrane-bound sac found in cells and used to transport materials around the cell.

Villi Folds in the wall of an organ or tissue that increase surface area.

Vitamin Chemical needed in small amounts for healthy metabolism. Some organisms can make them, some organisms have to obtain them in the diet.

Water potential (Ψ) A measure of the ability of water molecules to move freely in solution. Measures the potential for a solution to lose water – water moves from a solution with high water potential to one of lower water potential. Water potential is decreased by the presence of solutes.

Water vapour potential The potential energy of water vapour in a gas – it is used to indicate how much water vapour is present.

Xerophyte A plant specially adapted to living in dry areas.

Xylem A plant tissue containing xylem vessels (and other cells) that are used to transport water in a plant and provide support.

Zygote Diploid cell made from fusion of male and female gametes.

Index

Your Exam Café CD-ROM

In the back of this book you will find an Exam Café CD-ROM. This CD contains advice on study skills, interactive questions to test your learning, a link to our unique partnership with New Scientist, and many more useful features. Load it onto your computer to take a closer look.

Amongst the files on the CD are PDF files, for which you will need the Adobe Reader program, and editable Microsoft Word documents for you to alter and print off if you wish.

Minimum system requirements:

- Windows XP SP2 and later, Vista
- Pentium 4 or later, 1 GHz or faster (2 GHz required for Vista)
- 512 MB RAM (1 GB for Vista)
- 1 GB or 10% of hard disc capacity, whichever is the greatest
- Internet Explorer 7 or Firefox 3
- Flash Player version 9 or later
- Microsoft Office 2003 or later (Microsoft viewer applications may be used, or 100% compatible office suites)
- Adobe Reader version 8 or later

To run your Exam Café CD, insert it into the CD drive of your computer. It should start automatically; if not, please go to My Computer (Computer on Vista), click on the CD drive and double-click on 'start.html'.

If you have difficulties running the CD, or if there is no CD in this book, please contact the helpdesk number given below.

Software support

For further software support between the hours of 8.30–5.30 Monday to Friday (excluding bank holidays) please contact:
Tel: 0845 313 8888
Email: digital.support@pearson.com